AGROFORESTRY: SCIENCE, POLICY AND PRACTICE

FORESTRY SCIENCES

Volume 47

Agroforestry: Science, Policy and Practice

Selected papers from the agroforestry sessions of the IUFRO 20th World Congress, Tampere, Finland, 6–12 August 1995

Edited by

FERGUS L. SINCLAIR

School of Agricultural and Forest Sciences, University of Wales, Bangor, UK

Reprinted from *Agroforestry Systems*, Volume 30, Nos. 1–2 (1995)

Kluwer Academic Publishers

Dordrecht / Boston / London

Library of Congress Cataloging-in-Publication Data

A C.I.P. Catalogue record for this book is available from the Library of Congress.

ISBN 0-7923-3696-8

Published by Kluwer Academic Publishers,
P.O. Box 17, 3300 AA Dordrecht, The Netherlands

Kluwer Academic Publishers incorporates
the publishing programmes of
D. Reidel, Martinus Nijhoff, Dr. W. Junk and MTP Press.

Sold and distributed in the U.S.A. and Canada
by Kluwer Academic Publishers,
101 Philip Drive, Norwell, MA 02061, U.S.A.

In all other countries, sold and distributed
by Kluwer Academic Publishers Group,
P.O. Box 322, 3300 AH Dordrecht, The Netherlands.

Printed on acid-free paper

Printed in the Netherlands

Table of contents

Agroforestry Systems **30:** 1–3, 1995.

Preface

This issue comprises a collection of papers, selected from those presented during the agroforestry sessions of the IUFRO[1] 20th World Congress held in Tampere, Finland in August, 1995. They establish agroforestry as an interdisciplinary science focused on the practical imperative of assisting farmers, forest dwellers and landscape-level planners in achieving sustainable food, fuel and timber production into the twenty-first century.

Agroforestry as a practically-based subject focus in international research and development has come a long way in the five years since the last IUFRO World Congress in Montreal. Pedro Sanchez, the current director of ICRAF[2], now part of the CGIAR[3] system, points in his keynote address to the emergence of principles that can begin to be applied in agroforestry practice. These are seen as the result of process-oriented research in the natural sciences and he highlights the need for similar progress in understanding social processes and the consequent development of appropriate policy frameworks, if practical developments are to be realised in terms of farmers adopting sustainable agroforestry approaches. Louise Buck picks up this theme in her examination of policy developments in industrialised and less developed countries, revealing an unprecedented level of agroforestry policy assessment and planning activity, while Dean Current and Sara Scherr present a regional analysis of benefits accruing to farmers and the environment from agroforestry development in Central America and the Caribbean that yields specific policy recommendations. Echoing Sanchez's call for the development of agroforestry practices that put 'money in farmers pockets' and consider the 'value of environmental externalities', Colin Price outlines appropriate methods of economic analysis that exist to evaluate such perceived benefits of agroforestry development and urges that they be applied in the agroforestry sphere to determine the circumstances in which agroforestry practices offer quantitative benefits over alternative land use options. Both Sanchez and Price consider the implications of agroforestry in terms of environmental benefits at the landscape or watershed scale, and with respect to topical global concerns involving the maintenance of biodiversity and carbon sequestration.

There can be no doubt that agroforestry research is maturing, with speculation about perceived benefits of integrated land use practices being replaced by quantification of their impact and the development of a fundamental understanding of the interactive processes that combine to produce the net outcomes

[1] International Union of Forest Research Organisations.
[2] International Center for Research in Agroforestry.
[3] Consultative Group on International Agricultural Research.

of combining various plant and animal species. While there is still a long way to go before this understanding is complete or even sufficient in many areas to inform practice, significant milestones have been reached. In reviewing what is known about the contribution that trees can make to the nutrient requirements of intercropped plants, Cheryl Palm is able to provide general guidelines on which trees to use for this purpose, if not specific recommendations on particular species for specific environments. Similarly, Götz Schroth uses the practical options open to farmers to manipulate tree root development, as the framework for reviewing what is known about tree root characteristics and the below-ground interactions with crops in agroforestry practices, with a view to selecting tree species and appropriate system designs. Meine van Noordwijk and P. Purnomosidhi take the first step in developing simple methods that can be used to gauge tree root development non-destructively by employing relationships between root diameter at the stem base and functionally important root parameters such as the total length of fine roots. It is increasingly clear that there are now sufficient data about agroforestry for simulation models to be developed in order to test our developing understanding of how agroforestry systems function; to predict the likely productivity and sustainability of particular agroforestry practices and to explore alternative management regimes. Philippe de Reffye et al. show how research on tree architecture can be used to simulate the growth of individual trees appropriate to exploring alternative agroforestry scenarios while Gerry Lawson et al. report on attempts at coupling forest and crop process models to examine yield effects at the tree-crop interface. Robert Muetzelfeldt proposes a modular modelling approach as a means of improving the efficiency of simulation modelling in agroforestry and making it a more accessible tool for the interdisciplinary group of scientists involved in agroforestry research.

Another indication that agroforestry is maturing both scientifically and as a development imperative, is the way in which the over-optimistic development of the alley-cropping technology by some agroforestry researchers as a panacea over the past decade has been acknowledged as a misdirection of effort. While much has been learnt through alley-cropping research about tree-crop interactions and the specific attributes of some multipurpose tree species, it is now clear, as Sanchez underlines, that alley-cropping is only likely to be appropriate in certain very site-specific circumstances. However, as Walker et al. point out, while prefabricated technology designs such as alley-cropping have not enjoyed high adoption rates, farmers have shown interest in elements of these technologies. This suggests that supporting farmers in incorporating trees on their farms by providing components and decision support about how to use them may be more appropriate than attempting to design whole technologies. This is in keeping with evidence, such as that reported by Thapa et al. from Nepal, that in some circumstances farmers already possess a sophisticated understanding of tree-crop interactions and that targeting research on gaps in local knowledge may be the most effective means of supplying what they need to improve their agroforestry practice. Methods of appraising and using indigenous knowledge in agroforestry research are now being developed and appear to be complementary to process-oriented research in the natural and social sciences.

The agroforestry sessions were one of many themes in the twentieth IUFRO World Congress which set out to address the broad topic of 'Caring for the forest: research in a changing world'. Clearly, agroforestry research is at the centre of developing sustainable use of natural renewable resources in the coming century and is evolving to address changing needs. Consolidation of the scientific gains now being made in process-oriented research, coupled with the development of a policy framework that encourages adoption of sustainable land use practices, are now required. It can be anticipated that agroforestry practices have an important role to play in conserving forest resources by reducing the need to deforest and, if 'forest' is defined in a broad sense as tree cover, by increasing the proportion of woody biomass in farming landscapes.

Acknowledgements

A special debt is owed to the thirty or so scientists who anonymously reviewed the papers considered for this special issue of *Agroforestry Systems*. They not only did the job willingly, they did it within the strict time limits imposed by the need to have the publication available at the IUFRO World Congress. The School of Agricultural and Forest Sciences at the University of Wales, Bangor provided institutional support and Julia Hindle, Michele Jones, Dee Aebischer and Karen Cooper went well beyond the call of duty in assisting with the administrative work involved in producing the copy that was forwarded to the publishers.

Fergus L. Sinclair
School of Agricultural and Forest Sciences
University of Wales, Bangor

Agroforestry Systems **30**: 5–55, 1995.

Science in agroforestry

P. A. SANCHEZ
International Centre for Research in Agroforestry, P.O. Box 30677, Nairobi, Kenya

Key words: competition, complexity, profitability, sustainability

Abstract. Agroforestry research is being transformed from a collection of largely descriptive studies into more scientific approaches, based on process-oriented research. The development of agroforestry as a science should be based on four key features: *competition, complexity, profitability and sustainability.* Managing the *competition* between trees and crops for light, water and nutrients to the farmers' benefit is the biophysical determinant of successful agroforestry systems. Simultaneous agroforestry systems are more susceptible to competition than sequential ones. A tree–crop interaction equation helps quantify competition vs. complementary effects on fertility. Alley cropping, a simultaneous agroforestry system, has limited applicability because the competition factor usually exceeds the beneficial fertility effects. The *Faidherbia albida* parkland, another simultaneous system, is almost always beneficial since the reverse phenology of *F. albida* minimizes competition while enhancing the fertility effect. Sequential systems such as relay intercropping and improved fallows also minimize competition but the processes responsible for crop yield increases are largely unquantified. New methodologies for reliably measuring complex below-ground interactions are being developed.

Socioeconomic and ecological *complexity* are typical of agroforestry systems. Participatory, analytical and multidisciplinary characterization at different spatial scales is the required first step in effective agroforestry research. Diversity of products and services should be manipulated in a way that puts money in farmers' pockets. Domestication of indigenous trees with high-value products enhances *profitability*, particularly those that can be marketed as ingredients of several finished products. Policy research interventions are often necessary to help farmers during the initial years before trees become productive and exert their positive ecological functions. *Profitable* agroforestry systems are potentially *sustainable*, controlling erosion, enhancing biodiversity and conserving carbon, provided nutrient offtake is balanced by nutrient returns via litter and the strategic use of fertilizers, particularly phosphorus. A list of research gaps indicates where hard data are needed to provide a predictive understanding of the competition, complexity, profitability and sustainability aspects of agroforestry.

Introduction

Agroforestry is the traditional practice of growing trees on farms for the benefit of the farm family. It has been in use for at least 1300 years according to pollen records [Brookfield and Padoch, 1994], although tree domestication probably started much earlier [Simmonds, 1985]. Agroforestry was brought from the realm of indigenous knowledge into the forefront of agricultural research less than two decades ago, and was promoted widely as a sustainability-enhancing practice that combines the best attributes of forestry and agriculture [Bene et al., 1977; Steppler and Nair, 1987]. Growing trees along with crops and livestock was postulated to enhance crop yields, conserve soil and recycle nutrients while producing fuelwood, fodder, fruit and timber.

The potential of agroforestry was initially based on largely descriptive and anecdotal information, usually collected from high-potential environments [Nair, 1989]. Applied research was vigorously pursued with limited attention to the underlying biophysical or socioeconomic processes.

While such efforts raised awareness and helped institutionalize agroforestry in national programs throughout the world, the need for a critical, science-based approach gradually emerged. Many agroforestry innovations did not deliver what their promoters expected. The limited usefulness of alley cropping, perhaps the prototype agroforestry technology, contrasts sharply with the very high expectations of it during the 1980s. A recent review of process-oriented agroforestry research produced very little hard data [Anderson and Sinclair, 1993].

The transformation of a descriptive practice into a science happens when hypotheses are developed by systematic observation, description and interpretation. Such hypotheses, coupled with inductive reasoning and deductive experimentation, can lead to theories that provide a predictive understanding applicable to other situations. Hypotheses are continually challenged by testing, disproving and formulation of new ones, which often result in new ways of thinking, or paradigms. The fundamental sciences, mathematics, philosophy, physics and chemistry, developed in such a manner. Refinements, combinations and new ideas gave birth to other basic sciences such as biology, astronomy and psychology. Further combinations and developments produced botany, zoology, economics, ecology, statistics and other sciences. Applications to real-life situations resulted in new applied sciences in medicine, engineering and agriculture. Further specialization led to more focused applied sciences such as agricultural economics, plant genetics, plant physiology and soil science. These are all recognized as sciences because their key principles have been established through the methods previously described.

Agroforestry is not there yet. Its theoretical underpinnings are yet to be developed in a sufficiently rigorous manner to assure predictive understanding. At this point agroforestry is a field of study which involves the combined application of ecology, economics, anthropology, agronomy, forestry, soil science, animal science, tree genetics, biometrics and other applied sciences. Out of this cauldron a recognized science may emerge, and what an exciting one it may be, given its broad interdisciplinary nature.

The fundamentals of agroforestry are now being intensively investigated by many institutions throughout the world. Two key principles set agroforestry systems apart from agricultural or forestry systems: *competition* and *complexity*. They in turn determine two desirable properties: *profitability* and *sustainability*. The purpose of this paper is to outline these principles and properties as an approach to science-based agroforestry. All four combine biophysical and socioeconomic issues. This paper, however, is not a comprehensive review of the literature, so many of the references and examples used are the ones most accessible to the author. Comments on the drafts of this paper were received from 35 colleagues, many of whom provided new insights and valuable criticisms. Key substantive additions not supported by references are credited as personal communications in the text.

Competition

When plants grow in proximity to each other they interact in positive ways (complementarity) or in negative ways (competition). The biophysical bottom line of agroforestry is how to manage the interaction for light, water and nutrients between the tree component and the crop and/or livestock components for the benefit of the farmer. Much of the initial enthusiasm when agroforestry was brought into the research agenda was based on the assumption that when trees and crops are brought together they complement rather than compete with each other in their capture of growth resources.

The anecdotal evidence about the potential of agroforestry raised some suspicions in this author's mind when it was realized that most of the successful examples of agroforestry came from high-potential environments, where water or nutrients were not major limiting factors. A soil–agroforestry hypothesis was formulated out of this inquiry [Sanchez, 1987]. Anthony Young transformed it into ten hypotheses [Young, 1989b], and the number keeps growing [Van Noordwijk and Dommergues, 1990; Wilson, 1990; Ong et al., 1991a]. The current 17 biophysical hypotheses are listed in Appendix 1. Many of them remain to be tested or rejected in a sufficiently rigorous manner and under a wide enough range of conditions to be accepted as broadly applicable principles. All of them deal with competition or complementarity. Much of the work to test these hypotheses is recent (see references listed in Appendix 1), indicating the current impetus of process-oriented research. Science advances as much by disproving as by proving hypotheses. Many of the hypotheses shown in Appendix 1 are very general; they should be tested in specific agroforestry systems under specific biophysical and socioeconomic conditions, such as differing climate, soil, slope, cultural tradition and income levels.

Simultaneous vs. sequential agroforestry systems

The term agroforestry system is used when the tree and crop and/or animal components interact with each other and have discrete boundaries, separating the system from others on a farm. Although agroforestry systems have been classified in a myriad of different ways [Nair, 1989], there are only two functionally different types, simultaneous and sequential [ICRAF, 1994]. Simultaneous agroforestry is where the tree and the crop components grow at the same time and in close enough proximity for interactions to occur. Examples of this type are alley cropping (hedgerow intercropping), contour hedges, parklands, boundary plantings, homegardens and several silvopastoral systems.

In sequential agroforestry systems the maximum growth rates of the crop and the tree components occur at different times even though both components may have been planted at the same time and are in close proximity. Examples of this type are shifting cultivation, improved fallows, taungya, and some multistrata systems. Interactions between the crop and the tree components are minimized with time in sequential agroforestry.

Simultaneous agroforestry systems can be transformed into sequential agroforestry systems. This is the case when the trees in an alley-cropping system are allowed to grow into a fallow and cropping is discontinued. In the next cropping cycle the trees are severely pruned to minimize competition with crops, but they are allowed to grow when the crops are gone (B. Duguma, pers. comm.). Long-duration multistrata or complex agroforests start as sequential systems with crops and trees interacting, but develop into simultaneous systems when the various tree species fully develop.

The following section describes the biophysical interactions of some short-duration simultaneous and sequential agroforestry, focusing on the competition issue.

The alley-cropping experience

Alley cropping, also known as hedgerow intercropping and alley farming, is a simultaneous agroforestry system where trees, mainly leguminous, are grown in dense hedges between 'alleys' of specified width where short-cycle food crops are grown. The hedges are pruned periodically and the resulting mulch is placed on the alleys to provide nutrients and control weeds [Kang et al., 1981]. Interactions between the tree and the crop components are maximized by this spatial arrangement. Alley cropping should not be confused with contour hedges, another simultaneous system where the hedges are planted along contours on steep slopes but at variable widths and usually include grasses and trees that are not pruned.

The principles of plant interactions have been well studied in natural systems by plant ecologists [Harper, 1961] and in multiple-cropping systems by agronomists [Papendick et al., 1976; Willey, 1979]. For simultaneous agroforestry systems that have crop yields as the only product, Ong [in press] and Ong and Black [in press] proposed a tree–crop interaction equation for quantifying the crop and the tree component effects; it is modified as follows:

$$I = F - C \tag{1}$$

where

I = overall interaction, as a percentage of sole-crop yields free from interference with trees,

F = the fertility effect, i.e., the percentage of crop-yield increase caused by soil fertility improvements (nutrient inputs from tree prunings plus microclimate modifications such as less soil evaporation, lower soil temperatures), and

C = the competition effect, i.e., the crop-yield decrease caused by competition with trees for light, water and nutrients.

A positive I value in this equation means net complementarity, the desired outcome of agroforestry. A negative I value means net competition to the detriment of crop yields. The I value is based on total area, including that occupied by trees. Since the trees are used only as a source of mulch in this system there is only one product: crop production. In agroforestry systems that also produce poles, fruits, firewood or fodder from the trees as well as crops, the equation has to be modified to take these tree products into account as is done in intercropping research [Andrews and Kassam, 1976; Willey, 1979; Anderson and Sinclair, 1993; Van Noordwijk, in press]. This equation is still at the developmental stage. Other factors such as allelopathy and positive or negative effects of pest activities could be specified, although they are included as yield effects. The final judgement is the economic value of all products versus the sole crop, and this is done by converting the biophysical yield data into monetary values. The equation differs from the classic intercropping ones of land equivalent ratios or relative yield totals [Andrews and Kassam, 1976] in that the complementary and competition effects are separated.

The measurement of the two parameters in Eq. (1) can be accomplished with four treatments in an alley-cropping experiment [Ong, in press].

Co = sole crop
Cm = sole crop + mulch from pruned trees
Ho = crop + tree with mulch removed
Hm = crop + tree with its mulch

F can be measured as $Cm - Co$, that is the effect of tree prunings due to their nutrient input upon decomposition plus microclimatic changes on the soil surface or weed suppression prior to decomposition. F is therefore the fertility effect in its broadest sense, involving both nutrient availability and soil physical changes due to litter or mulch inputs.

C can be measured as $Hm - Cm$, or $Ho - Co$, that is, the effect of trees on crop yields separate from the fertility effect. Therefore a second equation makes the determination of crop-yield effects operational [Ong, in press]:

$$I = (Cm - Co) - (Hm - Cm) \qquad (2)$$

Unfortunately, there are very few alley-cropping experiments that have these four treatments arranged in a proper experimental design that assures no interference between treatments [Ong, in press]. A bias in favor of a positive alley-cropping effect is introduced when plots are so small that the sole-crop treatment (*Co*) is invaded by tree roots from adjacent *Hm* or *Ho* plots. In such cases there is unintended competition for below-ground resources which depresses yields in the sole-crop treatment and allows the trees from adjacent plots to capture additional water and nutrients. A review by Coe [1994] on this topic describes these pitfalls. The methodological solution is to use large plots, on the order of 10 × 20 m, and root barriers or trenches down to 1 m depth. Unfortunately, solid root barriers can be effective for as little as one season; after that tree roots grow under the barriers and up into the control plots [Singh et al., 1989; Hauser, 1993].

Ong [1994, in press; Ong et al., in press], Reynolds [1994] and Woomer et al. [in press] have analyzed the available literature, comprising over 100 alley-cropping experimental data sets. A summary of some long-term experiments considered to be reasonably devoid of interference between plots is shown in Table 1. Trials ranging from semi-arid to humid climates, in widely different soils are interpreted according to Eq. (2).

The effect of alley cropping was strongly negative (–58%) in a fertile semi-arid Alfisol in India [Rao et al., 1991]. In this environment the fertility effect was moderate (+19%) but the competition effect for water was very strong (–77%). A similar effect was observed in another semi-arid site with moderately fertile Alfisols at Machakos, Kenya [ICRAF, 1993]. Competition for moisture was very marked in a season with low rainfall, but a positive alley-cropping effect of +11% was observed at the same site during a season with high rainfall. The magnitude of the competition effect decreased from –34% to –8% with increasing rainfall. The fertility effect also increased with higher rainfall, because more tree biomass was produced and returned to the soil as mulch. These and other data sets [Reynolds, 1994] indicate that the likelihood of drought stress results in a strong probability that the alley-cropping effect will be negative in semi-arid areas.

Examples from the subhumid tropics show intermediate effects of fertility and competition. At Chipata, Zambia, the negative alley-cropping effect shown in Table 1 is typical of many results in the area [Akyeampong et al., in press]. The original alley-cropping trial at Ibadan, Nigeria [Kang et al., 1981], showed a tremendous positive fertility effect (+58%). Unfortunately this trial did not have a sole-crop control, impeding the estimation of *I* and *C*. Estimates of sole-crop maize yields from adjacent areas at the IITA station, however, suggest a *C* effect of –54% and therefore an overall alley-cropping effect of +4% (calculated from IITA, 1988].

Finally, in the humid tropics, where moisture is not expected to be limiting but fertility may be, trials still show a major competition effect, presumably because of competition for light and nutrients. In a fertile soil of volcanic origin from Costa Rica the overall alley-cropping effect was close to neutral (–3% with *Erythrina poeppigiana* and +1% with *Gliricidia sepium*, Kass [1987]). In a strongly acid and infertile soil at Yurimaguas, Peru, the overall effect was strongly negative (–30%) and attributed to competition for nutri-

Table 1. Examples of alley cropping results interpreted in terms of the tree-crop interaction equation. Values for *I* (overall interaction], *F* (fertility effect] and *C* (competition effect] are expressed as percentages of sole-crop yields. Compiled from information by C. K. Ong and M. van Noordwijk.

Location (original source)	Climate	Soil	Tree	Crop	*I*	*F*	*C*
Hyderabad, India [Rao et al., 1991]	Semiarid 550 mm	Alfisol pH 6.0	*Leucaena leucocephala*	Millet Peanut	–58	+19	–77
Machakos, Kenya (poor rainy season) [ICRAF, 1993]	Semiarid 230 mm this crop	Alfisol pH 5.6	*Senna siamea*	Maize	–31	+3	–34
Machakos, Kenya (good rains) [ICRAF, 1993]	Semiarid 417 mm this crop	Alfisol pH 5.6	*Senna siamea*	Maize	+11	+19	–8
Chipata, Zambia [Akyempaong et al., in press]	Subhumid 760 mm	Alfisol pH 4.6	*Leucaena leucocephala*	Maize	–20	+36	–56
Ibadan, Nigeria [Kang et al., 1981, 1985]	Subhumid 1400 mm	Sandy Entisol pH 6.2	*Leucaena leucocephala*	Maize	+4	+58	–54
Yurimaguas, Peru [Szott et al., 1991]	Humid 2200 mm	Ultisol pH 4.2	*Inga edulis*	Rice	–30	+28	–58
Turrialba, Costa Rica [Kass, 1987]	Humid 2650 mm	Volcanic (Tropept) pH 4.6	*Erythrina poeppigiana* *Gliricidia sepium*	Maize	–3	+52	–55
North Lampung, Indonesia [Van Noordwijk et al., 1995]	Humid 2700 mm	Ultisol pH 4.7	*Peltophorum dasyrachis*	Maize	+32	+58	–26

ents [Szott et al., 1991]. One major exception to this trend is alley cropping with *Peltophorum dasyrachis* in an acid Ultisol from Lampung, Indonesia, with an I value of +32% [Van Noordwijk et al., 1992, 1995]. This non-legume species grows as very dense, low hedgerows, producing a large mulch-to-shade ratio. Its F value is not unusually large (+58%), but its C effect is unusually small (–26%).

Many factors affect alley-cropping performance: the choice of tree species and crop species, alley width, biomass production, number of crop cycles, time and frequency of prunings, tillage, fertilization and weed dynamics. These are well described [Kang et al., 1981, 1985, 1990; Yamoah et al., 1986; Szott et al., 1991; Fernandes et al., 1993; Kang, 1993, in press; Salazar et al., 1993; Juo et al., 1994; Wendt et al., 1994; Palm et al., 1995b] but the bottom line appears to be that alley cropping will only work in limited and very site specific circumstances.

There is now a predictive understanding as to where alley cropping should work: at sites where the supply of water and nutrients is likely to be ample; that is, on fertile soils with a reliable and adequate rainy season. Immediately one wonders whether alley cropping should be the system of choice under such favorable environments.

In addition, alley cropping has major soil conservation and socioeconomic implications. When used as contour hedges on sloping lands, alley-cropping systems clearly decrease runoff and erosion [Lal, 1989; Young, 1989a; ICRAF, 1994; Kiepe and Rao, 1994]. As mentioned earlier, contour hedges are different from alley cropping in that the width of the hedge is determined by the slope and not by a specified alley width; also, the hedges can contain both grasses and trees, which are frequently used for fodder instead of for mulch, and include freely growing trees for fruit or timber [Garrity and Mercado, 1994]. The soil conservation effects of contour hedges are discussed in a later section. It should be clear, however that alley cropping on slopes is an effective way to control erosion, as long as the hedges on the contour are sufficiently well developed to provide a continuous barrier. This effect may in some instances override competition considerations.

Socioeconomic studies in western Kenya indicate that alley cropping requires about an 18% increase in maize yields above those for sole cropping to cover additional labor costs during establishment [ICRAF, 1993]. Swinkels and Franzel [submitted] found that pruning the trees increases labor requirements by only 5%. But there are two critical constraints on pruning labor. First, pruning is required during the farmers' peak periods of labor use, and delaying pruning can greatly decrease crop yields, because of increased competition. Second, women consider pruning too tedious and expect males to do it, who would rather be doing something else. Studies in Nigeria indicate farmers are unwilling to plant many trees in hedgerows unless they have clear ownership of the land [Lawry et al., 1994]. Therefore, alley cropping is most likely to work where the following conditions apply.

- Soils are fertile without major nutrient limitations.
- Rainfall is adequate during the cropping season.
- Land is sloping with erosion hazards.
- There is an ample supply of labor, coupled with a scarce supply of land.
- Land tenure is secure.

It is unfortunate that over 15 years of research has been devoted to this agroforestry technology without fully taking into account the principles of competition. As one of the scientists who got excited about alley cropping in the early 1980s and initiated research in the Amazon, this author shares this responsibility with others. He was also one of the first to question its suitability for low fertility soils [Sanchez, 1987].

In retrospect, the principal strategic mistake was to equate alley cropping with intercropping of annual crops, where yield increases often occur when C3 and C4 crops with drastically different architectures are grown in rows [Papendick et al., 1976; Ong et al., 1991b]. But trees differ from short-cycle crops in two fundamental ways: they grow for several years and they also develop extensive root systems that capture nutrients and water at a time when crop root systems are beginning to develop. Both traits give trees a major competitive advantage. Short-cycle food crops are in a way analogous to uptight city dwellers who are always looking at their watches: crops must be planted, flower, set fruit and die on schedule. Trees are more akin to laid-back people: if there is a drought they can stop growing and shed a few leaves; there is plenty of time to recuperate and grow rapidly when conditions are favourable. Which strategy wins in the long run?

Alley cropping is like playing with fire, because it has strongly positive and strongly negative interactions and thus requires precise fine tuning in every situation, according to Van Noordwijk's [in press] shade and mulch model. Such factors as alley width, avoiding fast-growing trees because they are strongly competitive, and supplemental fertilizer applications are among the key fine tunings. As a system to be promoted widely to farmers, it probably requires too much management. The challenge is how to manage the intrinsic comparative advantages of trees for the benefit of the farmer. In most circumstances, alley cropping is not the way to do it.

Nevertheless, much has been learned from the alley-cropping experience. The original hypothesis about how trees can improve crop yields by enhancing soil fertility has been replaced with more focused hypotheses, as shown by the more recent ones listed in Appendix 1. Improving hypotheses is a way for science to progress. An immense amount of hard data have also been obtained about the tree–crop interface, plant litter quality, biological nitrogen fixation, pruning regrowth, and the adaptation of trees to adverse soil conditions.[1] Alley-cropping research also produced practical means to control erosion biologically. All these results are valuable for the understanding and improvement of other agroforestry or intercropped systems. Therefore, the general failure of this technology from the development perspective carries with it the benefit of new and valuable scientific insights in terms of concepts and methods.

The albida effect

After the negative experience with alley cropping, one may wonder if there are any simultaneous agroforestry systems that work. There is one that does so remarkably well: the *Faidherbia (Acacia) albida* parklands in the Sahel [Charreau and Vidal, 1965; Vandenbelt, 1992]. Millet and sorghum crops growing within a 5 to 10 m radius around mature *F. albida* trees yield two to three times more than adjacent crops growing in the open [Bonkoungou, 1992]. This may represent a positive *I* value of 100 to 150% in the tree–crop interaction equation, without including two other important benefits, providing fuelwood and dry-season fodder. The necessary studies with the four treatments previously mentioned, however, are yet to be done.

The Sahelian environment is characterized by a nine month dry season and a short, erratic rainy season coupled with sandy, infertile soils very deficient in phosphorus and nitrogen [Sivakumar et al., 1991]. *F. albida* trees exhibit a reverse phenology, developing a green canopy during the dry season and dropping leaves at the onset of the rainy season [Joly, 1992]. They tend to establish themselves on patches of soil that have a slightly higher clay content than surrounding areas, which is correlated with a higher content of organic matter and nutrients in this environment [Geiger et al., 1992]. Their presence also seems to be associated with old termite mounds in the subsoil, capturing the generally higher nutrient status of termite nests [Brouwer et al., 1993]. Tap roots reach down as much as 30 m, capturing soil moisture at such depths during the dry season [Alexandre and Ouedraogo, 1992; Deans et al., 1993]. Therefore, *F. albida* has a comparative advantage in capturing nutrients and water in an environment where both resources are limiting.

F. albida foliage is used as fodder for ruminant animals during the dry season. Pods are also good feed for ruminants and are commonly sold at local markets. Animals enjoy the shade, urinate and defecate more often under the *F. albida* trees. Birds congregate on tree branches, adding nutrient inputs to the soil under the tree canopy. *F. albida* may also fix nitrogen during the dry season with nodules located on roots several meters below the soil surface, near to the water table [Sprent et al., 1989; Dupuy and Dreyfus, 1992].

Leaves of this species have a nutritive value similar to other African browse legumes but with higher crude protein (20%) and lower contents of lignin and soluble polyphenolics than Australian acacias [Reed et al., 1992]. These investigators also showed that rams fed with *F. albida* grew at rates similar to rams fed with *Leucaena leucocephala*, *Sesbania sesban* and *Medicago sativa* hay. The relatively low lignin : nitrogen ratio (4 : 3) and the moderate soluble polyphenolics : nitrogen ratio (11 : 3), calculated from data of Reed et al. [1992], suggest a rapid rate of leaf litter decomposition at the onset of the rainy season. These processes result in the formation of 'islands of fertility' as the trees grow in size. There is huge genetic variability in this species but most trees grow slowly, such that it is estimated that the full effect takes more than 10 years to develop. Such islands of fertility, however, are considered to be indicators of an advanced stage in the process of desertification [Schlesinger et al., 1990].

When the rains start, decomposition of the litter and manure around the *F. albida* trees occurs rapidly, resulting in high mineralization rates. For example, alluvial soils in Malawi mineralized 125 kg N ha^{-1} in four months under trees in comparison with 47 kg N ha^{-1} in the open [ICRAF, 1993].

In addition, the microclimate under the defoliated trees also improves. Branches provide about 50% shade to the crops below the tree canopy, ameliorating excessively high soil temperatures that can kill germinating cereal seedlings, plus lowering leaf temperatures and transpiration rates during the hottest parts of the day, while allowing full solar radiation during the morning and afternoon [Vandenbelt and Williams, 1992]. Consequently, *F. albida* influences the crop microclimate positively.

Table 2 shows the different site properties of two *F. albida* parklands in Burkina Faso [Depommier et al., 1992]. One site is dominated by old trees. In both cases there are modest increases in soil organic matter, total nitrogen, available phosphorus and potassium. Sorghum yields under the canopy averaged 2.1 times higher than in the open.

There is little question about the positive tree–crop interaction in *F. albida*

Table 2. Site differences under the canopy and 15 m away from *Faidherbia albida* trees in two parklands in Burkina Faso. Adapted from Depommier et al. [1992].

Parameter	Park 3		Park 4	
Site characteristic				
Total area (ha)	0		2.2	
F. albida trees (ha^{-1})	7		19	
Average canopy size (m^2)	140		88	
Ground cover by *F. albida* (%)	9.8		16.5	
	Under	Away	Under	Away
Soil differences (0–20 cm)				
pH (water)	5.9	5.8	6.1	6.0
%SOM	2.7	2.4	1.6	1.6
%Total N	0.18	0.13	0.12	0.09
P (–g g^{-1})	14.0	10.6	8.3	6.1
Ca (cmol kg^{-1})	10.0	8.9	5.2	4.6
Mg (cmol kg^{-1})	3.9	3.9	2.3	2.3
K (cmol kg^{-1})	0.7	0.4	0.6	0.3
Grain yield				
Sorghum (t ha^{-1})	1.6	0.7	1.8	0.9

parklands, but it does not always happen [Brouwer et al., 1993]. The processes involved need to be fully quantified at well-characterized sites. How densely can the *F. albida* trees be planted and still provide the positive effects? How long does it take for the trees to start exerting such positive effects? Even more intriguing is the evidence of similar positive effects with two non-nitrogen-fixing tree species also present in the Sahelian parklands, karité or shea butter tree (*Vitellaria paradoxa*) and neré (*Parkia biglobosa*). Such tree effects on crop growth are not, according to visual observations, as spectacular as those of *F. albida*. Kessler [1992] suggests that the fertility effect of karité and neré is due to nutrient trapping instead of N-fixation. These two species do not exhibit reverse phenology; therefore, several processes are taking place and these need to be disaggregated and quantified.

Water, temperature and nutrient budgets, therefore, need to be assembled at several key sites and under the three tree species to ascertain how far the *F. albida* effect can be extrapolated and to what extent it can be manipulated. The time course of development of the crop effects also needs to be established. Observations, however, indicate that the effect diminishes and eventually disappears as the subhumid tropics are approached. Quantification is badly needed.

Other simultaneous systems

Tree–crop interactions in other simultaneous agroforestry systems are probably somewhere in between the generally negative scenario of alley cropping and the generally positive scenario of the *F. albida* parklands. Homegardens, boundary plantings, live fences, complex agroforests and silvopastoral systems should be characterized in terms of the tree–crop interaction equation, taking into account that the I parameter will be the sum of the different products out of those systems and not simply crop yields.

Minimizing competition is the paramount objective in simultaneous agroforestry systems. Perhaps the simplest way to achieve this is to choose tree species that compete less with crops for growth resources. One obvious example is the choice of *Grevillea robusta* vs. *Eucalyptus* spp. for boundary plantings in the East African highlands. *G. robusta* is known by farmers as ‘friend of the crop’, because it does not affect maize yields much, while *Eucalyptus* spp. in this ecosystem certainly do. *Eucalyptus* spp., as well as neem (*Azadirachta indica*) in the Sahel, can devastate cereal crops growing close to them when resources become scarce, because of their extensive root systems.

Silvopastoral systems are widespread simultaneous systems throughout the tropics [Budowski, 1993; Dzowela and Kwesiga, 1994; Singh et al., 1994]. The knowledge base about them, however, is essentially at the descriptive stage. Quantification of these important systems is badly needed as they are believed to be a key component of pasture recuperation in the humid tropics of Latin America [Serrão, 1993]. Trees in pastures often provide the major source of dry-season fodder in subhumid and semi-arid areas. Live fence posts are an important component of silvopastoral systems, providing fodder and decreasing the rates of deforestation caused by the need to replace rotten fence posts [Budowski, 1993]. Process-oriented research on silvopastoral systems is a high-priority research area.

Sequential systems: a different story

When the tree and the crop components of an agroforestry system grow at different times or at widely different rates, competition is decreased simply because the peak demands for growth resources occur at different times for each component. This may be achieved either by planting the components at different times or by planting them at the same time when one component grows rapidly while the other grows slowly. The first case is called relay intercropping in the agronomic literature [Andrews and Kassam, 1976]. The second case, perhaps the most common in sequential agroforestry systems, is usually called mixed intercropping. They are both classified as sequential agroforestry, for lack of a better term. Figure 1 shows this situation in a maize–*Cajanus cajan* system in Trinidad [Dalal, 1974]. Both species in this trial were planted at the same time but their initial growth rates were so different that their nutrient uptake patterns were really in sequence; when maize approached maximum nutrient uptake, *Cajanus cajan* initiated its period of fast growth, resulting in minimum competition for nutrient resources.

Shifting cultivation is also a sequential agroforestry system. After slashing and burning the forest, crops are planted between tree stumps, many of which resprout but are kept small by the farmer's machete. When farmers plant what they expect to be their last crop, they let the trees regrow as a secondary forest fallow for 10 to 20 years. There is little competition between crops and trees in traditional shifting cultivation. Trees in the fallows, because they are perennial and have extensive root systems, accumulate large quantities of nutrients in their biomass, which upon slashing and burning are transformed into ash, which acts as fertilizer for the next cropping cycle [Sanchez, 1976]. The demand for resources between the uptight city person (the crop) and the laid-back hillbilly (the trees) is nicely spaced in time. Unfortunately, traditional shifting cultivation is now almost an anthropological curiosity, having been replaced by various forms of slash-and-burn systems in disequilibrium, primarily because of shortening the fallow period as a consequence of pressures on the land [Bandy et al., 1993].

The tree–crop interaction equation should be modified for sequential agroforestry systems. Such an equation must include the variety of products obtained: grain crop yields, fuelwood, fruit, fodder and timber. Developing such an equation requires conceptual refinement to account for the recycling of nutrients and the fact that products are not harvested every year. Like the original one, this equation would be in biophysical terms and must be converted into monetary values as well. Two examples of sequential agroforestry

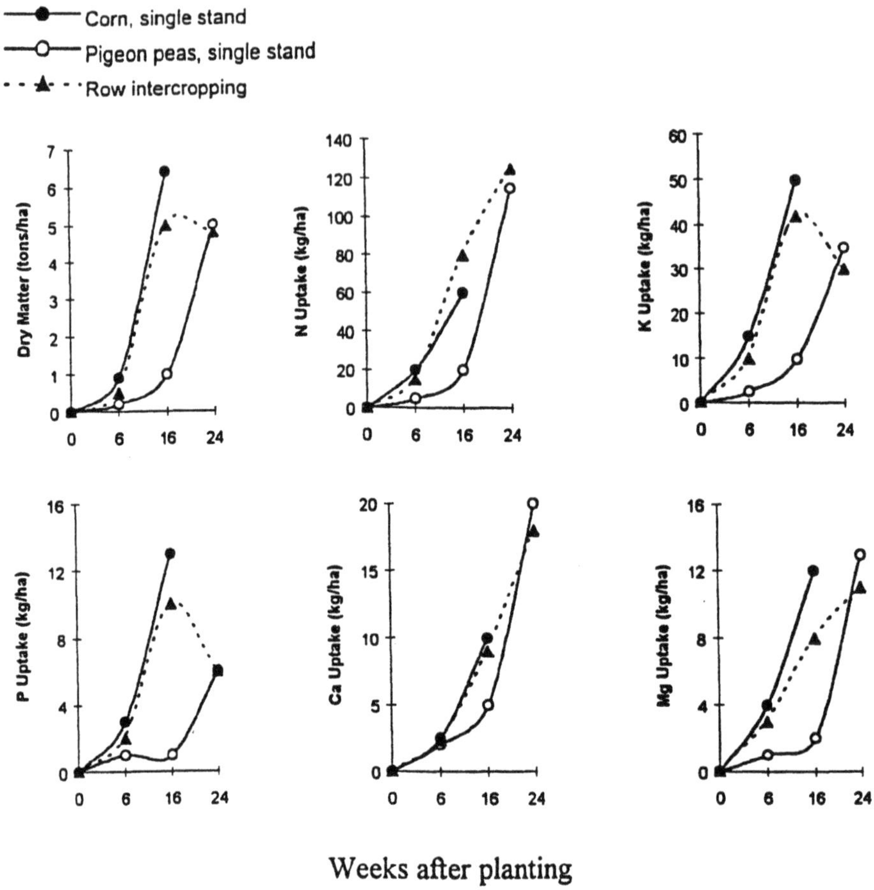

Fig. 1. Growth and nutrient uptake patterns of maize and pigeon peas (*Cajanus cajan*) in monoculture and row intercropping trial in Trinidad. *Source:* adapted from Dalal [1974].

systems merit attention because of their potential practicality: relay intercropping and improved fallows.

Relay intercropping in southern Africa

The vast subhumid plateau of southern Africa is a tropical savanna locally known as the 'miombo'. It is characterized by a rainy season four to six months long, followed by a long dry season, and by the dominance of trees of two leguminous genera, *Brachystegia* and *Julbernardia* [Huntley, 1982]. Average elevation is 1000 m and annual rainfall about 1000 mm; the soils, classified as Alfisols, Ultisols and Oxisols, are of moderate to low fertility. Typically, farmers grow maize during the rainy season and little is grown during the dry season. Maize responds strongly to nitrogen fertilizers, but most farmers have not been able to afford the recommended rates since the elimination of fertilizer subsidies. Frequent drought periods during the rainy season or delays in the start of the rains often decimate crop yields [ICRAF, 1993]. Diagnostic studies have indicated that low soil fertility, lack of fuelwood and lack of dry-season fodder are the main constraints perceived by farmers in the miombo [Dzowela and Kwesiga, 1994].

In densely populated areas of southern Malawi, farm size is extremely small, from 0.2 to 0.5 ha, precluding the possibility of fallow periods or cattle grazing. Nitrogen deficiency is the main fertility constraint. Given the doubtful on-farm results of alley cropping in this region, a sequential agroforestry alternative was sought to provide nitrogen inputs other than with fertilizers. *Sesbania sesban*, locally known as sesbania, grows fast and nodulates naturally in these soils. After considering the farmer labor calendar and in consultation with many farmers, the concept of relay intercropping maize with sesbania was developed.

Maize is planted in rows at normal spacing and population density, while sesbania seedlings are transplanted in rows at a rate of 7400 seedlings ha^{-1} shortly afterwards. Maize grows rapidly while sesbania establishes itself slowly, not competing much with the crop. After the first maize harvest, sesbania begins to grow rapidly and does so during much of the dry season when most fields are bare. To do this, sesbania taps water deep in the subsoil and fixes nitrogen, as evidenced by healthy nodules on its roots. *S. sesban* roots have reached a depth of 4 m in one year at Chipata, Zambia [Torquebiau and Kwesiga, submitted]. At Zomba, Malawi, sesbania trees accumulated about 1.7 t ha^{-1} of leaves, with very high nutrient concentrations (3.6% N and 0.37% P). Leaflets drop during the dry season, providing nutrient inputs in the order of 80 kg N ha^{-1} and 6 kg P ha^{-1} (J. A. Maghembe, unpub. data). Farmers then harvest about 1.7 t ha^{-1} of sesbania for fuelwood. At the onset of the next rainy season maize is planted and sesbania reseeds itself or is replanted, growing slowly and not competing much with the crop, because the crop for once has the competitive advantage. So far, this cycle has been repeated for five years.

The results of this system appear in Table 3, in which sesbania–maize intercropping is compared with sole maize to which no fertilizer is added and with sole maize to which fertilizer is added at the economically optimum rates of nitrogen and phosphorus [ICRAF, 1995]. There were no differences in maize yields during the first crop, because the trees were small and did not exert competitive or complementary effects. From the second year on, maize yields with sesbania intercropped were double those of the sole-crop maize, in both the presence and the absence of fertilizers. Drought was particularly acute in two years, 1992 and 1994, which resulted in widespread crop failures in southern Africa. Even in these years, maize yields doubled with sesbania intercropping in relation to the sole, unfertilized maize crop.

The positive effect of sesbania intercropping was also evident when the recommended fertilization was used. The five year average yields increased

Table 3. Performance of mixed intercropping system of maize and *Sesbania sesban* for five consecutive years (1989–1994) near Makoka, Zomba District, Malawi. The system is compared with sole maize with and without half the optimum fertilizer applications (28 kg N ha^{-1} and 5 kg P ha^{-1}). Maize grain yields and sesbania wood in tonnes ha^{-1}. Adapted from Maghembe, unpub.; ICRAF, 1993, pp 78–80; ICRAF, 1995.

Year	Seasonal rainfall (mm)	Sole maize no fert.	Sole maize + fert.	Maize + relay sesbania, no fert.	Maize + relay sesbania, no fert	SED	Wood[a]
1	1001	1.00	3.75	1.02	3.27*	0.40	2.08
2	1017	1.33	6.21	2.41*	6.13	0.60	0.05
3	551[b]	0.61	2.93	1.86**	3.54	0.39	2.24
4	962	1.02	4.79	2.33**	5.99*	0.51	3.01
5	522[b]	0.51	2.41	1.16*	3.29**	0.36	1.01
Mean	810	0.89	4.01	1.76	4.44		1.68

* Significant difference at 10%.
** Significant differences at 5%.
[a] Wood of *Sesbania sesban* relay-planted with unfertilized maize.
[b] Drought years (1992, 1994).

from 4.0 to 4.4 t ha^{-1} with sesbania intercropping plus fertilization. Fertilizing sole crops is still the most profitable option, but for those farmers who are unable to purchase chemical inputs, sesbania intercropping doubles yields during both good and bad years.

Fuelwood is another output of this system, which is currently valued at US$25 t^{-1} in Zomba District, Malawi. An average production of 1.7 t ha^{-1} yr^{-1} adds considerable value to sesbania intercropping.

The processes responsible for doubling maize yields in this sequential system have yet to be rigorously quantified. Nevertheless, on a yearly basis the I value of the tree–crop interaction equation must be over +100%, including the value of fuelwood. Growth and nutrient uptake curves of maize and sesbania need to be generated as well as quantification of the processes involved in nitrogen cycling.

Improved fallows in southern Africa

In less densely populated areas of the miombo woodlands, grass fallows of one to five years coexist with continuous maize cultivation, which no longer receives adequate fertilizers since the beginning of the structural adjustments. In a sense this is a form of shifting cultivation in tropical savannas. The strategy here was to use leguminous fallows to accumulate nitrogen in the biomass, smother weeds and improve soil physical properties [Kwesiga and Chisumpa, 1992]. Work done in Chipata, Zambia, demonstrated that two-year-old *S. sesban* fallows increased maize yields greatly compared with unfertilized maize [Kwesiga and Coe, 1994]. Results shown in Table 4 indicate that sesbania fallows doubled maize yields over a six year period, in comparison with continuous unfertilized maize production. This was accomplished in spite of two years without crop production while the sesbania was growing. Cost-benefit analysis also shows about twice the cumulative net benefit of the sesbania fallows over unfertilized maize (Table 4). This includes the value of sesbania wood. Sensitivity analysis indicates that two year fallows maintain their economic superiority over continuous, unfertilized maize under all realistic assumptions, including one or two drought years, fluctuations in maize and fuelwood prices, labor costs and discount rates [Place et al., 1994]. Therefore, this sequential system is also very promising; it is currently being tested extensively on-farm and is also being improved by farmers.

As in the case of relay intercropping, the processes responsible for such yield increases are yet to be quantified, but work has started on studies of litter decomposition, rooting depth, root productivity and changes in soil physical properties. Research is also in progress on alternative *Sesbania sesban* cultivars and on other species of *Sesbania*, cheaper establishment methods such as bare-root seedlings, and combining improved fallows with nitrogen fertilization to push yields to a higher plateau. Researchers are also examining farmer perceptions, local policies to protect the fallows from grazing and the overall adoption potential. A major constraint that remains to be resolved is

Table 4. Performance of two-year-old *Sesbania sesban* improved fallows in comparison with continuous maize production at Msekera, Chipata District, Zambia. *Sources*: Kwesiga and Coe [1994]; Place et al. [1994]; ICRAF [1995].

System	1988	1989	1990	1991	1992	1993	Sum of 6 yrs
Maize yield (t ha^{-1})							
Sesbania fallow	F	F	5.0	5.6	0.6	3.1	14.3
Continuous maize, no fert.	1.8	1.6	1.2	1.9	0.2	0.9	7.6
Continuous maize, 112 kg N ha^{-1} yr^{-1}	6.6	6.1	4.9	4.3	0.9	5.6	28.4
Net benefit US$ ha^{-1}							
Sesbania fallow[a]	–145	11	310	286	13	95	570
Continuous maize, no fert.	113	79	32	71	–10	7	289
Continuous maize, 112 kg N ha^{-1} yr^{-1}	460	343	200	134	–40	143	1240

[a] Includes wood production of 10 t ha^{-1} at harvest.

sesbania's susceptibility to root-knot nematodes (*Meloidogyne* sp.) as a host plant, which could harm sensitive crops like tobacco if they were planted in rotation. Screening for tolerance to nematode infestations is in progress in Malawi and Kenya.

Relay intercropping and improved fallows are promising alternatives for farmers in the miombo. The biophysical and socioeconomic processes involved must be fully quantified and understood before widespread adoption is promoted. These examples are described to emphasize the inherent advantage of sequential agroforestry systems over simultaneous ones in managing competition for the farmers' benefit. Such advantage, however, has not been satisfactorily quantified.

Complexity

Agroforestry research is full of complexity. The gist of the biophysical complexities has been described. In addition there are major socioeconomic and ecological complexities. Appendix 2 lists ten socioeconomic and ecological statements and hypotheses proposed by Izac [1994]; they will frame the discussions of the rest of this paper. When one adds to them the 17 biophysical hypotheses, complexity becomes obvious. The following section addresses some issues of socioeconomic, ecological and methodological complexity.

Socioeconomic complexity

The socioeconomic dimension of agroforestry is complex because of aspects related to temporal and spatial variability; scale factors; the multiplicity of products and services; the economic, social and ecological processes involved; the methods of characterization and diagnosis; and the diversity of institutions involved.

Agroforestry systems vary in space and time. Trees are seldom used uniformly across the farmer's land, and typically occupy discrete areas such as homegardens, contours, boundaries and fallow plots. Agroforestry frequently coexists with pure agricultural systems and pure forestry ones. Some sequential systems closely resemble agriculture in their early stages and forestry in their later stages.

The socioeconomic front is complex also because of the scale factor [Izac and Swift, 1994]. At one end is the farm household; at the other are the international pressures on national governments to 'internalize externalities' (assign monetary value to positive or negative consequences beyond the farm, such as siltation, loss of biodiversity and carbon sequestration). In between are two critical scales; one is the community or watershed scale with influences by local elders or their equivalents in the control of communal land. Community or watershed are seldom the same; one is a social demarcation while the other is geographical. The other critical scale is the national government where, for example, policies favoring timber exports may obliterate traditional community rights. The interactions between the different scales is indeed complex. Geographic information systems can help organize the data at the different spatial scales, aggregate or disaggregate them. This tool provides order to this difficult issue [Izac and Swift, 1994].

Another level of complexity is the multiplicity of products and services. Agroforestry systems not only produce food crops but also fruit, fodder, fuelwood, poles, timber and medicines; they also provide services: they conserve the soil and biodiversity and delineate boundaries. Most of the products have economic value, while most of the services have nonmonetary, environmental value.

Such systems, products and services are managed by different members of the household for their own purposes. The literature is full of descriptions of agroforestry systems at the household level [Nair, 1989], but rigorous analysis of the underlying socioeconomic processes involved is limited. Process-oriented research is not the monopoly of the biophysical sciences; it is equally important in the social sciences. The need to develop a predictive understanding on how farm households make decisions regarding land use is as essential as developing a predictive understanding of competition between tree and crop roots. In both cases there is a great deal of site specificity: cultural values, policies and income levels in one case; tree and crop species, climate and soil properties in the other. The general statements and hypotheses listed in Appendix 2 are primarily methodological; as the science evolves these hypotheses should become more process oriented.

Examples of socioeconomic processes are how households react to a diminishing resource base caused by nutrient depletion, how communities react to a different policy environment caused by increasing environmental consciousness by governments and how farmers intensify their land use in response to demographic pressure. Generalizations should be made about these processes to provide a predictive understanding beyond the specific cultural circumstances in which the data were obtained. This is what a science is all about.

The first step in agroforestry research, therefore, should be the characterization of the farmers' situations at a sufficiently detailed scale: watershed or regional. ICRAF pioneered the D&D (diagnosis and design) approach, which includes rapid rural appraisal and other methods. D&D is carried out by a team composed of at least an agronomist, a forester and a social scientist [Raintree, 1987; Avila and Minae, 1991]. Subsequent improvements added participatory approaches [Scherr, 1991], focus on gender issues [Kemerwa et al., 1994], and analytical (vs. descriptive) approaches based on quantitative and georeferenced data, combining both biophysical and socioeconomic parameters [Palm et al., 1995a].

Many of the diagnostic surveys are superficial in their analysis of constraints. It is not particularly useful to conclude that the main constraints facing farmers are low soil fertility, high soil erosion, lack of fodder, little fuelwood, unresponsive government policies, and limited access to credit and capital, because such conclusions apply to most tropical smallholder situations. Characterization and diagnosis should therefore be participatory, analytical and multidisciplinary. An example of this latter approach is the characterization methodology developed for the Alternatives to Slash-and-Burn Initiative [Palm et al., 1995a]. Such studies must be firmly based on relevant indigenous knowledge and also should be iterative. The latter is a characteristic of the social sciences that deals with the changing behavior of human beings (E. Zulberti, pers. comm.).

An operational dimension of additional complexity is the diversity of institutions involved in agroforestry research. Such institutions are attached to several public sectors such as agriculture, forestry, environment and soil conservation autofinance. Within a research institution or university, agroforestry cuts across many disciplinary or commodity-oriented departments. This is an integral part of natural resource management research.

Ecological complexity

Agroforestry systems exert two main ecological functions, productive and protective. The high total biomass of most agroforestry systems is responsible for the protective or buffering function [Kessler and Wiersum, 1993]. The scale and time factors of agroforestry add complexity from the ecological perspective [Fresco and Kroonenberg, 1992]. For example, the effects of a few trees on one farmer's field are different from those of thousands of trees on many farmer fields at the watershed scale. Ecological externalities such as stream sedimentation or, on the positive side, increases in plant species diversity, are seldom felt at the farm scale but they are certainly felt at higher spatial scales.

Ecological parameters interact with socioeconomic ones [Izac, 1994], building complexity upon complexity. This, however, is part of all natural resource management research, of which agroforestry is but one subject. Solid, scientifically sound approaches to natural resource management research are still in their infancy and must be fully developed.

In particular, next to nothing is known about the effects of different agroforestry systems, on the dynamics of pest–predator–disease interactions affecting food chains and life cycles that eventually determine sustainability and risk (R. Leakey, pers. comm.). The pantropical spread of the leucaena psyllid (*Heteropsylla cubana*) from its center of origin in the Caribbean to Hawaii in 1984, then to Asia and now in Africa since 1992 has caused major damage to *Leucaena leucocephala* [Vandenbeldt and Napompeth, 1992]. This, however, should not be viewed as an exceptional event. Exotic species leave their predators behind when they are moved away from their center of origin, but eventually predators catch up with them or new ones arise. Multipurpose tree species should not be expected to be pest or disease proof. Scientists should be prepared for such outbreaks by having a diverse genetic base available and they should alert the entomologists and plant pathologists associated with their research teams.

Methodological complexity

Methodological complexities are also part and parcel of agroforestry research. The challenge is to make research methods simple while remaining accurate and reproducible; experimental design and statistical analyses require extra care in agroforestry [Rao and Coe, 1992]. Oversimplification, however, has led to major shortcomings. One example is the inappropriate design of many alley-cropping trials, which has rendered them invalid as a result of root interference between plots [Coe, 1994]. There have been major methodological advances in measuring competition for light and water [Ong et al., in press]. Water-use measurement in agroforestry systems has been greatly simplified by straightforward methodologies that can follow water uptake of individual roots, branches and whole trees or crop plants [Ong et al., 1990; ICRAF, 1994].

The main methodological challenge is below ground, where straightforward ways to measure soil processes and root interactions are not well developed. Conventional soil fertility methods for assessing the availability of nitrogen and phosphorus are not sensitive enough to detect changes in low external input systems. New methods and approaches need to be developed [Sanchez, 1994a]. Relevant changes in soil organic matter content, for example, are seldom detected by conventional methods such as total soil carbon and total nitrogen determinations. Conventional phosphorus soil tests often fail to detect important changes in phosphorus availability. Recently, methods for evaluating soil biological processes have advanced much [Anderson and Ingram, 1993] but several parameters such as microbial biomass and labile organic phosphorus pools still lack replicability and threshold or critical levels. Simplifying root interactions is taking on a new dimension by the fractal branching approach proposed by Van Noordwijk et al. [1994] and Spek and Van Noordwijk [1994]. Methods for assessing the resilience and sustainability of agroforestry systems in economic, social and ecological terms remain to be developed.

Modelling in agroforestry is still in its infancy, in terms of both mechanistic and simulation models [Muetzelfeldt and Sinclair, 1993; Parton et al., 1994; Spek and Van Noordwijk, 1994; Ong et al., in press; Van Noordwijk, in press]. High priority should be given to developing robust models based on hard data, now that such data are becoming increasingly available.

Profitability

Agroforestry is also about putting money in farmers' pockets. This is in addition to the goals of providing food security, enhancing soil fertility, conserving soil and water, as well as increasing fodder and fuelwood production. Improved agroforestry systems should do all that, but to be sustainable, they must increase profitability in the medium term. Profitability is a necessary but not a sufficient condition of sustainability; in fact, many profitable systems are not sustainable in agronomic, social or ecological terms. Nevertheless, the most sustainable agroforestry systems are those that combine a high-value marketable product, such as coffee or cacao, with sound biophysical practices. The complex agroforests in Indonesia are perhaps the epitome of sustainability, with biodiversity levels approaching the original tropical rainforest they replaced after slash-and-burn [Michon and de Foresta, 1990; Michon et al., in press]. They provide farmers with a steady flow of income from jungle rubber, damar resins and fruits, all high-value–low-volume products that have assured markets and that are grown with little nutrient removal from the soil. Three research strategies to increase profitability in agroforestry are domestication, marketing and policy innovation.

Domestication

The first strategy is to introduce new products through the domestication of 'Cinderella' tropical tree species that have been largely overlooked by science [Leakey and Newton, 1994]. These include indigenous fruit trees and others that provide medicinal products and high-grade timber [Toledo, 1994]. Examples are the peach palm (*Bactris gasipaes*), indigenous to humid tropical Latin America, which produces food and animal feed from its fruits, 'heart of palm' from the apical meristem, medicines from its roots and wood for parquet floors [Clement, 1989; Szott et al., 1991; Mora Urpi et al., 1993; Villachica, 1994]. Other examples are camu-camu (*Myrciaria dubia*), a riverine Amazonian bush that produces berries with 40 times the vitamin C concentration of citrus [Peters and Vazquez, 1987; Peters, 1990; Ordoñez, 1994] and pygeum (*Prunus africana*), a tree from the African highlands whose bark produces an extract used as an ingredient in chemotherapy for prostate cancer [Cunningham and Mbenkum, 1993].

A wide variety of fruit trees are locally known but have limited international trade. They include the bush mango (*Irvingia gabonensis*) from the humid tropics of West Africa, which is traded regionally [Okafor, 1975]; cupuacú (*Theobroma grandiflora*) and arazá (*Eugenia stipitata*) from Amazonia [Clement and Villachica, 1994]; mahobohobo (*Uapaca kirkiana*), a source of wine in southern Africa [Maghembe et al., 1994], and karité (*Vitellaria paradoxa*), a Sahelian tree that produces oils used in the cosmetic industry [Bonkoungou, 1992].

Very little is known about how to convert most of these species into domesticated crops in agroforestry systems [Sinclair et al., 1994]. Research on reproductive biology and phenology is often in its infancy. Domestication in agroforestry consists of three stages [Leakey and Newton, 1994]. The first is the identification of potentially useful species through socioeconomic appraisal or ethnobotanical surveys [Franzel et al., 1995]. The second is the capture of germplasm by seed or vegetative propagation, followed by its conservation and selection, and the third is the incorporation into existing agroforestry systems. Instead of requiring many years as in traditional tree breeding, domestication can be done rapidly using the recently modified traditional horticultural techniques of vegetative propagation and clonal selection that maximize genetic diversity and minimize risk [Leakey, 1991; Leakey and Maghembe, 1995]. In the case of fruit trees, detailed physiological research is required to enable mature tissues to be easily propagated by cuttings [Leakey et al., 1994]. A list of species ICRAF is currently domesticating is shown in Appendix 3 as an example of the scope of such activities.

Domestication of indigenous fruit trees would in addition contribute to improved nutrition, particularly during the dry season when food supplies are limited and of low quality. Fruit of some of the species mentioned becomes available during the dry season (S. Minae, pers. comm.).

Marketing

Successful marketing of indigenous fruits and non-timber forest products is necessary for putting money in farmers' pockets [Toledo, 1994]. Market development of nontraditional products is best done by the private sector, but government policies can help remove key obstacles. An example is the relaxation of air freight regulations by the governments of Kenya and Zimbabwe, which opened European markets to horticultural products and flowers from these two countries [Tomich et al., 1994]. These issues were articulated at a recent workshop on the marketing of non-timber tree products (NTTPs), sponsored by the Alternatives to Slash-and-Burn Initiative (T. Tomich, pers. comm.). One conclusion was the need for policy-oriented research on the marketing of these products using existing methods rather than developing new methodologies. The research toolkits of agricultural economists and anthropologists are suitable for policy research on NTTP markets.

A second major conclusion was to focus on existing markets for NTTPs or on commodities for which NTTPs can be substituted. First, a demand must be created for new products, then the supply be created, not the other way around. The development of an international market for a totally new product, for example another kiwi fruit, is possible but risky. Perhaps the best approach is to expand on already existing national or regional markets, such as *Irvingia gabonensis* kernels in West Africa. Adoption is likely to be high when the products are already well known. It is easy, however, to saturate a local market by rapidly increasing production, causing prices to crash. Truly multipurpose trees also serve as a buffer against wide price fluctuations. *Bactris gasipaes* fruits can be used as animal feed when their price as food is low, or suckers can be allowed to grow from the base of the stem for heart-of-palm production. This is a major advantage over single-purpose trees like coffee.

Another approach is to focus on products that are ingredients of other products but not products themselves, for example, rubber or palm oil. The advantage is a wider market and less danger of market saturation. Environmentally friendly products can be successful, for example the 'green ivory' products of the tagua palm (*Phytelephas aequatorialis*) from the forests of Ecuador [Tangley, 1993]. Ingredients such as fruit concentrates or pulp and vitamin C are examples of this approach.

Regardless of approaches, research institutions should not undertake their own marketing initiatives to develop new NTTP markets. They should link with specialized NGOs and the private sector, who clearly have the comparative advantage, and back them up with policy research (T. Tomich, pers. comm.).

A way to promote long-term profitability is to ensure that processing or transformation facilities are located near the sources of production, so local communities benefit from the value added. This may be as rudimentary as simple drying and grinding facilities to transform products such as cassava roots, peach palm fruits and legume leaves into dried animal feed. The traditional on-farm coffee drying facilities of Colombia have been successfully adapted for such purposes in the Chapare region of the Bolivian Amazon. It does not make much sense to export water out of the humid tropics. Likewise pulp extraction of indigenous fruits and canning facilities for heart of palm exist in the Brazilian and Peruvian Amazon. Bark of the cinchona tree is no longer exported from Indonesia; a pharmaceutical facility at Bandung transforms it into quinine. These examples should gradually become the norm, as value-added products will strengthen rural communities.

Policy

Policy intervention is perhaps the most critical element for making agroforestry happen. Technology development is a necessary but not a sufficient condition for the widespread adoption of most agroforestry innovations. Policy research, therefore, is an essential component of science in agroforestry. Successful policy research in agroforestry must integrate two schools of thought, the traditional agricultural economics/agricultural anthropology approach with natural resource or environmental economics. When working in pure agriculture the author always enjoyed working with 'economists with chlorophyll', those who really understood farming. In agroforestry, they are much needed, but social scientists who are comfortable with the principles and jargon of soils, biodiversity and systems ecology are also necessary.

There are two broad kinds of policy research: policies to favor agroforestry's contribution to sustainable land use [Brooks et al., 1995], and those that aim to reverse existing perverse policies [Sanchez, 1994b]. Land and tree tenure, assistance to farmers during the time lag before trees become productive, and the valuation of environmental externalities are the three main policy research issues, each encompassing both kinds of policy research.

Tenure. Secure tenure, be it land tenure or tree tenure, is a prerequisite for planting trees and investing in agroforestry. Farmers are understandably reluctant to plant many trees if they are unsure about long-term access to their land. This is probably the first policy issue to consider [Gregersen et al., 1994]. There are often perverse policies related to land tenure; for example, no land titling without first deforesting the land; no rights to services when the land is classified as forest reserve even though farmers have used it for many generations [Sanchez, 1994b; Vosti and Witcover, in press]. Some common issues in Africa are the inability of women to plant trees, various disincentives of land fragmentation to tree planting and the conflicts between agroforestry plots and free cattle grazing [Place, 1994].

The time lag. One of the key policy issues is how to cope with the fact that economic returns from improved agroforestry systems are not likely to be realized during the first three to five years, before the trees become productive. What is the real discount rate farmers intuitively apply? Interventions to help farmers endure this difficult period before the trees exert their profitability and ecological functions is a major issue in agroforestry policy research. In some cases, farmers are perfectly willing to wait out this period, particularly when they have land tenure and fully value the importance of tree products or services. The challenge to policy research is to identify critical entry points and create appropriate incentives. One common constraint is access to planting materials. Incentives for tree nursery development by farmer groups with the use of the recommended germplasm is a high-priority policy research issue throughout the tropics. Recapitalization of soil fertility in nutrient-depleted lands is another.

Valuation of environmental externalities. Another major policy issue is how to reconcile the gap between the individual farmer's benefits and society's benefits arising out of agroforestry adoption, for example, improved water quality from erosion control. In such cases, society is likely to prefer more agroforestry than individual farmers may be willing to undertake. Policies can then be developed to provide additional incentives to farmers (F. Place, pers. comm.). The challenge to policy research is to identify the critical entry points, identify the benefits or costs external to the farm, assess their value and design mechanisms that create appropriate incentives for smallholders to incorporate environmental costs and benefits into their decisions (T. Tomich, pers. comm.). This challenge is not specific to agroforestry policy research; it applies to all natural resources research.

Sustainability

The concept of sustainability has profitability, productivity and environmental components, with a strong time dimension. Agroforestry systems that are profitable make farming systems more robust and less subject to perturbations. The diversity of outputs and the positive environmental attributes of trees on farms make them so. Although agroforestry interventions can make many agricultural systems more robust, the focus in this paper is on the less well-endowed areas where rainfall is erratic, the soil is not fertile, topography is difficult and farmers have limited land and capital resources. It is under such circumstances that agroforestry can show its mettle in terms of environmental sustainability, even with low use of external inputs. Sustainability can be described in terms of conserving the soil, enhancing biodiversity, conserving carbon in terrestrial ecosystems, and enhancing nutrient capture and retention.

Soil conservation

Keeping the soil resource in place is one of the major sustainability issues because it ensures site productivity and avoids negative downstream externalities such as siltation and eutrophication of surface waters and in some cases pollution of coastal marine resources. Several paradigm shifts are taking place in hillslope conservation management. Garrity and Van Noordwijk [1995] identified three: 1) the engineering approach has yielded to the biological approach; 2) the top-down watershed management approach is yielding to bottom-up approaches with a farmer or community focus; and 3) the pruned, leguminous alley-cropping concept of contour farming is diversifying toward a much wider array of contour hedgerow options.

Many agroforestry systems help keep the soil resource in place by biological instead of engineering means. Closely spaced trees on slopes reduce soil erosion by water through two main processes: first, as a physical barrier of stems, low branches, superficial roots and leaf litter against running surface water; second, as sites where water infiltrates faster because of generally better soil structure under trees than on adjacent land [ICRAF, 1994]. The scientifically acceptable evidence is overwhelming [Young, 1989a; Lal, 1989; Kiepe and Rao, 1994; ICRAF, 1994; Juo et al., 1994; and many others]. Trees, however, do not provide these functions until they are well established and have developed a litter layer [Sanchez et al., 1985]. Sequential agroforestry systems that include crops or ground cover while the trees are small can overcome this limitation.

Once established, most trees protect the soil constantly, provided they are healthy and the litter layer is not removed. Biomass transfer of tree leaf litter to cropped fields undermines this process [Nyathi and Campbell, 1993]. The perennial nature of tree root systems provides a dependable source of carbon substrate for micro-organisms in the rhizosphere; microbial mucilage binds soil particles into stable aggregates, which results in improved soil structure [Tisdall and Oades, 1982]. This important process takes place continually in agroforestry systems instead of temporarily, as in annual crop systems.

Some trees, however, exert a strong allelopathic effect on the underlying vegetation to the point that they actually promote erosion when their own litter layer is sparse. Black wattle (*Acacia mearnsii*) is one example; farmers in Tanzania regard it as an erosion-prone tree on slopes (K. Snyder, pers. comm.).

Trees also help control wind erosion by providing a physical barrier. Windy areas in many parts of the world have tree windbreaks. Two issues merit emphasis: one, the effect is not instantaneous; trees must develop first. Second, the key height against wind erosion is the first meter above the soil surface. Scientifically acceptable evidence about the effect of boundary plantings with trees in reducing wind erosion is available in the tropics [ICRISAT, 1990]. Windbreaks, however, can reduce crop production in adjacent areas when tree root systems extend into cropped fields. Very competitive trees such as neem (*Azadirachta indica*) and *Eucalyptus* spp. can decimate crop yields several meters away. Less competitive trees such as *Grevillea robusta* do not.

There is very clear evidence that contour hedges and alley cropping effectively control erosion as long as the hedges provide an effective barrier [Lal, 1989; Kiepe and Rao, 1994; Banda et al., 1994; ICRAF, 1994; Fujisaka et al., in press]. Erosion control can also be handled by other means, such as terraces, grass strips and ditches along the contour of slopes [Young, 1989a; Kiepe and Rao, 1994]. Contour hedges, often composed of fodder trees and grasses, also provide additional products such as fodder for farm animals, which is produced in areas that do not compete with crops. Fruit or timber trees grown as an upperstory in such contours also produce additional services and income [Garrity, 1993; Garrity and Mercado, 1994]. Contour hedges do require management although certainly less maintenance than earth terraces; the key is that they become a productive niche in the farm in addition to conserving the soil [Kiepe and Rao, 1994].

Controlling soil erosion biologically has an additional advantage: the slope between the hedges becomes less steep and even flat in some cases [Garrity, in press; Kiepe and Rao, 1994]. These 'biological terraces' are produced by taking advantage of the erosion process within the contour hedges, with the tree growth keeping up with the higher soil surface at the lower end, something nonbiological terraces cannot do. Scouring of the topsoil at the upper end of these terraces occurs, with detrimental effects on crop yields [Garrity, in press]. This is usually the result of hoeing downwards. Reversing uneven terrace productivity is a high-priority research issue in steeplands.

Soil conservation, although usually expressed in terms of physical sediment transport, has major fertility implications. For example, erosion and runoff accounts for about half of the negative nitrogen and phosphorus balance in typical Rwandan farms [Smaling, 1993]. This process, carried over decades of cultivation, is one of the main causes of the depletion of soil fertility in the uplands of Africa. Contour hedges can put a definite and lasting stop to that. Erosion control through agroforestry is a prerequisite to nutrient replenishment efforts. The use of phosphate rock as a capital investment in phosphorus-depleted soils of Africa [World Bank, 1994] must be accompanied by erosion control measures.

Enhancing biodiversity

Agroforestry systems are more diverse than crop or forest plantation monocultures. But the most important way agroforestry helps conserve plant and animal biodiversity lies in its preventive value, the savings in current biodiversity attained by reducing further clearing of tropical forests with the use of viable alternatives to slash-and-burn agriculture [Sanchez, 1994b; Schroeder, 1994; Vosti, 1995]. Farmers at the forest margins clear additional lands when their fields are no longer productive. One such alternative is the multistrata or complex agroforests where annual food crops are planted along with trees that eventually occupy different strata and produce high-value products such as fruits, resins, medicinals and high-grade timber. Food crops cover the ground quickly and grow until they are shaded out by the faster growing trees. There are many such multistrata systems throughout the humid tropics [Nair, 1989; Michon and de Foresta, 1990; Subler and Uhl, 1990; Sanchez, 1994b; Toledo, 1994]. For every hectare farmers put into such systems, several hectares of rainforest can be saved every year from the shifting cultivator's axe in the farmer's search for a livelihood. Precise estimates of these substitution values do not exist for agroforestry systems, although figures of 7.1 and 11.5 have been reported [Schroeder, 1993].

Agroforestry plays a major role in the reclamation of degraded and abandoned lands and is generally considered the most workable approach to mimic natural succession and increase biodiversity [Anderson, 1990]. Hard data on increasing biodiversity in degraded lands through agroforestry, however, are practically non-existent [Sanchez et al., 1994]. This is an important research imperative.

Agroforestry systems are more biodiverse above and below ground than monocultures, at least in the humid tropics. Plant diversity is in the order of 300 species per hectare in the mature complex rubber agroforests of Sumatra, Indonesia [E. Penot, pers. comm.). The level of plant biodiversity attained in such long-term agroforestry systems by far exceeds that of rubber plantations (5 species ha^{-1}) but is still less than that of adjacent undisturbed forests, with 420 plant species ha^{-1}. The richness of bird species in 50-year-old damar agroforests is 70% that of the original rainforest and almost all mammal species are present in the agroforest [Michon et al., in press]. This is possible because such agroforests, composed of hundreds of small plots managed by individual families, occupy contiguous areas of several thousand hectares in Sumatra. Tracks of the rare Sumatran rhino (*Dicerorhinus sumatrensis*) were recently discovered in one of these rubber agroforests, implying that they may provide a habitat similar to the natural rainforest [Sibuea, 1995]. Such high biodiversity levels, however, cannot be expected of shorter duration agroforestry systems, such as improved fallows or in less geographically extensive systems [Sinclair et al., 1994].

Below-ground biodiversity is higher in agroforestry systems than in crop monocultures, approximating the levels of the natural forest in the Amazon

[Lavelle and Pashanasi, 1989] as well as in Indonesia [Michon et al., in press]. Soil macrofauna and microflora are key regulators of the basic decomposition processes that provide nutrients to higher plants and animals, but methodological difficulties in their quantification remain. Microbial biomass is concentrated in the rhizosphere and in fertility 'hot spots', which soil analysis does not detect, but plant roots certainly do. It is difficult, however, to apply molecular biology methods to soil microbial biomass, because of the current inability of isolating DNA from soil organisms in sufficient quantity and purity [Coleman et al., 1994]. Overcoming these obstacles will enhance our understanding of microbial biomass diversity and its role in soil processes. While not as attractive as 'furry and feathered creatures', soil communities are a major and largely ignored component of biodiversity conservation and ecosystem functioning. Quantifying the effect of agroforestry interventions on below-ground biodiversity is a research imperative.

Maintaining carbon in the terrestrial ecosystem

Agroforestry systems play an important role in keeping carbon in the terrestrial ecosystem and out of the atmosphere. This is accomplished by preventing further deforestation and by accumulating biomass and soil carbon [Schroeder, 1994]. Few hard data, however, are available on the magnitude of these processes.

As in the case of biodiversity conservation, the main contribution of improved agroforestry systems to terrestrial carbon conservation comes from its preventive effect, that is, the area of natural forests that will not be cleared because farmers can make continuous use of already cleared land through improved agroforestry systems [Schroeder, 1993; Unruh et al., 1993; Sanchez, 1994b]. Recent estimates indicate that tropical deforestation contributes approximately 20% to global warming and that this percentage is expected to increase in the coming decades [Dale et al., 1993]. One hectare of humid tropical forests contains on average 160 t ha^{-1} of carbon in the above-ground biomass [Houghton et al., 1987]. When it is slashed and burned, most of the carbon is emitted to the atmosphere either immediately during the burn or gradually through the decomposition of unburned logs and branches. Keeping this carbon resource (some 96 billion tonnes in the remaining humid tropical forest biomass) in situ is of critical importance.

Complex agroforestry systems of long duration such as the jungle rubber and damar agroforests of Sumatra and multistrata systems throughout the humid tropics can sequester carbon in their tree biomass which remains sequestered for decades. Again, quantification is lacking.

Soil organic matter can act both as a source and a sink of carbon. Contrary to popular opinion, total soil organic matter contents are not uniformly low in the tropics [Sanchez and Buol, 1975]. Soils contain as much carbon as the above-ground vegetation in humid tropical forest ecosystems, an average of 145 t ha^{-1} of carbon [Sombroek et al., 1993]. Deforestation increases topsoil carbon emissions to the atmosphere because of higher soil temperature, which leads to accelerated organic matter decomposition [Swift et al., 1979].

Topsoil carbon decreases of 25 to 40% are commonly measured in the top 25 cm during the first two to three years after slash-and-burn following conversion to agriculture, pastures or tree plantations [Sanchez et al., 1983, 1985]. Afterwards soil carbon will either reach equilibrium, increase or decrease, depending on the quantity and quality of organic inputs, including crop residues and the decomposition rate of soil organic matter [Sanchez et al., 1989]. In the cases of continuous cropping and plantation forestry, evidence indicates real decreases in topsoil carbon [Sanchez et al., 1983, 1985]. In the case of improved pastures, however, increases have been recorded [Serrão and Toledo, 1990; Serrão, 1993]. Similar increases also happen in mature secondary forest fallows that approximate the biomass level of primary forests [Nye and Greenland, 1960]. There are, however, no hard data about topsoil carbon dynamics in agroforestry systems.

Subsoil carbon is a major pool in tropical soils because of the considerable depth of the dominant soils of the humid and subhumid tropics. Amounts of subsoil carbon in the soil layer at 25–200 cm depth are about twice as large as the amounts found in the top 25 cm of the dominant Oxisols and Ultisols of the Amazon. This is because the greater volume of the subsoil more than compensates for its lower carbon concentration [Sombroek et al., 1993]. Subsoil carbon contents generally do not decrease upon deforestation and can actually increase with new land-use systems that bring extensive root systems to depths that were previously essentially devoid of roots. This is the case with the introduction of aluminum-tolerant grasses in the Llanos Orientales of Colombia. Fisher et al. [1994] report that grass–legume pastures of *Andropogon gayanus–Stylosanthes capitata* and *Brachiaria humidicola–Arachis pintoy* sequestered an additional 34 t C ha^{-1} in the subsoil layer at 40–100 cm depth in 9 to 13 years of grazing. They attribute the additional carbon to root turnover of the grasses that developed deep root systems in these aluminum-toxic Oxisols.

Trees certainly develop deep root systems in semi-arid areas but not so much in the humid tropics, where water is more available and where nutrients are concentrated in the topsoil. Thus, trees may perform a carbon-sequestration function similar to that of the aluminum-tolerant tropical pastures. Root studies in Pucallpa, Peru, indicate that peach palm (*Bactris gasipaes*) has as extensive but deeper rooting pattern than the aluminum-tolerant grass *Brachiaria dictyioneura* [Riesco and Ara, 1994]. The deep-rooting pattern of leguminous fallows in southern Africa could also result in carbon sequestration. The accumulation of root dry matter in the top 100 cm of an Alfisol in Zimbabwe by two year *Sesbania sesban* and *Acacia angustissima* fallows was 3 and 6 t ha^{-1}, respectively [B. Dzowela, unpub. results]. Data should be gathered on this important topic in agroforestry systems, in both humid and drier areas.

Soil organic carbon consists of a series of functional pools or fractions that perform different functions [Parton et al., 1994]. The more active fractions are responsible for nutrient release processes while the slow and passive fractions are more responsible for soil structure development and maintenance. The decomposed products of the organic input are distributed in the different pools, depending on litter quality, soil texture and clay mineralogy. The effect of agroforestry systems on the distribution of newly fixed carbon into the different pools is just beginning to be studied [Barrios et al., in press]. For carbon sequestration, the aim is to increase the size of the slow and passive soil carbon pools, since most of the carbon added to the active pools will decompose in less than two years.

In most agroforestry systems, carbon sequestration should be considered a byproduct rather than the principal objective. The magnitude of carbon sequestration by agroforestry, however, is considered among the highest of land-use systems by climate change researchers. Unruh et al. [1993] performed complex calculations of agroforestry systems in Africa, their biomass accumulation and their potential distribution using GIS techniques. Their results suggest a huge amount of carbon can be sequestered, ranging from 8 to 54 Gt (billion tonnes) of C in a total of 1.55 billion hectares where agroforestry could potentially be practised. Above-ground biomass would sequester 3–15 t ha^{-1} of C, while soil and root biomass would sequester an additional 1.3–6.5 t ha^{-1} of carbon. Another estimate by Schroeder [1993, 1994] suggests above-ground accumulation rates of 9 t C ha^{-1} in five years for the semi-arid tropics, 21 t C ha^{-1} in eight years for the subhumid tropics and 50 t C ha^{-1} in five years for the humid tropics but gave no below-ground estimates. Above- and below-ground carbon sequestration values, therefore, need to be generated locally, taking into account the duration of each agroforestry system, and extrapolated geographically in a realistic fashion, based on actual rates of agroforestry adoption.

Enhancing nutrient capture and retention

Soil nutrients are considered among the least resilient components of sustainability [Fresco and Kroonenberg, 1992]. A fundamental principle of sustainability, therefore, is to return to the soil the nutrients removed from it through harvests, runoff, erosion, leaching, denitrification and other loss pathways [Sanchez, 1994a]. Nutrient budgets are a tool for determining whether there is an adequate balance or whether nutrients are being depleted. In natural rainforest ecosystems the nutrient balance is adequate, while in intensively fertilized cropping systems of the temperate region a large, positive balance results in nitrate and phosphorus pollution [Sanchez, 1994a]. In the small farms of Africa, the net balance of nitrogen, phosphorus and potassium is negative, which is responsible for the widespread nutrient depletion throughout tropical Africa [Smaling, 1993]. Can agroforestry help? Early results are encouraging, particularly for nitrogen.

Nitrogen. Trees can provide additional nitrogen inputs via two processes: biological nitrogen fixation (BNF) and deep nutrient capture. Although the magnitude of BNF is methodologically difficult to quantify, some studies give ample proof of it [Ladha et al., 1993]. Empirical evidence, such as presence of active nodules and nitrogen uptake of leguminous species of the Papilionaceae and Mimosaceae families, suggest that BNF can supply considerable nitrogen inputs to crops via litter, in soils sufficiently well supplied with phosphorus. This is a definite nutrient input.

Deep nutrient capture is the uptake of nutrients by tree roots at depths where crop roots are not active. It can be considered an additional nutrient input in agroforestry systems, because such nutrients are leached as far as the crop is concerned. They become an input upon being transferred to the soil via tree litter decomposition. Evidence of this process in *Faidherbia albida* parklands has been described in earlier sections of this paper. An exciting dimension has recently been discovered in western Kenya for subsoil nitrate in nitrogen-deficient Nitisols (red Alfisols and Oxisols with high iron content). Hartemink et al. [in press] detected nitrate levels in the order of 120 kg N ha^{-1} in the 50–200-cm layer. They also found that *Sesbania sesban* fallows depleted this pool, thus capturing a resource that was unavailable to maize. The source of this nitrate pool is believed to be the result of the mineralization of topsoil organic nitrogen, which is relatively high in these soils; subsequent nitrate leaching from topsoil layers, particularly at the start of the rainy season; and nitrate sorption by positively charged subsoil clay surfaces. What the trees have done in effect is to expand the volume of soil used by a considerable amount [Cahn et al., 1992].

Nitrogen-fixing trees may have been oversold in comparison with non-fixing trees, particularly those of the Caesalpiniaceae family [Garrity and Mercado, 1994]. There is ample evidence that non-fixing trees, including several species of *Cassia (Senna)*, accumulate as much or more nitrogen in their leaves than nitrogen-fixing legumes, presumably because of their greater root volume [Szott et al., 1991; Garrity and Mercado, 1994].

Can agroforestry, through BNF and deep nutrient capture, eliminate the need for application of nitrogen fertilizers to crops? The answer depends largely on crop yield-level expectation. A typical maize crop in smallholder African farms yields 1 t ha^{-1} of grain and requires a plant accumulation of about 40 kg N ha^{-1}. A crop of 4 t ha^{-1} requires 80 kg N ha^{-1}, and a crop of 7 t ha^{-1} requires 200 kg N ha^{-1} [Sanchez, 1976]. Topsoil nitrogen mineralization provides an average input of about 30 kg N ha^{-1}, which approximates to the needs of a maize crop of 1 t ha^{-1}. Nitrogen inputs from 4 t ha^{-1} of leguminous leaf mulch dry matter range from 60 to 150 kg N ha^{-1} [Palm 1995, this issue]. This mulch input rate is a realistic one and the one actually recorded in the *Sesbania seban* improved fallows in Chipata, Zambia, discussed previously, which added 128 kg N ha^{-1} to the maize crop, resulting in a 4 t ha^{-1} maize yield. The fertilizer recommendation in that area is 112 kg N ha^{-1} , about the same as the nitrogen added by the *S. sesban* fallow.

Since the recovery of leguminous leaf nitrogen incorporated into the soil is about the same as fertilizer N (about 20 to 50%, Palm [1995, this issue]), agroforestry can replace fertilizer nitrogen applications at the grain-yield level for maize of 4 t ha^{-1}. At high yield levels comparable to commercial farms in the industrialised world, say 7 t ha^{-1}, organic nitrogen inputs are likely to be insufficient and must be supplemented by inorganic fertilizers [Szott and Kass, 1993]. The interaction between organic and inorganic sources of nutrients is essentially a new subject of research in the tropics. Very little is known about it, because previous research has mainly compared one source against the other.

Organic inputs, however, have an important advantage over inorganic fertilizers in sustainability terms. Much of the remaining 50 to 80% of the applied organic nitrogen not utilized by crops is incorporated into active pools of soil organic matter, because these mulches also provide a carbon source needed as energy for microbial immobilization [Palm, 1995, this issue]. Readily available or soluble carbon is often low in nutrient-depleted soils, even though the total organic matter contents appear normal. Soil micro-organisms need a carbon substrate to form organic nitrogen. Inorganic fertilizers do not contain such carbon sources; therefore, most of the fertilizer nitrogen not used by crops is subject to leaching and denitrification losses, while much of the nitrogen released from organic inputs and not utilized by crops will be 'saved' as soil organic nitrogen.

The slow accumulation of soil organic nitrogen with agroforestry organic inputs is likely to make a difference in terms of long-term sustainability. This strategy is not new and has been used for centuries in temperate-region agriculture, with crop rotations and winter leguminous cover crops. What is new is the potential to do something similar in the tropics with low-input systems that fit with farmer perspectives. The potential for improved nitrogen management needs to be quantified in agroforestry systems by measuring processes such as mineralization, immobilization, denitrification, volatilization and leaching, along with changes in the soil organic nitrogen pools in systems combining organic and inorganic sources of nitrogen.

Phosphorus. Agroforestry, however, cannot supply most of the phosphorus inputs required by crops. Deep capture of phosphorus is likely to be negligible because of the very low concentration of available phosphorus in the subsoil. Improved fallows and other sequential systems do accumulate phosphorus in their biomass and return it to the soil via litter decomposition. But this is recycling and does not constitute an input from outside the system. This situation is compounded by the physiological fact that most of the phosphorus accumulated by cereal crops and grain legumes is in the grain and is thus removed from the soil at harvest. The proportion of phosphorus recycled back to the soil in basic grain crops, assuming complete crop residue return, is in the order of 38%. Comparable figures for other nutrient elements are 54% for nitrogen, 89% for potassium, 95% for calcium and 70% for magnesium

[Sanchez and Benites, 1987]. Therefore, phosphorus is often the key critical nutrient in agroforestry and other low external-input systems.

Leguminous mulches applied at a realistic rate of 4 t ha^{-1} provide 8 to 12 kg P ha^{-1}, about half the phosphorus requirements of a maize grain crop of 4 t ha^{-1}, which accumulates 18 kg P ha^{-1} [Palm et al., 1991; Palm, 1995, this issue]. Therefore, inorganic sources of phosphorus must be applied to agroforestry systems in soils depleted of this element. The strategy is to utilize at maximum all the available organic sources first, including manures, and supplement the difference with phosphorus fertilizers [Sanchez, 1994a]. Interactions between organic and inorganic sources of phosphorus also need to be quantified.

In soils with high phosphorus fixation capacity by iron and aluminium oxides (which can usually be identified as having red, clayey topsoils) a different strategy is possible. Large applications of indigenous rock phosphates may replenish the phosphorus supply of these soils, the phosphates being gradually released by desorption from the oxide clay surfaces to plants for the next five to ten years. This is being considered as a new approach, investing in natural resource capital [World Bank, 1994].

The soil conservation dimension of many agroforestry systems ensures that nutrient inputs added through biological nitrogen fixation, deep nutrient capture or phosphorus fertilizers are not lost by runoff and erosion. A phosphorus investment program that does not include contour hedges or other erosion control technologies is likely to do more harm than good.

Nutrient budgets. Individual nutrient budgets at the field scale are useful, but whole-farm nutrient budgets are better. Small farms in Kenya often mix crop with milk production in zero-grazing systems, with several nutrient compartments and flows between them. Examples are the cropped fields, patches of napier grass (*Pennisetum purpureum*), contour hedges, boundary plantings, the cattle enclosure (boma) and a compost pile where some crop residues and household waste are added. Shepherd et al. [in press] modelled nutrient budgets for conventional practices and improved agroforestry systems, which consisted of fodder legumes as contour hedges and under large trees along boundaries for livestock feed, and nitrogen and phosphorus fertilizers for the grain crops. The results, presented in Table 5, show the different compartments that were measured or estimated as an example of the type of data that should be collected. Agroforestry interventions decreased the net negative nitrogen balance from –76 to –45 kg N ha^{-1} a^{-1} but transformed a net negative phosphorus balance of –3.8 kg P into a positive 9.4 kg P ha^{-1} a^{-1}. Although the results are far from ideal, they suggest that agroforestry systems can improve nutrient balance.

One way to balance nutrient budgets is with fertilizer applications in agroforestry systems. Fertilizer response has been most commonly reported in alley-cropping systems and perennial shade systems [Szott and Kass, 1993]. Nutrient balances in alley cropping are usually negative, particularly with

Table 5. Annual soil N and P budgets for the field compartment in an existing farm system and an improved agroforestry system in Western Kenya. *Source:* Shepherd et al., in press.

Components	kg N ha^{-1} a^{-1}		kg	
	Existing	Improved	Existing	Improved
Rock weathering	–	–	0.1	0.1
Deep uptake	0	69	–	–
Biological fixation	6	69	–	–
Asymbiotic fixation	4	4	–	–
Atmospheric deposition	6	6	1.0	1.0
Fertilizer application	0	115	0	35.0
Sum inputs	16	263	1.1	36.1
Faeces deposited	3	0	0.5	0
Urine deposited	2	0	–	–
Manure from boma	3	31	0.4	6.7
Urine from boma pit	0	31	–	–
Compost	6	116	2.0	27.0
Sum transfers in	14	178	2.9	33.7
Denitrification	22	39	–	–
Volatilization	2	0	–	–
Leaching	39	69	–	–
Sum outputs	63	108	0	0
Tree leaf fed to animals	0	48	0	2.4
Fuelwood	0	4	0	0.3
Crop harvest, fodder	0	224	0	45.1
Crop harvest, grain	26	75	3.5	9.9
Crop residues, fodder	4	27	0.4	2.7
Crop residues, fuel	2	0	0.2	0
Erosion to hedgerow	11	0	3.7	0
Sum transfers out	43	378	7.8	60.4
Balance	–76	–45	–3.8	+9.4

phosphorus [Palm et al., 1991; Woomer et al., in press], so fertilizer responses are common. Intensive fertilization is the norm in the very successful multistrata systems practised by descendants of Japanese settlers in the Amazon of Brazil, commonly referred to as the paragon of sustainability in that region [Subler and Uhl, 1990].

Conclusions

Agroforestry is well on its way to becoming a specialized science at a level similar to that of crop science, but is perhaps more exciting because of its multidisciplinary mix. The advances in process-oriented research since the beginning of this decade are impressive, but there is still a long way to go. The major gaps are research dealing with complexity, profitability and sustainability (both biophysical and socioeconomic) and in the development of testable hypotheses in socioeconomic aspects. The following research needs were outlined in the four sections of this paper and are summarized below.

Competition

- The biophysical research hypotheses listed in Appendix 1 must be thoroughly tested in specific agroforestry systems under a representative range of biophysical and socioeconomic conditions.
- The tree–crop interaction equation needs further conceptual development, particularly in relation to sequential systems.
- Long-term trials with the four key treatments of this equation need to be conducted in agroforestry systems, including the *Faidherbia albida* parklands.
- Particular care needs to be given to experimental design in agroforestry to avoid root interference between experimental plots.
- Process-oriented research on silvopastoral systems remains a major gap in work among the types of agroforestry systems.
- The processes responsible for major crop-yield increases in promising sequential systems and their socioeconomic implications must be thoroughly understood and quantified before they are recommended for wide-scale adoption.

Complexity

- Process-oriented research on socioeconomic complexities should be strengthened, particularly research on how households or communities react to policy environments.
- Characterization and diagnosis studies should be *participatory, analytical and multidisciplinary*, with due attention to indigenous knowledge and gender issues.
- Geographic information systems help organize data at the different spatial scales needed in agroforestry research.
- Research on the dynamics of pest-predator-disease interactions needs to be systematically initiated.
- Reliable and straightforward ways for measuring soil processes and root interactions need further development.
- High priority should be given to developing robust agroforestry models based on hard data, now that such data are becoming increasingly available.

Profitability

- Domestication of tropical tree species that produce high-value products such as fruits, medicines and high-grade timber is a research approach that will increase money in farmers' pockets and in some cases improve human nutrition.
- Market development of new, non-timber forest products is best done by NGOs or the private sector.
- Policy research on market development should focus on existing markets, on ingredients rather than totally new products and on local transformations for value-added production.
- Policy research is an essential component of science in agroforestry and a prerequisite for adoption, in most circumstances. Priority research areas are land or tree tenure, assistance to farmers during the time-lag before trees become productive, and the valuation of environmental externalities.

Sustainability

- Soil conservation is undergoing several paradigm shifts that require modifications in the research agenda: the engineering approach has yielded to the biological approach; the top-down watershed approach is yielding to bottom-up farmer- or community-focused approach; and the alley-cropping concept of contour farming is diversifying into an array of contour hedge options.
- Estimates of substitution values of agroforestry in preventing additional deforestation need to be developed to assess positive externalities of biodiversity conservation and prevention of greenhouse gas emissions.
- Hard data need to be collected on the increases in plant and animal biodiversity upon reclaiming degraded lands through agroforestry.
- The effects of agroforestry interventions on below-ground biodiversity need to be quantified.
- Above- and below-ground carbon sequestration values need to be generated locally, taking into account the duration of each agroforestry system, and extrapolated geographically based on realistic rates of agroforestry adoption.
- Several agroforestry systems can meet the nitrogen requirements of moderate yields of grain crops. Quantification is needed on factors affecting components of the nitrogen cycle, including the magnitude of inputs from biological N fixation, deep nitrate capture, litter decomposition and the fate of nitrogen not taken up by the crop but used to build soil organic nitrogen capital.
- As agroforestry is unable to meet about half of the phosphorous requirements of such crops, research should focus of the interaction between

organic and inorganic phosphorus inputs, using rock phosphates as an investment in nutrient capital.

The alley-cropping experience has done considerable harm to the reputation of agroforestry research, but scientists have learned much from it in terms of research approaches and methodologies. Care must be taken not to oversell the new and exciting technologies currently being developed, and to ensure that the biophysical and socioeconomic processes involved are satisfactorily understood. It is hoped that as confidence increases with time, new, scientifically solid results will become the norm rather than the exception in agroforestry research.

Acknowledgements

Many colleagues have reviewed drafts of this paper and provided helpful suggestions. The author would like to acknowledge their help and thank the following who made particularly insightful or extremely thorough comments: Edmundo Barrios, Roland Buresh, Hubert de Foresta, Steve Franzel, Sam Fujisaka, Dennis Garrity, Anne-Marie Izac, Bert Janssen, Roger Leakey, Susan Minae, Chin Ong, Cheryl Palm, Frank Place, Keith Shepherd, Fergus Sinclair, Paul Smithson, Thomas Tomich and Meine Van Noordwijk.

Note

1. Kang et al., 1981, 1985; Yamoah et al., 1986; Mulongoy and Van der Meersch, 1988; Lal, 1989; Atta-Krah, 1990; Kang et al., 1990; Lawson and Kang, 1990; Palm and Sanchez, 1990, 1991; Ong et al., 1991b; Szott et al., 1991; Hariah et al., 1992; Fernandes et al., 1993; Haggar et al., 1993; Hauser, 1993; Kang, 1993; Rao et al.,1993; Salazar et al., 1993; Jabbar et al., 1994; Juo et al., 1994; Kass et al., in press; Sanginga and Mulongoy, in press; Palm, 1995, this issue, and many others.

Appendix 1

Biophysical agroforestry hypotheses. Adapted from the following sources: Hypotheses 1–10 from Young [1989b]; 11 from Wilson [1990]; 12 from Van Noordwijk and Dommergues [1990]; 13–18 from Ong et al. [1995].

Hypothesis	Status and references
1. Agroforestry systems can control soil erosion.	1. Proven in contour hedgerows and multistrata systems [Roose, 1970; Lal, 1989; Young, 1989b; Alegre and Fernandes, 1991; Banda et al., 1994; ICRAF, 1994: 80–88; Juo et al., 1994]; Kiepe and Rao, 1994.

Appendix 1 (Continued)

Hypothesis	Status and references
2. Agroforestry systems can maintain soil organic matter (SOM] at levels satisfactory for soil fertility.	2. Not proven. Too simplistic, since there are no reliable SOM levels related to satisfactory soil fertility. SOM increases have been detected temporarily in sandy soils under alley cropping [Lal, 1989; Kang et al., 1990;] but not for other soils [Rao, in press]. Relevant proof should be in terms of functional SOM pools in relation to system nutrient uptake and overall productivity.
3. Agroforestry systems maintain more favorable soil physical properties than agricultural systems.	3. Partially proven, for soils under contour hedgerows in relation to adjacent cultivated fields [Van Noordwijk et al., 1992; ICRAF, 1994: 85–86].
4. Nitrogen-fixing trees can substantially augment nitrogen inputs in agroforestry systems.	4. Proven [Ladha et al., 1993]. Limited quantification of N-fixation by legume species and subsequent biomass N accumulation and return to the soil via litter.
5. Trees in agroforestry systems provide deep nutrient capture from subsoil layers that are inaccessible to crop roots.	5. Proven for deep nitrate capture in oxic subsoils with positive charge [Hartemink et al., in press; ICRAF, 1995]. Not yet proven widely and unlikely to be relevant in other infertile subsoils.
6. Agroforestry systems can lead to more closed nutrient cycling and to more efficient use of nutrients and less leaching losses.	6. Not quantitatively proven. Highly probable in many systems, but the data are not there.
7. The cycling of bases accumulated by trees in agroforestry systems and returned to the soil as litter can help reduce soil acidity.	7. Proven for litter high in calcium and magnesium in non-agroforestry systems [Sanchez et al., 1985]. Decomposition of leaf litter produces metabolic organic products that temporarily complex aluminum in the soil solution thus decreasing soil acidity, for short time periods [Davelouis et al., 1991; Wong et al., in press].
8. Agroforestry is a useful component of systems for the reclamation of degraded soils.	8. Too general, probably will be proven in many circumstances. Proven for saline and alkaline soils [Singh et al., 1994] Proven for nitrogen-depleted soils of Eastern Zambia with sesbania fallows [Kwesiga and Coe, 1994].
9. The role of tree roots is as important as that of above-ground biomass in soil fertility maintenance.	9. Not proven. An important research topic.
10. Shade from tree canopy improves soil biological activity and nitrogen mineralization.	10. Proven [Wilson, 1990; ICRAF, 1993: 58–60].

Appendix 1 (Continued)

Hypothesis	Status and references
11. Roots of N-fixing trees have more nodules when in close contact with roots of non N-fixing plants. This may lead to direct N transfer to the non-nodulating plant.	11. Not proven. Highly controversial.
12. Annual crops are unable to use all water stored in the soil.	12. Proven for shallow rooted crops [Ong et al., in press].
13. The combination of trees and crops greatly enhances rainfall water use.	13. Proven [ICRAF, 1994: 67–73; Ong et al., in press].
14. Because agroforestry systems use more water than annual cropping systems, they should increase primary productivity.	14. Being tested; depends on competition.
15. There is less competition between tree and crop species that develop canopies at different times.	15. Proven [Dalal, 1974].
16. Competition for water in agroforestry systems can be reduced by modifying the spatial arrangement of trees.	16. Proven [ICRAF, 1994: 67–73].

Appendix 2

Socioeconomic and ecological guiding principles and hypotheses in agroforestry research, developed by ICRAF Programme 1 staff. Adapted from Izac [1994], and A.-M. Izac, pers. comm.

1. The identification of key driving socioeconomic and ecological processes within land-use typologies permits the spatial delineation of target and recommendation domains for agroforestry interventions.
2. Defining land use typologies across the landscape (based on parameters such as soils, climate, population density, level of market integration, policy level and the policy environment), is a prerequisite to the development and targeting of effective agroforestry interventions.
3. The adoptability of a new agroforestry practice is determined by five principal components: the farmers' natural resource base, their resource endowment, degree of market integration, cultural preferences and perceived benefits.
4. At the farming systems scale, agroforestry adoption has different impacts for different classes of farmers such as their gender.
5. Farmers are principally interested in the on-farm profitability and risk buffering capacity of agroforestry systems, while society is principally interested in the resilience, sustained productivity and biodiversity of these systems on the watershed and regional scales.

Appendix 2. (Continued).

6. Adoption of agroforestry practices usually necessitates policy changes that reconcile individual and social interests and benefits. The private benefits of agroforestry are restricted by policy and market constraints. Policy reform addressing such constraints will enhance the adoption of agroforestry.
7. Adoption of new agroforestry practices is principally a strategy for increased household welfare; adoption can be measured in terms of increased productivity, decreased risk and uncertainty, diversification of output, increased nutrient cycling efficiency, and securing land tenure.
8. Agroforestry interventions can lead to ecologically sustainable systems which at the same time are sufficiently productive and profitable to decrease rural poverty.
9. Policy interventions focusing on the maintenance of biodiversity at the landscape patch level can reduce pest outbreaks and enhance adoption of agroforestry.
10. The positive ecological and social impacts of agroforestry are superior in marginal soils, steep slopes, degraded lands, forests and desert margins compared to impacts on fertile lands.

Appendix 3

Summary of ICRAF's current domestication activities.

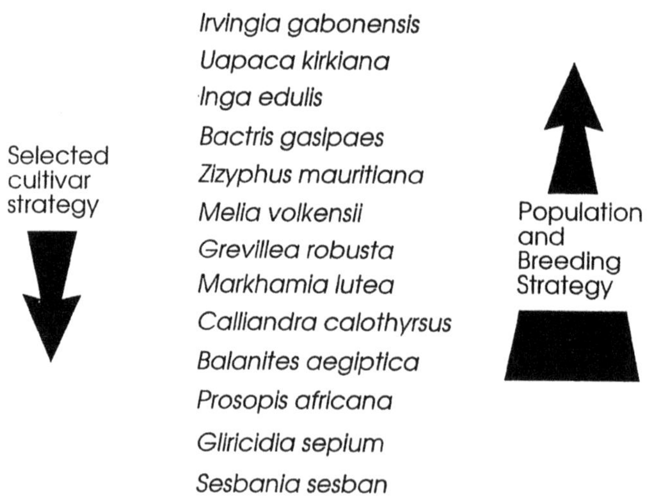

References

Akyeampong E, Duguma B, Heineman AM, Kamara CS, Kiepe P, Kwesiga F, Ong CK, Otieno HK and Rao MR (1995) A synthesis of ICRAF's research on alley cropping. In: Kang BT (ed) Proceedings Second International Alley Farming Workshop. IDRC, Ottawa (in press)
Alegre JC and Fernandes ECM (1991) Runoff and erosion losses under forest, low input agriculture and alley cropping on slopes. In: TropSoils technical report, 1988–1989, pp 227–228. North Carolina State University, Raleigh, NC, USA
Alexandre DY and Ouedraogo SJ (1992) Variations in root morphology of *Faidherbia albida* in relation to soil and agronomic effects. In: Vandenbelt RJ (ed) *Faidherbia albida* in the West African Semi-Arid Tropics, pp 107–110. ICRISAT, Patancheru, India
Anderson AB (ed) (1990) Alternatives to deforestation: Steps toward sustainable use of the Amazon rainforest. Columbia University Press, New York, USA, 281 pp
Anderson JM and Ingram JSI (eds) (1993) Tropical Soil Biology and Fertility. A Handbook of Methods. 2nd Ed. CABI, Wallingford, UK, 221 pp
Anderson LS and Sinclair FL (1993) Ecological interactions in agroforestry systems. Forestry Abstracts 54: 489–523 and Agroforestry Abstracts 6(2): 57–91
Andrews DJ and Kassam AH (1976) The importance of multiple cropping in increasing world food supplies. ASA Special Publication 27: 1–10. American Society of Agronomy, Madison, WI, USA
Atta-Krah AN (1990) Alley farming with leucaena: Effect of short grazed fallows on soil fertility and crop yields. Experimental Agriculture 26: 1–10
Avila M and Minae S (1991) Diagnosis and design: ICRAF develops a methodology for planning agroforestry research. Agroforestry Today 3(3): 9–11
Banda AZ, Maghembe JA, Ngugi DN and Chome VA (1994) Effect of intercropping and closely spaced leucaena hedgerows on soil conservation and maize yield on a steep slope at Ntcheu, Malawi. Agroforestry Systems 27: 17–22
Bandy DE, Garrity DP and Sanchez PA (1993) The world-wide problem of slash and burn agriculture. Agroforestry Today 5(3): 2–6
Barrios E, Buresh RJ and Sprent JI (1995) Organic matter in soil particle size and density fractions from maize and legume cropping systems. Soil Biology and Biochemistry (in press)
Bene JG, Beall HW and Cote A (1977) Trees, Food and People. IDRC, Ottawa, Canada
Bonkoungou JE (1992) Sociocultural and economic functions of *Acacia albida* in West Africa. In: Vandenbelt RJ (ed) *Faidherbia albida* in the West African Semi-Arid Tropics, pp 1–8. ICRISAT, Patancheru, India
Brookfield H and Padoch C (1994) Agrodiversity. Environment 36(5): 7–11, 37–45
Brooks KN, Gregersen HM and Folloitt PM (1995) Policy measures to increase agroforestry's contribution to sustainable land use. Draft policy brief. EPAT/MUCIA/USAID, University of Minnesota, St. Paul, MN, USA, 3 pp
Brouwer J, Fussell LK and Herrmann L (1993) Soil and crop microvariability in the West African semi-arid tropics: A possible risk reducing factor for subsistence farmers. Agriculture, Ecosystems and Environment 45: 229–238
Budowski G (1993) The scope and potential of agroforestry in Latin America. Agroforestry Systems 23: 121–132
Cahn MD, Bouldin DR and Cravo MS (1992) Nitrate sorption in the profile of an acid soil. Plant and Soil 143: 179–183
Charreau C and Vidal P (1965) Influence de l'*Acacia albida* sur le sol: Nutrition minerale et rendements des mils *Pennisetum* au Senegal. Agronomie Tropicale 6–7: 600–626
Clement CR (1989) The potential use of the pejibaye palm in agroforestry systems. Agroforestry Systems 7: 201–212
Clement CR and Villachica H (1994) Amazonian fruits and nuts: Potential for domestication in various ecosystems. In: RRB Leakey and Newton AC (eds) Tropical Trees: Potential for Domestication and the Rebuilding of Forest Resources, pp 230–238. HMSO, London, UK

Coe R (1994) Through the looking glass: 10 common problems in alley-cropping research. Agroforestry Today 6: 9–11

Coleman DC, Dighton J, Ritz K and Giller KE (1994) Perspectives on the compositional and functional analysis of soil communities. In: Ritz K, Dighton J and Giller KE (eds) Beyond the Biomass, pp 261–271. Willey-Sayce, Chichester, UK

Cunningham AB and Mbenkum FT (1993) Sustainability of harvesting *Prunus africana* bark in Cameroon: A medicinal plant in the international trade. People and Plants Working Paper 2. UNESCO, Paris, France, 28 pp

Dalal RC (1974) Effect of intercropping maize with pigeon peas on grain yields and nutrient uptake. Experimental Agriculture 10: 83–90

Dale VH, Houghton RA, Grainger A, Lugo AE and Brown S (1993) Emissions of greenhouse gases from tropical deforestation and subsequent uses of the land. In: Sustainable Agriculture and the Environment in the Humid Tropics, pp 215–260. National Academy Press, Washington, USA

Davelouis JR, Sanchez PA and Alegre JC (1991) Green manure incorporation and soil acidity amelioration. In: TropSoils technical report 1988–1989, pp 286–289. North Carolina State University, Raleigh, NC, USA

Deans JD, Lindley DK and Munro RC (1993) Deep beneath the trees in Senegal. ITE Annual Report for 1993: 1–3. Institute of Terrestrial Ecology, UK

Depommier D, Jadonet E and Oliver R (1992) *Faidherbia albida* parks and their influence on soils and crops at Watinoma, Burkina Faso. In: Vandenbelt RJ (ed) *Faidherbia albida* in the West African Semi-Arid Tropics, pp 111–116. ICRISAT, Patancheru, India

Dupuy N and Dreyfus B (1992) Presence of bradyrhizobia under *Acacia albida*. In: Vandenbelt RJ (ed) *Faidherbia albida* in the West African Semi-Arid Tropics, pp 145–148. ICRISAT, Patancheru, India

Dzowela B and Kwesiga F (1994) The potentials and limitations of agroforestry for improving livestock production and soil fertility in southern Africa. In: Craswell ET and Simpson J (eds) Soil Fertility and Climatic Constraints in Dryland Agriculture. ACIAR Proceedings 54: 19–25. Canberra, Australia

Fernandes ECM, Davey CB and Nelson LA (1993) Alley cropping on an acid soil in the upper Amazon: Mulch, fertilizer and hedgerow pruning effects. ASA Special Publication 56: 77–96. American Society of Agronomy, Madison, WI, USA

Fisher MJ, Rao IM, Ayarza MA, Lascano CE, Sanz JI, Thomas RJ and Vera RR (1994) Carbon storage by introduced deep-rooted grasses in the South American savannas. Nature 371: 236–238

Franzel S, Hitimana L and Akyeampong E (1995) Farmer participation on-station tree species selection for agroforestry: A case study from Burundi. Experimental Agriculture 31: 27–38

Fresco LO and Kroonenberg SB (1992) Time and spatial scales in ecological sustainability. Land Use Policy (July): 155–167

Fujisaka S, Mercado J and Garrity DP (1995) A case of farmer adaption and adoption of contour hedgerows for soil conservation. In: Kang BT (ed) Proceedings Second International Alley Farming Workshop. IDRC, Ottawa (in press)

Garrity DP (1993) Sustainable land use systems for the sloping lands of Southeast Asia. ASA Special Publication 56: 41–66. American Society of Agronomy, Madison, WI, USA

Garrity DP (1995) Tree–crop interactions in slopes. In: Ong CK and Huxley PA (eds) Tree–Crop Interactions, a Physiological Approach. CABI, Wallingford, UK (in press)

Garrity DP and Mercado AR Jr (1994) Nitrogen fixation capacity in the component species of contour hedgerows: How important? Agroforestry Systems 27: 241–258

Garrity DP and Van Noordwijk M (1995) Research imperatives in conservation farming and environmental management of sloping lands: An ICRAF perspective. ICRAF, Bogor, Indonesia (manuscript), 5 pp

Geiger SC, Vandenbelt RJ and Manu A (1992) Variability in growth of *Faidherbia albida*: The soils connection. Soil Science Society of America Journal 58: 227–231

Gregersen H, Belcher B and Spears J (1994) Creating policies to contain unproductive

deforestation. Policy Brief No. 8. EPAT/MUCIA, University of Wisconsin, Madison, WI, USA, 6 pp
Haggar JP, Tanner EVJ, Beer JW and Kass DCL (1993) Nitrogen dynamics of tropical agroforestry and annual cropping systems. Soil Biology and Biochemistry 25: 1363–1378
Hariah K, Van Noordwijk M, Santoso B and Syekhfani MS (1992) Biomass production and root distribution in eight trees and their potential for hedgerow intercropping on an Ultisol in Southern Sumatra. Agrivita 15: 54–68
Harper JL (1961) Approaches to the study of plant competition. In: Milthorpe FL (ed) Mechanisms in Biological Competition. Journal of Experimental Biology 15: 1–39
Hartemink AE, Buresh RJ, Jama B and Janssen BH (1995) Soil nitrate and water dynamics in sesbania fallows, weed fallows and maize. Soil Science Society of America Journal (in press)
Hauser S (1993) Root distribution of *Dacyladenia (Acioa) barteri* and *Senna (Cassia) siamea* in alley cropping on Ultisol. 1. Implications for field experimentation. Agroforestry Systems 24: 111–121
Houghton RA, Boone RD, Fruci JR, Hobbie JE, Melillo JM, Palm CA, Peterson BJ, Shaver GR, Woodwell GM, Moore B, Skole DL and Myers N (1987) The flux of carbon from terrestrial ecosystems to the atmosphere in 1980 due to changes in land use: Geographic distribution of the global flux. Tellus 39B: 122–139
Huntley BJ (1982) Southern African savannahs. In: Huntley BJ and Walker BH (eds) Ecology of Tropical Savannas, pp 101–119. Springer Verlag, Berlin, Germany
ICRAF (1993) Annual Report for 1992, pp 47–49. ICRAF, Nairobi, Kenya
ICRAF (1994) Annual Report for 1993, pp 80–87. ICRAF, Nairobi, Kenya
ICRAF (1995) Annual Report for 1994. ICRAF, Nairobi, Kenya
ICRISAT (1990) Soil, Crop and Water Management in the Sudano-Sahelian Zone, ICRISAT, Patancheru, AP, India, 384 pp
IITA (1988) Annual Report for 1987. IITA, Ibadan, Nigeria
Izac A-MN (1994) Characterization and impact, a concept paper for ICRAF's Research Programme 1. ICRAF, Nairobi, Kenya, 17 pp
Izac A-MN and Swift MJ (1994) On agricultural sustainability and its measurement in small-scale farms in sub-Saharan Africa. Ecological Economics 11: 105–125
Jabbar M, Larbi A and Reynolds L (1994) Profitability of alley farming with and without fallow in southwest Nigeria. Experimental Agriculture 30: 319–327
Joly H (1992) The genetics of *Acacia albida* (syn. *Faidherbia albida).* In: Vandenbelt RJ (ed) *Faidherbia albida* in the West African Semi-Arid Tropics, pp 53–62. ICRISAT, Patancheru, India
Juo ASR, Caldwell JO and Kang BT (1994) Place of alley cropping in sustainable agriculture in the humid tropics. Transactions 15th World Congress of Soil Science (Acapulco, Mexico) 7a: 98–109
Kang BT (1993) Alley cropping: Past achievements and future directions. Agroforestry Systems 23: 141–155
Kang BT (ed) (1995) Proceedings Second International Alley Farming Workshop. IDRC, Ottawa, Canada (in press)
Kang BT, Grimme H and Lawson TL (1985) Alley-cropping, sequentially cropped maize and cowpea with Leucaena on a sandy soil in southern Nigeria. Plant and Soil 85: 267–277
Kang BT, Reynolds L and Atta-Krah AN (1990) Alley farming. Advances in Agronomy 43: 315–359
Kang BT, Wilson GF and Sipkens L (1981) Alley cropping maize and *Leucaena leucocephala* Lam. in Southern Nigeria. Plant and Soil 63: 165–179
Kass DLC (1987) Alley cropping of food crops with wood legumes in Costa Rica. In: Beer J, Fassbender JH and Heuveldop J (eds) Advances in Agroforestry Research, pp 197–208. CATIE, Turrialba, Costa Rica
Kass DL, Araya JS, Sanchez JO, Pinto LS and Ferreira P (1995) Ten years experience with alley farming in Central America. In: Kang BT (ed) Proceedings Second International Alley Farming Workshop. IDRC, Ottawa, Canada (in press)

Kemerwa A-M, Peden D and Bamwerinde W.1994) Balancing gender to make agroforestry work: Two-Wings agroforestry groups in Uganda. Agroforestry Today 6(1): 15–17

Kessler JJ (1992) The influence of karité (*Vitellaria paradoxa*) and neré (*Parkia biglobosa*) trees on sorghum production in Burkina Faso. Agroforestry Systems 17: 97–118

Kessler JJ and Wiersum KF (1993) Ecological sustainability of agroforestry in the tropics. Entwicklung + Ländicher raum 27(5): 8–11

Kiepe P and Rao MR (1994) Management of agroforestry for the conservation and utilization of land and water resources. Outlook on Agriculture 23(1): 17–25

Kwesiga F and Chisumpa S (1992) Multipurpose trees of eastern Zambia: An ethnobotanical survey of their use in farming systems. AFRENA Report No 49. ICRAF, Nairobi, Kenya

Kwesiga F and Coe R (1994) The effect of short rotation *Sesbania sesban* planted fallows on maize yields. Forest Ecology and Management 64: 199–208

Ladha JK, Peoples MB, Garrity DP, Capuno VT and Dart PJ (1993. Estimating dinitrogen fixation of hedgerow vegetation using the nitrogen-15 natural abundance method. Soil Science Society of America Journal 57: 732–737

Lal R (1989) Agroforestry systems and soil surface management of a tropical Alfisol. 2. Water runoff, soil erosion and nutrient loss. 3. Changes in soil chemical properties. Agroforestry Systems 8: 97–132

Lavelle P and Pashanasi B (1989) Soil macrofauna and land management in Peruvian Amazonia. Pedobiologia 33: 283–291

Lawry S, Steinbarger D and Jabbar MA (1994) Land tenure and the potential for adoption of alley farming in West Africa. Outlook on Agriculture 23(3): 183–187

Lawson TL and Kang BT (1990) Yield of maize and cowpea in an alley cropping system in relation to available light. Agricultural and Forest Meteorology 53: 347–357

Leakey RRB (1991) Toward a strategy for clonal forestry; some guidelines based on experience with tropical trees. In: Jackson JE (ed) Tree Breeding and Improvement, pp 27–42. Royal Forestry Society of England, Wales and Northern Ireland, Tring, UK

Leakey RRB and Maghembe JA (1995) Domestication of high-value trees for agroforestry: An alternative to slash and burn agriculture. ICRAF Position Paper. ICRAF, Nairobi, Kenya, 12 pp

Leakey RRB and Newton AC (1994) Domestication of 'Cinderella' species as a start of a woody plant revolution. In: Leakey RRB and Newton AC (eds) Tropical Trees: Potential for Domestication and the Rebuilding of Forest Resources, pp 3–6. HMSO, London, UK

Leakey RRB, Newton AC and Dick JP (1994) Capture of genetic variation by vegetative propagation: Processes, determining causes. In: Leakey RRB and Newton AC (eds) Tropical Trees: Potential for Domestication and the Rebuilding of Forest Resources, pp 72–83. HMSO, London, UK

Maghembe JA, Kwesiga F, Ngulube M and Malaya FM (1994) Domestication potential of indigenous fruit trees of the miombo woodlands of southern Africa. In: Leakey RRB and Newton AC (eds) Tropical Trees: Potential for Domestication and the Rebuilding of Forest Resources, pp 220–229. HMSO, London, UK

Michon G and de Foresta H (1990) Complex agroforestry systems and conservation of biological diversity. 1. Agroforestry in Indonesia, a link between two worlds. Malaysian Nature Journal (Golden Jubilee Issue) 45: 457–473

Michon G, de Foresta H and Arif Aliadi (1995) Appropriation of forest resources by local communities: An agroforestry strategy in Sumatra. Fourth International Congress of Ethnobiology, Lucknow, India (in press)

Mora Urpi J, Szott LT, Murillo M and Patiño VM (eds) (1993) IV congreso internacional sobre biolog ía, agronomía, e industrialización del pijuayo. Editorial Universidad de Costa Rica, San José, 492 pp

Mulongoy K and Van der Meersch MK (1988) Nitrogen contribution by *Leucaena leucocephala* prunings to maize in an alley cropping system. Biology and Fertility of Soils 6: 282–285

Muetzelfeldt RI and Sinclair FL (1993) Ecological modelling of agroforestry systems. Agroforestry Abstracts 6(4): 207–247 and Forestry Abstracts 54(12): 1103–1143

Nair PKR (ed) (1989) Agroforestry Systems in the Tropics. Kluwer, London, 664 pp
Nyathi P and Campbell BM (1993) The acquisition and use of miombo litter by small-scale farmers in Masvingo, Zimbabwe. Agroforestry Systems 22: 43–48
Nye PH and Greenland DJ (1960) The soil under shifting cultivation. Commonwealth Agricultural Bureaux Technical Communication 51, Harpenden, UK, 156 pp
Okafor JC (1975) Varietal delimitation in *Irvingia gabonensis*. Bulletin du Jardin Nationale de Belgique 45: 211–221
Ong CK (1994) Alley cropping, ecological pie in the sky? Agroforestry Today 6 (3): 8–10
Ong CK (1995) The 'dark side' of intercropping: Manipulation of soil resources. In: Ecophysiology of Tropical Intercropping. Institut National de la Recherche Agronomique, Paris, France (in press)
Ong CK and Black CR (1995) Complementarity in resource use in agroforestry systems. In: Kang BT (ed) Proceedings Second International Alley Farming Workshop. IDRC, Ottawa, Canada (in press)
Ong CK, Black CR and Marshall FM (1995) Principles of resource capture and utilisation of light and water. In: Huxley PA and Ong CK (eds) Tree–Crop Interactions, a Physiological Approach, Chapter 4. CABI, Wallingford, UK (in press)
Ong CK, Corlett JE, Singh RP and Black CR (1991b) Above and below ground interactions in agroforestry systems. Forest Ecology and Management 45: 45–47
Ong CK, Odongo JCW, Marshall F and Black CR (1991a) Water use by trees and crops: Five hypotheses. Agroforestry Today 3: 7–10
Ong CK, Singh RP, Khan AAH and Osman M (1990) Recent advances in measuring water loss through trees. Agroforestry Today 2 (3): 7 –9
Ordoñez J (1994) Desarrollo de mercados para productos amazónicos no tradicionales. In: Toledo JM (ed) Biodiversidad y desarrollo sostenible de la Amazonía en una economía de mercado, pp 175–183. FUNDEAGRO, Lima, Peru
Palm CA (1995) Contribution of agroforestry trees to nutrient requirements of intercropped plants. Agroforestry Systems (this issue)
Palm CA, Izac A-M and Vosti SA (1995a) Guidelines for characterization of Alternatives to Slash-and-Burn sites. ICRAF, Nairobi, Kenya (in press)
Palm CA, McKerrow AJ, Glazener KM and Szott LT (1991. Agroforestry systems in the lowland tropics: Is phosphorus important? In: Tiessen H, López-Hernandez D and Salcedo IH (eds) Phosphorus Cycles in Terrestrial and Aquatic Ecosystems, Regional Workshop 3: Central and South America, pp 134–141. Saskatchewan Institute of Pedology, Saskatoon, Canada
Palm CA, Salazar AA, Szott LT and Fernandes ECM (1995b) Long-term performance of alley cropping on an acid soil of the Amazon basin. In: Kang BT (ed) Proceedings Second International Alley Farming Workshop. IDRC, Ottawa, Canada (in press)
Palm CA and Sanchez PA (1990) Decomposition and nutrient release patterns of the leaves of three tropical legumes. Biotropica 22: 330–338
Palm CA and Sanchez PA (1991) Nitrogen release from the leaves of some tropical legumes as affected by their lignin and polyphenolic contents. Soil Biology and Biochemistry 23: 83–88
Papendick RI, Sanchez PA and Triplett GB (eds) (1976) Multiple cropping. ASA Special Publication 27. American Society of Agronomy, Madison, WI, USA, 378 pp
Parton WJ, Woomer PL and Martin A (1994) Modelling soil organic matter dynamics and plant productivity in tropical ecosystems. In: Woomer PL and Swift MJ (eds) The Biological Management of Tropical Soil Fertility, pp 171–188. Wiley, Chichester, UK
Peters CM (1990) Population ecology and management of forest fruit trees in Peruvian Amazonia. In: Anderson AB (ed) Alternatives to Deforestation; Steps toward Sustainable Land Use of the Amazon Rainforest, pp 86–98. Columbia University Press, New York
Peters CM and Vázquez A (1987) Estudios ecológicos de camu-camu (*Myrciaria dubia*). 1. Producción de frutas en poblaciones naturales. Acta Amazónica 16/17: 161–173
Place F (1994) The role of land and tree tenure on the adoption of agroforestry technologies: A summary and synthesis. ICRAF, Nairobi, Kenya (manuscript), 88 pp

Place F, Mwanza S and Kwesiga F (1994) A cost-benefit analysis of improved fallows in Eastern Province, Zambia. ICRAF, Nairobi, Kenya (manuscript), 14 pp

Raintree JB (1987) The state of the art on agroforestry diagnosis and design. Agroforestry Systems 5: 219–250

Rao MR (1995) Agroforestry for sustainable soil management in humid and subhumid tropical Africa. In: Strategies for the Management of Upland Soils of Humid and Subhumid Africa. IBSRAM, Abidjan, Côte d'Ivoire (in press)

Rao MR and Coe R (1992) Evaluating the results of agroforestry research. Agroforestry Today 4(1): 4–9

Rao MR, Muraya P and Huxley PA (1993) Observations of some tree root systems in agroforestry intercrop situations and their graphical representations. Experimental Agriculture 29: 183–194

Rao MR, Ong CK, Pathak P, and Sharma MM (1991) Productivity of annual cropping and agroforestry systems on a shallow Alfisol in semiarid India. Agroforestry Systems 15: 51–63

Reed JD, Rittner U, Tanner J and Wiegand O (1992) Nutritive value of leaves and fruits of *Faidherbia albida* and their use for feeding ruminants. In: Vandenbelt RJ (ed) *Faidherbia albida* in the West African Semi-Arid Tropics, pp 43–52. ICRISAT, Patancheru, India

Reynolds L (1994) A review of the biophysical and socioeconomic basis of alley farming and its adoption potential. Consultant report to AFNETA. International Agricultural Consultancy, Leicester, UK, 90 pp

Riesco A and Ara M (1994) Perspectivas de la integraci6n de sistemas agrosilvopastoriles. In: Toledo JM (ed) Biodiversidad y desarrollo sostenible de la Amazonía en una economía de mercado, pp 83–103. FUNDEAGRO, Lima, Peru

Roose E (1970) Importance relative de l'érosion, du drainage oblique et vertical dans la pedogenèse d'un sol ferralitique de moyenne Côte d'Ivoire. Cahiers ORSTOM Serie Pedologique 8: 469–482

Salazar A, Szott LT and Palm CA (1993) Crop–tree interaction in alley cropping systems on alluvial soils of the upper Amazon basin. Agroforestry Systems 22: 67–72

Sanchez PA (1976) Properties and Management of Soils in the Tropics. Wiley, New York, USA, 618 pp

Sanchez PA (1987) Soil productivity and sustainability in agroforestry systems. In: Steppler HA and Nair PKR (eds) Agroforestry: A Decade of Development, pp 205–226. ICRAF, Nairobi, Kenya

Sanchez PA (1994a) Tropical soil fertility research, towards the second paradigm. Transactions 15th World Congress of Soil Science (Acapulco, Mexico) 1: 65–88

Sanchez PA (1994b. Alternatives to slash and burn: A pragmatic approach for mitigating tropical deforestation. In: Anderson JR (ed) Agricultural Technology, Policy Issues for the International Community, pp 451–480. CABI, Wallingford, UK

Sanchez PA and Benites JR (1987) Low input cropping for acid soils of the humid tropics. Science 238: 1521–1527

Sanchez PA and Buol SW (1975) Soils of the tropics and the world food crisis. Science 188: 598–603

Sanchez PA, Palm CA, Davey CB, Szott LT and Russell CE (1985) Trees as soil improvers in the humid tropics? In: Cannell MGR and Jackson JE (eds) Trees as Crop Plants, pp 327–358. Institute of Terrestrial Ecology, Huntingdon, UK

Sanchez PA, Palm CA, Szott LT, Cuevas E and Lal R (1989) Organic input management in tropical agroecosystems. In: Coleman DC, Oades JM and Uehara G (eds) Dynamics of Soil Organic Matter in Tropical Ecosystems. University of Hawaii Press, Honolulu, HI, USA

Sanchez PA, Villachica JH and Bandy DE (1983) Soil fertility dynamics after clearing a tropical rainforest in Peru. Soil Science Society of America Journal 47: 1171–1178

Sanchez PA, Woomer PL and Palm CA (1994) Agroforestry approaches for rehabilitating degraded lands after tropical deforestation. In: Rehabilitation of Degraded Forest Lands in the Tropics–Technical Approach. JIRCAS International Symposium Series 1: 108–119. JIRCAS, Tsukuba, Japan

Sanginga N and Mulongoy K (in press) Increasing biological nitrogen fixation and the contribution to the nitrogen cycle in alley cropping systems. In: Kang BT (ed) Proceedings Second International Alley Farming Workshop. IDRC, Ottawa, Canada

Scherr SJ (ed) (1991) Methods for Participatory On–Farm Agroforestry Research. ICRAF, Nairobi, Kenya, 72 pp

Schlesinger WH, Reynolds JF, Cunningham GL, Huenneke LF, Jarrell WA, Virginia RA and Whitford WG (1990) Biological feedbacks in global desertification. Science 247: 1043–1048

Schroeder P (1993) Agroforestry systems: Integrated land use to store and conserve carbon. Climate Research 3: 53–60

Schroeder P (1994) Carbon storage benefits of agroforestry systems. Agroforestry Systems 27: 89–97

Serrão EAS (1993) Brazil. In: Sustainable Agriculture and the Environment in the Humid Tropics, pp 263–351. National Academy Press, Washington, USA

Serrão EAS and Toledo JM (1990) The search for sustainability in Amazonian Pastures. In: Anderson AB (ed) Alternatives to Deforestation; Steps toward Sustainable Land Use of the Amazon Rainforest, pp 195–214. Columbia University Press, New York, USA

Shepherd KD, Ohlsson E, Okalebo JR, Ndufa JK and David S (1995) A model of nutrient cycling in mixed farms in the highlands of western Kenya to show the impact of improved management. International Conference on Livestock and Sustainable Nutrient Cycling in Mixed Farming Systems of Sub-Saharan Africa. ILCA, Addis Ababa, Ethiopia (in press)

Sibuea TTH (1995) Short notes on the Sumatran rhino (*Dicerorhinus sumatrensis*) in the agroforest areas (damar garden) in Krui, Lampung. AWB-Indonesia, Bogor, 3 pp

Simmonds NW (1985) Perspectives on the evolutionary history of tree crops. In: Cannell MGR and Jackson JE (eds) Trees as Crop Plants, pp 3–12. Institute of Terrestrial Ecology, Huntingdon, UK

Sinclair FL, Verinumbe I and Hall JB (1994) The role of tree domestication in agroforestry. In: Leakey RRB and Newton AC (eds) Tropical Trees: Potential for Domestication and the Rebuilding of Forest Resources, pp 124–136. HMSO, London, UK

Singh G, Singh NT and Abrol IP (1994) Agroforestry techniques for the rehabilitation of degraded salt-affected lands in India. Land Degradation and Rehabilitation 5: 223–242

Singh RP, Ong CK and Saharan N (1989) Above and below ground interactions in alley cropping in semiarid India. Agroforestry Systems 9: 259–274

Sivakumar MVK, Wallace JS, Renard C and Giroux C (eds) (1991) Soil Water Balance in the Sudano-Sahelian Zone. IAHS Publication 199. Institute of Hydrology, Wallingford, UK, 628 pp

Smaling E (1993) An agroecological framework for integrating nutrient management, with special reference to Kenya. Doctoral thesis. Agricultural University, Wageningen, the Netherlands, 250 pp

Sombroek WG, Nachtergaele FO and Hebel A (1993) Amounts, dynamics and sequestration of carbon in tropical and subtropical soils. Ambio 22(7): 417–426

Spek LY and van Noordwijk M (1994) Proximal root diameters as predictors of total root system size for fractal branching models. 2. Numerical model. Plant and Soil 164: 119–127

Sprent JI, Sutherland JM and de Faria SM (1989) Structure and function of root nodules from woody legumes. In: Stirton CH and Zarucchi JL (eds) Advances in Legume Biology. Missouri Botanical Garden Monographs on Systematic Botany 29: 559–578

Steppler HA and Nair PKR (eds) (1987) Agroforestry: A Decade of Development. ICRAF, Nairobi, Kenya, 335 pp

Subler S and Uhl C (1990) Japanese agroforestry in Amazonia: A case study in Tome-Açú, Brazil. In: Anderson AB (ed) Alternatives to Deforestation; Steps toward Sustainable Land Use of the Amazon Rainforest, pp 152–166. Columbia University Press, New York, USA

Swift MJ, Heal OW and Anderson JM (1979) Decomposition in Terrestrial Ecosystems. University of California Press, Berkeley, CA, USA, 372 pp

Swinkels R and Franzel S (submitted) Adoption potential of hedgerow intercropping systems in the highlands of western Kenya. 2. Economic and social evaluation. ICRAF, Nairobi, Kenya, 19 pp

Szott LT and Kass DCL (1993) Fertilizers in agroforestry systems. Agroforestry Systems 23: 157–176

Szott LT, Palm CA and Sanchez PA (1991) Agroforestry in acid soils of the humid tropics. Advances in Agronomy 45: 275–301

Tangley L (1993) Marketing biodiversity products, the taungya initiative. Lessons from the field 2. Conservation International, Washington, USA

Tisdall JM and Oades JM (1982. Organic matter and water-stable aggregates in soils. Journal of Soil Science 33: 141–163

Toledo JM (ed) (1994) Biodiversidad y desarrollo sostenible de le Amazonía en una economía de mercado. FUNDEAGRO, Lima, Peru, 295 pp

Tomich T, Roemer M and Vincent J (1994) Development from a primary export base. In: Lindauer DL and Roemer M (eds) Asia and Africa: Legacies and Opportunities in Development, pp 151–193. ICS Press, San Francisco, CA, USA

Torquebiau E and Kwesiga F (submitted) The Role of Roots in an Improved Fallow Agroforestry System in Eastern Zambia. ICRAF, Nairobi, Kenya

Unruh JD, Houghton RA and Lefebvre PA (1993) Carbon storage in agroforestry: An estimate for sub-Saharan Africa. Climate Research 3: 39–52

Vandenbelt RJ (ed) (1992) *Faidherbia albida* in the West African semi-arid tropics. ICRISAT, Patancheru, India, 206 pp

Vandenbelt RJ and Napompeth B (1992) The leucaena psyllid comes to Africa. Agroforestry Today 4(4): 11–12

Vandenbelt RJ and Williams JH (1992) The effect of soil surface temperature on the growth of millet in relation to the effect of *Faidherbia albida* trees. Agricultural and Forest Meteorology 60: 93–100

Van Noordwijk M (1995) A simple model to quantify mulch and shade effects. In: Huxley PA and Ong CK (eds) Tree–Crop Interactions, a Physiological Approach, Chapter 3. CABI, Wallingford, UK (in press)

Van Noordwijk M and Dommergues YR (1990) Agroforestry and soil fertility: Root nodulation – the twelfth hypothesis. Agroforestry Today 2(2): 9–10

Van Noordwijk M, Hairiah K, Sitompul SM and Syekhfani MS (1992) Rotational hedgerow intercropping + *Peltophorum pterocarpum* = new hope for weed-infested soils. Agroforestry Today 4(4): 4–6

Van Noordwijk M, Sitompul SM, Hariah K, Listyarini E and Syekhfani M (1995) Nitrogen supply from rotational and spatially zoned inclusion of Leguminosae for sustainable maize production on an acid soil in Indonesia. Proceedings Third International Symposium on Plant-Soil Interactions at low pH. Brisbane, Australia

Van Noordwijk M, Spek IY and De Willigen P (1994) Proximal root diameters as predictors of total root system size for fractal branching models. 1. Theory. Plant and Soil 164: 107–118

Villachica H (1994) Investigación y el desarrollo de sistemas sustentables para frutales nativos amazónicos, el caso pijuayo. In: Toledo JM (ed) Biodiversidad y desarrollo sostenible de la Amazonía en una economía de mercado, pp 109–135. FUNDEAGRO, Lima, Peru

Vosti S (1995) The role of agriculture in saving the rainforest. 2020 Vision Brief 9, IFPRI, Washington, DC, USA, 2 pp

Vosti SA and Witcover J (1995) Slash and burn agriculture: Household perspectives. Agriculture, Ecosystems and Environment (in press)

Wendt JW, Jones RB and Itimu OA (1994) The potentials and limitations of agroforestry for improving livestock production and soil fertility in southern Africa. In: Craswell ET and Simpson J (eds) Soil Fertility and Climatic Constraints in Dryland Agriculture. ACIAR Proceedings 54: 19–25. Canberra, Australia

Willey RW (1979) Intercropping–its importance and research needs. Part 1. Competition and yield advantages. Field Crop Abstracts 32: 1–10

Wilson JR (1990) Agroforestry and soil fertility: The eleventh hypothesis-shade. Agroforestry Today 2(1): 14–15

Wong MTF, Akyempaong E, Nortcliff S, Rao MR and Swift RS (1995) Initial response to maize and beans to decreased aluminium activity with organic residues on an Oxisol in Burundi. Plant and Soil (in press)

Woomer P, Bajah O, Atta-Krah AN and Sanginga N (1995) Analysis and interpretation of alley farming data from Tropical Africa. In: Kang BT (ed) Proceedings Second International Alley Farming Workshop. IDRC, Ottawa, Canada (in press)

World Bank (1994) Feasibility of Phosphate Rock Use as a Capital Investment in Subsaharan Africa: Issues and Opportunities. Africa Technical Division, World Bank, Washington, DC, USA, 72 pp

Yamoah CF, Agboola AA and Wilson GF (1986) Decomposition, nitrogen release and weed control by prunings of alley cropping shrubs. Agroforestry Systems 4: 239–246

Young A (1989a) Agroforestry for Soil Conservation. CABI, Wallingford, UK, 276 pp

Young A (1989b) Ten hypotheses for soil–agroforestry research. Agroforestry Today 1: 13–16

Agroforestry Systems **30**: 57–73, 1995.

Agroforestry policy issues and research directions in the US and less developed countries: insights and challenges from recent experience

L. E. BUCK
CIIFAD/CIFOR, Department of Natural Resources, Fernow Hall, Cornell University, Ithaca, NY 14850, USA

Key words: participatory research, policy research, strategic research

Abstract. Efforts to improve the performance of agroforestry systems, and to expand the land area and number of people able to benefit from this integrative approach to agriculture and natural resource management, are constrained throughout the world by non-supportive land use policies. A growing sense of urgency that policy change is needed to enable agroforestry to flourish has contributed during the past two years to an unprecedented level of agroforestry policy assessment and planning activity.

In the US, agroforestry has emerged from academia, where it has incubated since the mid-1980s, into the professional resource management arena. A multi-organizational agroforestry evaluation process has driven national policy and program formation to the forefront of the agenda of the agroforestry community, as it seeks to influence the 1995 Farm Bill. Internationally, the Consultative Group on International Agricultural Research and collaborators fostered a sequence of policy issue identification activities as a basis for setting strategic research priorities for forestry and agroforestry.

Following a brief review of forces driving agroforestry development in industrialized and less developed countries, the paper highlights recent policy assessment initiatives in each sphere. Observations on the issues driving and the priorities emerging from these processes are offered, to lend perspective to the critical challenges facing the agroforestry policy research community. An explanation for pervasive constraints and inconsistencies in policy effectiveness is then explored, from which a promising approach to research intervention is forwarded.

It is argued that social scientists might influence agroforestry policy most favorably at this critical juncture, as perceptions of inter-dependence increase among different stakeholders in the policy system, by employing interventionist, actor-oriented perspectives and participatory methods to facilitate policy innovation and evaluation. The approach is consistent with participatory technology design processes that earlier helped to establish agroforestry as a prototype for sustainable development.

Introduction

Agroforestry in tropical and sub-tropical regions has been recognized by the international scientific community for almost 20 years, while scientific engagement with temperate zone agroforestry became evident about a decade later. During this time agroforestry practice in the South has proven its ability to address some of the region's pressing land use problems, while the science has become a strategic integrating discipline that cuts across the boundaries of traditional land use disciplines [Nair, 1993]. The expansion of agroforestry activity in industrialized country settings is more recent and the pace of development slower. Evidence suggests, however, that the temperate zone branch of the field may soon catch up.

Recently, the agroforestry community in both spheres has noticeably broadened its focus of attention from technology to policy dimensions of agroforestry development and research. Reasons differ between the two regions as a function of the stage of development of agroforestry and associated expectations for its advancement. There is universally growing appreciation, however, that agroforestry as a prototype of sustainable development is constrained by policies and institutions that were codified to support more conventional, industrial models of agriculture, forestry and rural development. To realize the potential which bio-technical and socio-economic research on agroforestry suggests is possible for enhancing agricultural productivity, while conserving natural resources and protecting the environment, policy constraints must be removed and replaced by a consistent constellation of incentives [Kidd and Pimentel, 1992].

The impetus for this growing sentiment is explored in this paper, together with the recent assessment and planning activity that it has spawned, and implications for agroforestry policy research. First, however, the driving forces behind agroforestry development in less developed and industrialized countries are briefly reviewed to help differentiate agroforestry in the two domains, and to appreciate how these factors affect perceived policy needs.

Driving forces in agroforestry development

Several authors have compared the emergence and application of agroforestry in developing and developed country contexts [Ehrenreich et al., 1994; Long and Nair, 1991; Lassoie and Buck, 1991]. Key issues that motivate the agroforestry communities in these overlapping spheres to seek the benefits of integrating trees, annual crops and livestock in complex, perennial-based production systems are summarized below. The US is used as an example of an industrialized country.

Less developed countries

Poverty alleviation and rural welfare was the leading justification for investment in agroforestry by the international aid community throughout most of the 1980s. By integrating selected trees back into rural systems, burdens faced by the poor in meeting their requirements for forest products from marginal and degraded areas could be reduced, often while improving soil conditions and plant productivity.

Throughout much of the South, agriculture remains the primary economic sector in terms of GDP, the main source of employment and often the principal source of foreign exchange. Thus, maintaining the quality of the natural resource base upon which commercial agricultural productivity depends is a critical priority for many countries. Agroforestry can play an important resource conserving role for this purpose.

As land is cleared by expanding populations for agriculture, environmentally unique or sensitive areas come under increasing threat. International organizations invest in agroforestry to help buffer threatened areas from the impact of agricultural encroachment. The influence of the environmental agenda on key development sectors has increased appreciably in recent years. While in the past this influence was discernable almost exclusively through NGOs, bi-lateral and international aid programs are becoming more visibly focused on the agricultural – environmental nexus [World Bank, 1994; World Resources Institute, 1992].

While the foregoing distinctions are of some consequence in assessing policy impact and planning interventions, agroforestry is widely appreciated for its capacity to perform any or all of these functions, depending on how technologies are designed and applied. Similarly, it is becoming less fruitful to differentiate among agroforestry research, extension and development activities. The participatory, systems approach commonly employed in technology development, utilizing the 'Diagnosis and Design' methodology, has essential features of each, which can be heightened or diminished depending on the objectives and capabilities of the implementing organization [Raintree, 1990; Buck, 1990].[1] In sum, investment in one domain of agroforestry is likely to bear fruit in others.

United States

In the United States the rediscovery of traditional agroforestry, and efforts to develop new practices, stem from concern about the persistence and high cost of natural resource degradation associated with industrialized agriculture. Public pressure is growing to limit soil erosion and water contamination and to restore wildlife habitat. This, together with changes in rural economics and demographics associated with the on-going restructuring of agriculture, as well as growing national and international markets for high quality forest products, are combining to influence land use decisions in new ways [Garrett and Buck, 1994]. Notably in these countries, the influence of agriculture on the national economy continues to shrink in terms of GDP, employment and export earnings.

Recent moves to make greater use of agroforestry practices in temperate agriculture have been driven by the proven or perceived ability of agroforestry to meet the following needs: i) product diversification, ii) environmental impact mitigation, iii) land rehabilitation, iv) land use conversion (from annual to timber crops), v) increased or decreased food production, vi) sustainable use or retirement of marginal or fragile land, vii) habitat enhancement, and viii) aesthetic appreciation.

Less urgency is associated with the agroforestry movement in industrialized countries than in less developed ones, as problems amenable to this approach are not viewed to be as threatening to lives or livelihoods. Appreciation of climatic differences and associated growing conditions also limits expectations for the immediacy of problem relief in temperate environments. It is often argued that agroforestry is unlikely to make as significant and immediate a difference to the land and lives of the North as it can in the South.

Nonetheless, the benefits appear real and important to growing numbers concerned with rural vitality, environmental protection and resource conservation. From a global perspective, agroforestry is seen to have a role in the North's transition to sustainable development as it strives to meet its environmental and social responsibilities to the South. As such justifications mount, the 'relevance distinction' between temperate and tropical agroforestry may be less 'relevant' in the future than it has been in the past.

Agroforestry policy

Policy has been defined as any governing principle, plan or course of action [Webster, 1988]. Examining policy from this broad perspective, it is not surprising that the domain of agroforestry policy seems unwieldy and subject to a variety of interpretations. Indeed, scholars of government and political science attest to the numerous schemes and perspectives that have emerged to study and understand the complex concept of public policy (Uphoff, pers. comm.).

To address the question of how policy affects agroforestry through a particular theoretical construct, thus, is unlikely to be productive. Instead, attention is focused on issues arising from the community of researchers, agency specialists, program and project managers, local organizations and donors concerned with advancing effective agroforestry practice. What do they perceive as key policy issues affecting the development and spread of agroforestry land use? I shall draw from recent organized efforts to elicit such views.

Policy assessment initiatives

During the past two years there has been a noticeable surge of activity within the overlapping agroforestry communities, North and South, to assess the role of policy in realizing the potential agroforestry has for sustainable development. A host of interrelated meetings, workshops and conferences in each sphere reveal strong convictions about the significant unmet potential of agroforestry to address mounting ecological and economic crises worldwide, and the important role of policy support in unleashing it. Some of the most visible and potentially influential of these initiatives are identified below.[2]

Less developed countries

Between 1992–1994, the Consultative Group on International Agricultural Research (CGIAR), in collaboration with the International Union of Forestry Research Organizations (IUFRO) and the Food and Agriculture Organization (FAO) sponsored a series of policy research workshops throughout the developing world. The aim was to better understand forestry and agroforestry problem areas that could benefit from strategic policy research. Views from several hundred policy leaders and scientists were elicited. Background materials for the workshops identified agroforestry policy issues from a variety of different perspectives [Gregersen et al., 1992]. A synthesis report was developed to help focus and coordinate the efforts of key centers and their partners, especially the International Center for Research in Agroforestry (ICRAF), the Center for International Forestry Research (CIFOR) and the International Food Policy Research Institute (IFPRI), to conduct strategic policy research in the closely interrelated forestry and agroforestry domains [CIFOR, 1994].

Among the initiatives this activity spawned was a research methodology workshop in 1994 on Policies for Local Organizations in Natural Resources Management [IFPRI et al., 1994]. Social scientists from policy research organizations throughout the world met to consider research priorities and methodologies for improving policy support to local organizations concerned with agroforestry and related land use systems.

A complementary workshop series was held in 1993–1994, bringing together members of the CGIAR Centers and the North American university research system, with selected National Agriculture Research Systems (NARS) and foundations to formulate the Global Research on the Environmental and Agriculture Nexus (GREAN) initiative [Cornell University and University of Florida, 1994]. The GREAN initiative aims to generate a second Green Revolution, to create livelihood and food security for the poorest people in the developing world on an environmentally sustainable basis. It calls for political will in developed and developing countries alike, to ensure that sufficient resources are applied to global research on reconciling sustainability with productivity growth. Predictably, there is a role for agroforestry in the proposed $ 100 million strategy for research collaboration.

United States

In 1993–1994 the USDA Soil Conservation Service (SCS), recently named the Natural Resources Conservation Service (NRCS), commissioned a comprehensive assessment of US agroforestry as a component of the Resource Conservation Act (RCA) appraisal [Garrett et al., 1994]. It was the first organized effort in the US to systematically review the scope of agroforestry activity and to assess its potential for further development. The report was researched and prepared by a select group of comparatively long-standing members of the agroforestry research community. It presents a vision that links land use policy and practice in the US to global concerns for meeting basic human needs and containing ecological disaster.

Following a positive reception of the RCA report by the United States Forest Service (USFS), Cooperative Extension Service (CES) and other concerned agencies, a multi-organizational initiative was launched to develop a framework for agroforestry development in the US [AFTA, 1994]. A consultation and workshop process involving numerous new and known actors in the agroforestry community resulted in a widely distributed 'white paper' for use in formulating national level agroforestry policy proposals. The circulation of this report throughout land grant universities, and to numerous government agencies, societies and membership groups has noticeably heightened awareness and interest in agroforestry activity (AFTA Steering Committee, pers. comm.).

In association with the foregoing activities, the visibility of the Association for Temperate Agroforestry (AFTA) has increased, as has the membership of this growing network organization.[3] Late in 1994 the organization facilitated the development of an agroforestry action team, an ad hoc committee charged with formulating an agroforestry policy, strategy and action plan for the 1995 Farm Bill. The action team is comprised of eleven politically astute members with diverse backgrounds and a common interest in agroforestry. It met in December, 1994 to distill information and priorities from the foregoing sources, and draft a succinct agroforestry policy statement to help influence members of Congress to introduce or support a package of measures aimed at advancing agroforestry.

Driving forces in agroforestry policy evaluation

It is not justified to compare agroforestry policy issues and priorities arising from the foregoing sets of activities on a uniform set of criteria. Aside from a common purpose to elicit resources for agroforestry research and development, their foci on agroforestry policy stem from different needs and concerns. Instead, the author's observations from participation in the international and the US activities described, offer different perspectives to draw from in evaluating the issue of agroforestry policy needs and research priorities.

Less developed countries

Three key issues appear to be driving the current international focus on agroforestry policy. First, limitations of project approaches to agroforestry development have been widely encountered throughout the South. Within the comparatively artificial system of incentives that it is possible to create in a project environment, efforts to influence institutional and farmer behavior in favor of agroforestry have often been rewarded. Attempts to ensure the sustainability of project activities and impacts following the withdrawal of external support, however, have repeatedly failed [Annis, 1988]. Similarly, strategies to 'scale-up' agroforestry project successes in order to benefit wider regions from experience gained, have often been unsuccessful [Josiah and Gregersen, 1994].

Second, once local organizations became widely appreciated as effective intermediaries for agroforestry development [Buck, 1990; Korten, 1990; Esman and Uphoff, 1984], obstacles began to surface in efforts to reproduce the innovative learning processes that made external interaction with certain local groups initially so rewarding. It was discovered that non-supportive regulations, administrative leader opinion, competing socio-cultural norms and financial disincentives often combined to limit tendencies for local group formation, action and networking.

Third, the scope and persistence of policy constraints to agroforestry land use are pervasive. They include insufficient institutional support, non-conducive systems of land and resource tenure, gender biases in development priorities, and the absence of sufficient economic incentives for tree growing. The latter are often due to policy distortions in the form of subsidies for substitutes.

Mounting frustration with these structural constraints has reinforced the need to address them at their roots. The agroforestry policy environment, however, is complex, subject to various perspectives and interpretations, and uneven with respect to power relations. Thus, expanded investment is warranted in strategically designed research to generate appropriate knowledge needed to penetrate these barriers and help induce the necessary change.

United States

The RCA assessment of US agroforestry was commissioned as part of a broader effort to evaluate the effectiveness of provisions in the Conservation Title of the 1985 and 1990 Farm Bills [Food Security Act of 1985; Food, Agriculture, Conservation and Trade Act of 1990]. With the benefit of the study, the construction of the 1995 Farm Bill provided a critical legislative opportunity to design a comprehensive constellation of measures aimed at integrating the research, technical assistance and training activities needed to advance agroforestry efficiently.

A widely recognized need to reform the agriculture sector's costly Conservation Reserve Program (CRP) [Soil and Water Conservation Society, 1994] provided an immediate impetus to invoke agroforestry. Groups of professional resource managers, politicians, farmers and environmentalists in the State of Missouri, for example, have become uniquely united through a common view that agroforestry holds promise for eliciting greater farmer enrollment, on a more 'voluntary' basis, to establish more permanent conservation practices. This is mainly a function of agroforestry's comparative economic advantages over more commonly employed farmland conservation practices (H.E. Garrett, pers. comm.). In Missouri, the capacity of black walnut (*Juglans nigra*) intercropping systems to produce marketable forage, nuts and high value timber in the short, medium and long term respectively, limits an operator's reliance on public subsidies in choosing a conservation-oriented land use practice.

At a broader level, the US agroforestry policy initiative has been fueled by sporadic and unevenly distributed, but noticeable changes in the country's dominant system of resource management values. These foster the separation of land and resources among uses, and pose rigid distinctions between public and private resource ownership [Jacobs and Harper, 1993]. While the tradition of segregation prevails upon our rural and urban landscapes, and the specialized institutions that support it, public sentiment is growing for more integrated, ecologically sound approaches.

International policy research priorities

Much of the recent focus on agroforestry policy research in developing countries is interrelated with forest policy issues. Of the following five main problem areas identified through the CGIAR-sponsored workshop series, the first four provide scope for priority-level, agroforestry policy research: i) the negative effects of tropical deforestation; ii) undervaluation of forest benefits; iii) insufficient knowledge and understanding of the effects of institutional reforms aimed at ensuring greater local participation in the policy dialogue and in the conservation and management of forest resources; iv) insufficient knowledge and understanding about the effect of policies aimed at improving the role of trees in rural welfare; and v) inefficiency and unprofitability of forest industries, and insufficient intensity and productivity of forest management.

Of 21 priority research topics falling within the four problem areas of interest, 15 concern agroforestry directly, as indicated by their assignment through this process to ICRAF's research domain. The majority of these fall within the mandates, also, of CIFOR or IFPRI, while certain topics associated more directly with these two centers are linked, in turn, with agroforestry. Problems iii and iv, above, are most centrally concerned with agroforestry. Distinct from the others, these statements indicate that policy reform efforts are evident, but the effects are not well understood or appreciated. Eight of the 15 agroforestry research priority areas are associated with institutional reform, and roles-of-trees-in-rural-welfare.

Space precludes an elaboration of the numerous statements of relationship that bear examination within the four agroforestry-relevant problem areas.[4] It is instructive, however, to develop an impression of the phenomena and categories of variables that are encompassed in the high priority policy research profile:

> macroeconomic and sectoral policy linkages and interactions . . . incentives for deforestation and tree/forest management . . . agricultural policies . . . population migration . . . forest department practices . . . slash and burn farming . . . forest buffer zones . . . land tenure/tree ownership . . . costs

and benefits of tree/forest goods and services . . . externality benefits of trees . . . value of non-traditional forest products . . . tree product markets . . . carbon emissions/carbon sequestration value of trees . . . North/South concessional transfers . . . household decisions about tree management . . . local organizations, agroforestry adoption . . . local participation . . . human welfare . . . institutional change . . . tree and forest productivity . . . soil and water protection . . . risk reduction . . . rural income generation . . . nutrition . . . food security . . . traditional agroforestry systems . . . household agroforestry needs . . . degraded land reclamation . . . agronomic zoning.

A similar constellation of factors, bearing urgent research attention, emerged from the Workshop on Policies for Local Organizations in Natural Resources Management, albeit with reference to this more specific domain of action. Finally, the GREAN initiative stresses that policy and institutional analysis should be incorporated into each of its four priority research programs, e.g., i) enhanced productivity, food security and nutrition, ii) soils, water, forests and fisheries, iii) conservation of biological diversity, and iv) coping with an uncertain and fluctuating climate. While GREAN's scope is much broader than that of agroforestry alone, the roots of the two concepts in conservation and sustainable development are consistent and overlapping.

In sum, while an emerging agroforestry research focus on policy is laudable, the magnitude of the need represented by these statements of priority presents a daunting challenge to an international policy research community which itself professes to be woefully underdeveloped.[5] Indeed, an overriding theme of each workshop concerned the need to overcome severe weaknesses in current forest and agroforestry policy research methodology, as well as national institutional capacity for policy analysis.

United States policy development priorities

In the US experience presented, policy research was employed to design a set of coordinated policy recommendations for advancing the development of agroforestry. While the SCS-commissioned study was exploratory and descriptive in its approach, it was an important first step in mapping out the programmatic and institutional domain for agroforestry in order to locate strategic intervention opportunities. An underlying objective of the policy evaluation team, and the broader-based agroforestry planning group, was to prevent perceptions from developing that agroforestry is the domain of a particular sector or activity, thus undermining its value as an integrating concept.

The strategy focusses on the need to get agroforestry 'on the ground', so the feasibility and benefits of respective systems become more widely recognized and appreciated [AFTA, 1994]. The action team's policy recommendations identify specific institutional niches where agroforestry is likely to gain a foothold at an early stage. These include land retirement programs, specifically the Conservation Reserve Program (CRP), the Wetlands Reserve Program (WRP) and others, as well as private forest improvement programs, including the Stewardship Incentive Program (SIP), the Forest Improvement Program (FIP) and others. More than ten federally supported program areas are identified as currently or potentially supportive of agroforestry activity. The strategy emphasizes the need for technology development research, to hasten the emergence and spread of innovative agroforestry practices.

The land grant system of agricultural colleges and universities is considered well suited to assume scientific leadership for agroforestry technology development, despite the widely discussed need for operational reforms [Lassoie and Buck, 1991]. A needs-driven, competitive, open grants program that encourages multi-disciplinary teamwork and partnership (among scientists, resource management agencies, membership groups, landowner associations, and farmers) constitutes the major operational vehicle and requirement for funds within the overall strategy. Furthermore, the program would be most effective if region-specific, and responsive to diverse socio-ecological conditions throughout the country. The success of each region's efforts and their contributions to agroforestry development nationally, thus will depend on effective coordination within and between the consortia of stakeholders that are emergent in several regions already.

Predictably, a key issue facing the agroforestry policy action team concerns where, specifically, within the array of candidate institutions and programs to center agroforestry. Each possibility must be considered in terms of its political implications for passage of agroforestry support measures within Congress, as well as that organization's legacy and future prospects for assuming the networking and power-sharing roles that will be vital to the long term success of agroforestry. The focus of the current debate concerns whether to broaden the mandate and geographic scope of the USFS's region-specific agroforestry center, to play a national coordinating role, or to lodge this function with the more agriculturally-oriented NRCS.

Policy research can play an important role in evaluating the potential effectiveness of respective choices. It can also be used to evaluate and help minimize the inevitable transaction costs of coordination and collaboration. Once agroforestry begins to develop the momentum that is hoped for, further challenges will arise to guide the emergence of innovative, institutional configurations that can effectively link the relevant public and private spheres of endeavor.

As in the arena of agroforestry technology development, there is much to be learned from nearly two decades of experience in less developed countries. In order to draw relevant lessons from this experience, however, and to constructively tackle the priority policy research lists that are emerging within this sphere, it is instructive to develop a common perspective concerning key causes of agroforestry policy failure.

Why is policy failing agroforestry?

Ultimately, most agroforestry practice stems from household decisions to maintain or incorporate, and manage trees within a farming enterprise. Simple as this sounds, from a policy perspective critical household decisions about trees constitute elements of a highly complex process of guided, or misguided action.

Factors affecting household land use decisions can be attributed to three broad domains and objectives: i) individual – personal welfare maximization; ii) community – social obligation and commitment; and iii) government – legal conformity. Corresponding roughly with these decision domains are the following three instruments of policy influence and action: i) market modifiers; ii) institutions; and iii) laws. Incentives or disincentives for certain decisions and behavior emanate from these instruments of policy with various degrees of force. This is a function, partly, of the comparative influence of three interacting development objectives: i) economic – national growth and development; ii) social – human rights and welfare; and iii) environmental – conservation, preservation, restoration. Key features of this simplified scheme are depicted in Fig. 1.

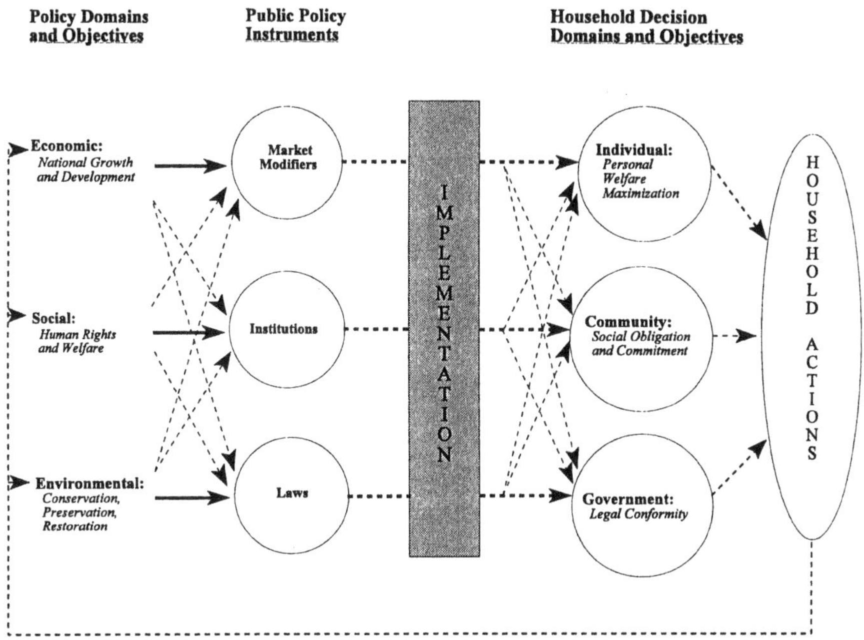

Fig. 1. Land use decision domains and influences. Broken and solid lines suggest the strength of conventional relationships among respective domains. Market modifiers such as taxes, subsidies and direct public investment are employed principally to serve economic development and personal welfare maximization objectives, for example. These affect social welfare and environmental protection goals, but to a lesser extent.

Drawing from this construct, for land use policy to foster a predictable course of acceptable action, signals from respective policy instruments must adequately convey development objectives from corresponding policy domains. These signals, in turn, must be consistent and mutually reinforcing in their effect on household decision-making. For the resulting action to positively affect the condition of land and resources, the farm enterprise, and the vitality of the region as a whole, it is important to calibrate policy signals to household decision-makers on the basis of sound knowledge of these relationships over time.

In this scheme, agroforestry policy problems may be viewed as a function of insufficient correspondence between signals emanating from the instruments of policy (market modifiers, institutions and laws) and respective household decision-making and action domains. Policy-makers may anticipate for example, that granting legal title to land will induce tree management as a function of the prospective owners' welfare maximization objective. In reality, however, customary tenure institutions based on norms of reciprocity and sanction might have an overriding influence. Local norms may not only mitigate against intensive tree management therefore, but leadership may incite social resistance to land titling procedures!

Indeed, examples of 'signal dysfunction' in the record of agroforestry development efforts abound. These can be attributed either to 'gaps' or to 'conflicts' between signals and decision domains. Thrupp [1994a] refers to gaps as delinkages or divergences, while conflicts entail explicit contradictions.

'Gaps' and 'conflicts' in turn, are partially attributable to inconsistent or competing policy objectives. Simply explained, countries everywhere are undergoing dramatic economic, social and environmental reorientations as a function of i) pressures to reform national economies through structural adjustment and market liberalization measures, ii) initiatives to respond to sustainable development mandates for reconciling economic growth with environmental and social equity objectives, and iii) associated political movements for democratization and decentralization. As new actors enter the policy domain to meet these challenges, and market mechanisms, institutional configurations and legal frameworks take shape to operationalize their respective objectives, competition for conceptual and operational space within which to influence land users' decisions intensifies.

By this argument, land and resource tenure emerge as persistent constraints to agroforestry development as a function of competing claims on property rights as an instrument of both economic development and social welfare. When these two objectives conflict, defenders of either domain are reluctant to cede control over this critical resource. Similarly, direct public investment in infrastructure, afforestation schemes, processing capacity, extension education, etc., may be conceived and deployed as instruments of economic, social or environmental policy. Because they are not adequately tailored to meet corresponding household security objectives however, they fall short of their intended purpose. The effects of such contradictions and delinkages can range from inefficiencies in policy or program implementation, to development malpractice as a function of misguided intervention.

Challenges: constructing agroforestry policy pathways

From the perspective of land use policy systems as a whole, that is, the configurations of actors engaged in various policy processes to influence the outcome of land use decisions, it has been shown that they are vastly too complex to predict the influence of a particular intervention with meaningful precision, or to control it. This is particularly so in many developing countries where colonialism, and its subsequent, erratic demise permanently disrupted historical patterns of socio-political management.

Thus, it is useful to disaggregate the system into its constituent stakeholders. To further focus this effort prominence is given to strategic clusters of stakeholder groups, recognizing that these fluctuate, depending on the context and issues at hand. Enough has been learned about agroforestry, and sustainable resource management in general, to appreciate the central role of the resource user and manager in technology innovation. It is also known that this capacity is multiplied when the individual's efforts are facilitated within motivated groups [Engel, 1995]. It is assumed, therefore, that these forces of self-interest and innovative capacity warrant the focus of our attention at this level.

It is appreciated that policy frameworks, or systems of rules and incentives imposed from higher levels of authority, can be essential for creating a supportive context for agroforestry. There is concern that uniformly constructed and applied measures may be inconsistent with local conditions, and thus disabling. Systems of rules and incentives should, therefore, be tailored to characteristic socio-ecological regions.

It is recognized that moving from the domain of agroforestry technology to policy involves confronting issues of inter-dependence. These arise when a number of actors recognize that they hold stakes in the same scarce resource, and can anticipate a collective impact from their actions [Roling, 1994]. Furthermore, that collective perceptions of inter-dependence are an essential condition for managing 'conflicts' and 'gaps' among stakeholders in the development of a land use policy system.

Experience with agroforestry technology and rural development suggests that participatory research mobilizes, empowers and induces technical innovation [Rocheleau, 1991; Buck, 1993; Scherr, 1991]. It is reasonable to surmise, therefore, that as agroforestry advances more emphatically into the realm of policy, it will be appropriate to draw from prior experience to fashion participatory methods and collaborative research strategies, suitable for building bridges and negotiating conflict among inter-dependent stakeholders within the system.

Indeed, during the period of the agroforestry policy assessment initiatives described herein, a profusion of effort has been witnessed to bring these methods to bear on linking local stakeholders with wider planning and policy-making processes, through collaborative research and development activities [Thrupp, 1994b]. Recent syntheses of this experience convey important successes and an expanding trend toward multiple, convergent research and development processes that combine the complementary strengths of organizations with different sectoral expertise and mandates, as well as distinct types of institution [Rocheleau, 1994; Thrupp et al., 1994]. The facilitation of multiple knowledge networks is at the heart of this process.

In the US, the theory and practice of participatory research in sustainable agriculture is just coming to light [Dlott et al., 1994]. By recent accounts the approach will be instrumental in advancing the science and application of agroforestry. As in developing countries, social scientists can help to serve the development effort by expanding their repertoire of participatory research methods to anticipate and meet the new policy challenges. These will surely arise for example, as agroforestry is applied to address watershed or landscape level conservation and development issues.

Conclusions

A recent surge of interest and activity in policy dimensions of agroforestry development throughout the world suggests growing recognition that further innovation may be advanced most effectively by tackling the structural barriers that appear to constrain this good idea from taking off! This perception serves, for a while, to put social scientists in the limelight. It also challenges them to provide a focus through which to penetrate the complex policy environment that characterizes land use decision-making and action. The conventional use of 'global', predictive models must begin to give way to decentralized policy analysis and design.

The coupling of agroforestry with forestry policy issues in less developed countries provides insight into the important phenomena of perceived interdependence among actors with a common stake in the management of natural resources. The condition generates opportunities for joint action, and suggests the critical roles of bridge-building and conflict management to facilitate the development of this perception among stakeholders. In the domain of agroforestry this insight can be usefully applied to foster joint action among economic, social and environmentally-oriented interests and objectives for land use. It serves also to link practitioners, technical advisors and scientific researchers in various configurations, to critically reflect on the policy environment in which they operate, and how they might act upon it.

The strategy places the researcher in the role of facilitator. He or she becomes concerned with generating knowledge and sharing information that will inform and motivate actors in respective knowledge networks to appreciate their interdependence. The researcher focuses also on linking networks to enhance their effectiveness. Participatory research is instrumental in this process.

Strategic research can contribute importantly through its use to evaluate the conditionality and the effectiveness of dispersed policy reform efforts. It should aim to deepen understanding of processes involved in the construction of alternative policy frameworks and mechanisms. It should be employed also to identify key attributes of the socio-ecological environments which give rise to such initiatives. Ultimately, analysts would seek to draw generalizable conclusions concerning conditions and opportunities for strategic policy interventions, designed to reduce gaps and manage conflicts within complex land use decision-making environments.

Methodologically, strategic agroforestry policy research should be concerned with understanding the mechanisms involved in facilitative networking. Improved knowledge about this form of social organization for innovation will help to multiply the efforts of those public and private organizations, variously configured, that are combining strengths to improve agroforestry and sustainable land use policy. This genre of collaborative action policy researchers may be wise, also, to link their efforts with more conventional, sector or policy-specific research activities. The latter can serve, when data are available, to test the feasibility and impact of specific propositions for policy reform that emerge from facilitated networks. Such unions, in turn, should help to develop data bases from which increasingly robust models of land use decision factors and processes eventually can be constructed.

Notes

1. Diagnosis and Design, or 'D&D', was developed at the International Center for Research in Agroforestry (ICRAF) in the mid-1980s as agroforestry's analogue to farming systems analysis. A key distinguishing feature of D&D is its focus on meeting the broad array of basic human needs from the agricultural landscape and farming enterprise. The methodology germinated from researchers' observations of rural development project experience with agroforestry. It was brought to fruitition as an interdisciplinary framework for scientific research, then re-interpreted as a participatory methodology for rural development planning and intervention.
2. In addition, the literature reports numerous case studies of field-based encounters with policy constraints, and efforts to overcome them. See for example, Follis and Nair [1994] and Josiah [1994].
3. AFTA has grown from about 100 members in 1993 when its first newsletter was published, to approximately 400 in January, 1995. Membership is open to scientists, professionals and practitioners of any nationality who have an interest in temperate zone agroforestry. It is coordinated by an elected president, based at Michigan State University, and a seven person steering committee.
4. The list of topics associated with each research problem is specified in CIFOR [1994], pp. 7–9.
5. The issue of institutional capacity for policy research is discussed in CIFOR [1994], pp. 7–9.

References

AFTA (1994) Agroforestry for sustainable development: a national strategy to develop and implement agroforestry. Proceedings of a Workshop to Develop a Framework for a Coordinated National Agroforestry Program, June 29–30, 1994, Nebraska City, NE. Department of Forestry, Michigan State University, MI, USA

Annis S (1988) Can small-scale development be large-scale policy? In: Annis S and Hakim P (eds) Direct to the Poor: Grassroots Development in Latin America, pp 209–218. Lynne Rienner Publications, Boulder, CO, USA

Buck L (1990) Planning agroforestry extension projects: the Care International approach in Kenya. In: Budd WW, Duchhart I, Hardesty LH and Steiner F (eds) Planning for Agroforestry, pp 101–131. Elsevier, Amsterdam, The Netherlands

Buck LE (1993) Development of participatory approaches for promoting agroforestry: Collaboration between the Mazingira Institute, ICRAF, Care-Kenya, KEFRI and the Forestry Department (1980–91). In: Wellard K and Copestake JG (eds) Non-Governmental Organizations and the State in Africa: Rethinking Roles in Sustainable Agricultural Development, pp 118–135. Routledge, London, UK

CIFOR (1994) A review of tropical forestry and agroforestry problem areas and policy research needs and the planned response of the CGIAR system. Working paper no. 5, Bogor, Indonesia

Cornell University and the University of Florida (1994) Global Research on the Environmental and Agricultural Nexus (GREAN) for the 21st Century: A Proposal for Collaborative Research among US Universities, CGIAR Centers and Developing Country Institutions. Report of the Taskforce on Research Innovations for Productivity and Sustainability, funded by the Ford and Rockefeller Foundations. University of Florida, Gainesville, FL, USA

Dlott JW, Altieri MA and Masumoto M (1994) Exploring the theory and practice of participatory research in US sustainable agriculture: A case study in insect pest management. Agriculture and Human Values 11(2–3): 126–139

Ehrenreich JH, Keerio G and Khan GS (1994) Opportunities and problems in agroforestry: A comparison between developed and lesser developed countries. In: Schultz RC and Colletti JP (eds) Opportunities for Agroforestry in the Temperate Zone Worldwide: Proceedings of the Third North American Agroforestry Conference, August 15–18, 1993, pp 317–318. Ames, IA, USA

Engel PGH (1995) Facilitating innovation: An action-oriented approach and participatory methodology to improve innovative social practice in agriculture. PhD thesis, Agricultural University, Wageningen, The Netherlands

Esman MJ and Uphoff NT (1984) Local Organizations: Intermediaries in Rural Development. Cornell University Press, Ithaca, NY, USA

Follis M and Nair PK (1994) Policy and institutional support for agroforestry: An analysis of two Ecuadorian case studies. In: Schultz RC and Colletti JP (eds) Opportunities for Agroforestry in the Temperate Zone Worldwide: Proceedings of the Third North American Agroforestry Conference, August 15–18, 1993, pp 319–326. Ames, IA, USA

Food, Agriculture, Conservation and Trade Act of 1990, United States Public Law 101–624, Title XIV: Conservation

Food Security Act of 1985, United States Public Law 99-198, Title XII: Conservation

Garrett HE and Buck LE (1994) Forest management practices in temperate zone agroforestry: silvicultural and policy aspects in the United States. International Symposium: Agroforestry and Land Use Change in Industrialized Nations, May 30–June 2, Berlin, Germany

Garrett HE, Kurtz WB, Buck LE, Gold MA, Hardesty LH, Lassoie JP, Pearson HA and Slusher JP (1994) Agroforestry: an integrated land-use management system for production and farmland conservation. Prepared for USDA SCS 68-3A75-3-134, Resource Conservation Act Appraisal

Gregersen H, Oram P and Spears J (1992) Priorities for Forestry and Agroforestry Policy Research: Report of an International Workshop. IFPRI, Washington, DC, USA

IFPRI, ODI, CIFOR, IICA and ISNAR (1994) Policies on Local Organizations for Natural

Resource Management: Towards an Interdisciplinary Research Agenda. Workshop Proposal. IFPRI, Washington DC, USA

Jacobs H and Harper S (1993) The changing nature of settlement policy in the USA: A theoretical and case study review. In: Harper S (ed) The Greening of Rural Policy: International Perspectives, pp 135–150. Belhaven Press, New York, NY, USA

Josiah SJ (1994) Implementing large-scale agroforestry projects through umbrella NGO's. In: Schultz RC and Colletti JP (eds) Opportunities for Agroforestry in the Temperate Zone Worldwide: Proceedings of the Third North American Agroforestry Conference, August 15–18, 1993, pp 349–360. Ames, IA, USA

Josiah SJ and Gregersen HM (1994) Expanding the impacts of social forestry programs in developing countries. The Environmental and Natural Resources Policy and Training Project Policy Brief. EPAT/MUCIA, University of Minnesota, St. Paul, MN, USA

Kidd CV and Pimentel D (eds) (1992) Integrated Resource Management: Agroforestry for Development. Academic Press, Inc, San Diego, CA, USA

Korten D (1990) Getting to the 21st Century: Voluntary Action and the Global Agenda. Kumarian Press, West Hartford, CT, USA

Lassoie JP and Buck LE (1991) Agroforestry in North America: New challenges and opportunities for integrated resource management. In: Garrett HE (ed) Proceedings of the Second Conference on Agroforestry in North America, August 18–21, 1991, pp 1–19. Columbia, MO, USA

Long AJ and Nair PK (1991) Agroforestry system design for the temperate zones: Lessons from the tropics. In: Garrett HE (ed) Proceedings of the Second Conference on Agroforestry in North America, August 18–21, 1991, pp 133–139. Columbia, MO, USA

Nair PKR (1993) State-of-the-art of agroforestry research and education. Agroforestry Systems 23: 95–119

Raintree JB (1990) Theory and practice of agroforestry diagnosis and design. In: MacDicken KG and Vergara NT (eds) Agroforestry: Classification and Management, pp 58–97. John Wiley & Sons, New York, NY, USA

Rocheleau DE (1994) Participatory research and the race to save the planet: questions, critique, and lessons from the field. In: Thrupp LA (ed) Agriculture and Human Values 11(2–3): 4–25

Rocheleau DE (1991) Participatory research in agroforestry: Learning from experience and expanding our repertoire. Agroforestry Systems 15: 111–137

Roling NG (1994) Platforms for decision-making about ecosystems. In: Fresco LO, Stroosnijder L, Bouma J, Keulen J (eds) The Future of the Land: Mobilizing and Integrating Knowledge for Land Use Options, pp 385–393. John Wiley & Sons, New York, NY, USA

Scherr SJ (ed) (1991) Special issue: On-Farm Agroforestry Research. Agroforestry Systems 13(2–3): 91–302

Soil and Water Conservation Society (1994) When Conservation Reserve Program Contracts Expire: The Policy Options, Conference Proceedings, February 10–11, 1994, Arlington, VA, USA

Thrupp LA (1994a) Building bridges for people-centered sustainable agriculture: Conflicts and convergences in the agriculture-environmental interface. Draft. World Resources Institute, Washington, DC, USA

Thrupp, LA (ed) (1994b) Special issue: Participation and Empowerment in Sustainable Rural Development. Agriculture and Human Values 11(2–3)

Thrupp LA, Cabarle B and Zazueta A (1994) Participatory methods in planning and political processes: Linking the grassroots and policies for sustainable development. Agriculture and Human Values 11(2–3): 75–84

Webster's Dictionary (1988) Second edition. Simon & Schuster, New York, NY, USA

World Bank (1994) Facing the Global Environmental Challenge: A Progress Report on World Bank Global Environmental Operations. The World Bank, Washington, DC, USA

World Resources Institute (in collaboration with UNEP and UNDP) (1992) World Resources: A Report. World Resources Institute, New York, NY, USA

Agroforestry Systems **30**: 75–86, 1995.

Economic evaluation of financial and non-financial costs and benefits in agroforestry development and the value of sustainability

C. PRICE
School of Agricultural and Forest Sciences, University of Wales, Bangor, Gwynedd LL57 2UW, UK

Key words: cost-benefit analysis, environmental values, futures, non-market effects

Abstract. Financial appraisal of agroforestry is not different in kind from appraisals applied to pure forestry. Existing techniques for valuing non-market effects may also be applied. Effects on hydrological and carbon fluxes might be less favourable than those of pure forests. In the confusing field of biodiversity valuation too, only limited and specific claims can be made for agroforestry. Amenity valuation techniques developed for trees and woodlands encounter additional problems of interpretation in an agroforestry context. Agroforestry may offer a quantifiable value in short-term sustainability, and might meet recent economic definitions of long-term sustainable development, without encountering the full dangers to future generations embodied in conversion of natural to human capital.

Introduction

By referring to financial and non-financial effects, to costs and to benefits, and (through sustainability) to the long term as well as the short, the title of this paper excludes nothing of relevance to agroforestry. Hence it must be extremely selective, or extremely superficial in the material treated. On the other hand, the number of economic studies of agroforestry which stray beyond discounted cash flow (DCF) appraisal of financial costs and benefits is pitifully small. Even those which evaluate environmental effects such as soil conservation [Ehui et al., 1990] or nitrogen fixing [Stone et al., 1993] address on-site financial benefits, rather than classical economic externalities. Some papers about non-financial benefits quote results from pure forestry valuations. However, it is in the interactive nature of agroforestry, that its effects cannot be compiled as a weighted mean of agricultural and forestry components.

Hence this paper takes a broad, superficial view, summarizing economists' current thinking about valuation of non-financial effects, and about sustainability. It presents no new evidence on biophysical effects, but exemplifies how valuation techniques might be applied to the known and expected effects of agroforestry. Where appropriate, comparison with the effects of gross scale land use mixtures is made.

Most reference is to temperate conditions, but where relevant – and particularly in relation to sustainability – tropical conditions are mentioned too.

Financial values

Financial evaluations of agroforestry based on projected or actual yields are becoming more abundant (many recent articles in *Agroforestry Systems*). Such DCF calculations require no innovations of economic technique, but merely a more elaborate application of protocols which have been central to forest economics for 150 years [Stone et al., 1993]. Nor are the results they deliver surprising. For example, in comparisons with pure forestry or pure agriculture on similar sites, a low discount rate favours pure forestry, and a high rate pure agriculture. But agroforestry may give greater net benefit than either pure land use in the following circumstances.

- There are supplementary or complementary productivity effects between crop components [Wannawong et al., 1991].
- Agricultural prices are falling relative to forestry prices: thus the agricultural component is sold when its price is relatively high, and the forestry component when its price is relatively high [Willis et al., 1993].
- Forestry revenues are higher per unit of photosynthetic potential utilized, but the discount rate is high enough to outweigh this for food output at the beginning of the production cycle (C. Price, unpub.). (This effect does not seem to have been widely noted.)
- Some non-market benefit arises from mixing of land uses, either physically within the site (for example, provision of edge habitats for wildlife), or within the economic enterprise (for example, stabilization of cash flow).

Cases of the last kind are mentioned below. Opposite cases, where the mixing of crops creates negative interactions, also receive speculative evaluation.

Non-financial values

Recent interest in environmental economics has brought numerous new classifications of techniques for monetizing extra-market effects. An exhaustive and clear-cut taxonomy recognises eight approaches to valuation [Price, 1989, Chapter 24].

Approach 1 – Market benefits are created or destroyed for other agents in the economy.

Approach 2 – Financial costs are imposed on or undertaken by other agents, in order to protect existing benefits or prevent future damage.

Approach 3 – Similar commodities have a price in other markets.

Approach 4 – Causes and campaigns produce voluntary contributions.

Approach 5 – Respondents to questionnaires state a willingness to pay for benefits or to accept compensation for costs.

Approach 6 – Expert judgements of value are formulated and subjected to discussion and refinement.

Approach 7 – Costs have previously been willingly incurred in order to favour non-market benefits.
Approach 8 – Financial costs are incurred to gain access to non-market benefits.

No systematic review of the merits and problems of these approaches is offered here: the interested reader is referred elsewhere [Hanley and Spash, 1993]. But at least one relevant example of each approach is given in the following sections.

Biophysical impacts and production function approaches

Approaches 1 and 2 (which include variants known severally as *production function, replacement cost, defensive expenditure, environmentally compensating expenditure*) recognise that particular land uses have effects which, though not themselves marketed, nonetheless hold financial implications for other land users, locally, regionally and even globally. The economic elements of these approaches are, in principle, straightforward: it is in obtaining reliable biophysical relationships that most of the problems arise.

Hydrological impact

It has long been known that conifer crops, under temperate oceanic conditions, evaporate significantly more water than grassland on equivalent sites [Calder and Newson, 1979]. They are also more successful in 'scavenging' particulates (by dry deposition from aerosols) and acidic solutions (from clouds) [Gee and Stoner, 1988].

Both ionic and sediment loads are increased by cultivation associated with afforestation, and by the gross intervention of forest harvesting [Leeks and Roberts, 1987; White, 1990].

Effects on water quality and differential evaporation bring various financial implications. Unexpected extra costs may be incurred, such as liming of catchments to counteract the acidification due to the scavenging of acid mists by conifers [Whitehead et al., 1990]. In one celebrated incident the sediment load running off a 90-ha plantation into Crai Reservoir in Wales led to £45,000 being spent on temporary alternative sources of water.

Discounted costs may increase because engineering works are brought forward in time. Advanced investment in new water storage capacity, due to increased evaporation from forest canopies, could incur a cost of the same order of magnitude as the cost of planting the forest [Collett, 1970]. The cost of advancing water treatment investment as a result of the Crai Reservoir incident was put at £317,000, giving a total imposed cost of about £4000 per hectare of forest.

Economically quantifiable *loss of benefits* may be incurred too, including reduced revenues from fisheries suffering acidification [Milner and Varallo, 1990], and reduced electricity sales where hydroelectric catchments are affected [Barrow et al., 1986]. In both cases, the effects may determine whether forestry is worthwhile or not.

The above effects are due partly to high moisture retention on large leaf surface areas, and partly to the aerodynamic roughness of tree canopies. In both respects, open-grown tree crops would be expected to exert an effect which is more-than-proportional to the volume of timber grown. Recent research [Teklehaimanot et al., 1991] has indeed shown greater rainfall interception and evaporation rates for widely spaced trees. In small woods greater losses were found at woodland edges than in interiors [Harding et al., 1992], while edges also scavenge mists and atmospheric particulates more efficiently, potentially contributing more to watercourse pollution. Thus greater costs may be incurred by arranging a given size and number of trees in agroforestry configurations, rather than as conventional woodlands.

Cultivation effects depend on the specific agroforestry system employed [White, 1990]. Where there is an arable component, and where terrain varies in erodibility, a *separation* of crop components may minimize the cultivation of susceptible sites, and thus soil losses.

Climate change

At least eight approaches to costing carbon fluxes are found in the literature [Cline, 1992]. They generally refer to losses of production consequent on global warming and sea-level rise, or to responsive/compensating expenditures undertaken to avoid or mitigate damage. Cost estimates have ranged widely, from 20p to £200 per tonne of carbon [Price and Willis, 1993], with typical estimates for loss and damage around £10, depending on the discount rate used and on whether some arbitrary time horizon is imposed.

Inserting the £10 value in a model for poplar grown at a wide, agroforestry spacing gives the estimated value of carbon fixing as about £470 per ha, or 30% of combined timber/carbon benefit. For a conventional forestry spacing, carbon is worth £960 per ha, or 43%. Assuming no differential carbon storage by ground vegetation and soil, the difference in favour of conventional spacing seems significant. It is attributable to the smaller volume per ha under agroforestry. Growing an equal volume of timber would give similar carbon values under either regime.

Willingness to pay for biodiversity

Biodiversity has become a fashionable field for economic research, full of unresolved debates about terminology and concepts, and much intertwined with philosophical issues [Randall, 1991]. An intensive and high-productivity cultural system as often used in agroforestry is based on numerously replicated genes. Expansion of the system thus has marginal direct benefit to the gene pool. On the other hand, compared with intensive conifer forest (low levels of light reaching the forest floor) and intensive agriculture (chemical application regimes designed to promote monoculture) agroforestry does offer an improved spectrum of niches for colonization.

To evaluate biodiversity benefits requires a deep enquiry into *why* diversity should be considered valuable. Biodiversity values are conveniently treated under the headings of instrumental, interest and intrinsic values [Price, 1993a], these referring respectively to:

- substances produced or functions served which promote the material cause of humanity;
- pleasure given to humanity by the existence of 'nature';
- the good supposed to reside in the existence of 'nature' in itself.

Instrumental values

Intimate mixing of crops promotes system function in numerous ways: fixing of nitrogen; soil stabilization; provision of nesting and foraging sites for crop-pollinating species [Saville, 1993]. These benefits of agroforestry are, however, mostly captured by financial analysis based on the enhanced productivity of individual enterprises: whether the long term implications are adequately addressed is touched on later in relation to sustainability.

Interest values

On-site enjoyment of wildlife has traditionally been valued by analysis of travel costs (approach 8). Using this method, the annual wildlife benefits of Britain's state forests have been estimated at about £15 million [Benson, 1994]. The method, however, fails to distinguish the extra value attributable to the particular land use, and is unlikely to allow separate valuation for particular species [Price, 1987]. The emphasis has now shifted to questionnaire research (approach 5), where, despite a plethora of problems associated with sample and questionnaire design, the contingent valuation method (CVM) has become a fashionable tool [Mitchell and Carson, 1989].

A particular recent focus of CVM studies has been on what are called non-use, or more properly passive use values, which are experienced without actual visits to site. While these values are susceptible to shameless exploitation as a means of boosting the value ascribed to any land use, they are normally important only for primeval ecosystems or ancient cultural landscapes, rather than for new land use systems. Thus agroforestry – and particularly agroforestry *development* – is unlikely to show any measurable passive use value in an unbiased survey of *habitat*. Colonizing *species* in an open-structured and rather dynamic agroforestry system are likely to be opportunists, which are not of the greatest public interest (S. Hinsley, pers. comm.): current conservation priorities in Britain are with species of open moorland, undrained wetland, and the deep interior of ancient semi-natural woodland.

This rather unenthusiastic assessment is less applicable to conditions, especially tropical ones, where agroforestry represents a richer successor to a multilayered forest ecosystem than any pragmatic alternative would do.

CVM studies of passive use should on the whole be treated with caution, for several reasons in addition to those normally associated with hypothetical questionnaires. It is, firstly, suggested [Minter, 1994] that willingness to pay is meaningful only after respondents have received adequate information about the nature and functions of species or habitats. However, such information both makes the respondent highly unrepresentative of the population generally, and risks double-counting instrumental values, which are appropriately addressed through other valuation approaches.

Secondly, the very fact of a questionnaire's being administered raises the profile of a particular species or habitat. It suggests that something important is at stake, and at the same time (temporarily) displaces other environmental concerns. The value derived is an artefact of the questionnaire, and overstates the long-term impact of a particular species or habitat on the interest value derived from nature conservation taken as a whole [Price, 1995].

Thirdly, it has been shown that CVM studies tend to elicit a 'symbolic value' [Blamey and Common, 1994] or a value for the 'moral satisfaction' [Kahneman and Knetsch, 1992] associated with a general support for the environment. Such values could be reallocated to other convenient symbols, without significant loss.

Money subscribed to wildlife campaigns (approach 5) is potentially even more likely to misrepresent values. Subscriptions can scarcely be interpreted in terms of rational consumers buying a conservation good, since the probability that an individual contribution will affect the outcome of the campaign is very small. Instead, contributions represent the purchase of image and self-image, and possibly altruistic motives which are likely to entail multiple counting of benefits.

Intrinsic values

Multiple definition of intrinsic value has led to a deep confusion of interpretation. It has been equated with some aspects of passive use value [Pearce and Turner, 1990]. It has also been suggested that 'the value we place may include our valuation of an animal's own existence'. Truly, however, intrinsic value is not a part of human satisfaction, and the only means to interpret it is via judgements about what sacrifices of profit *ought to be made*, to improve conditions for other species. Past 'ethical' conservation decisions might be thought to reflect such valuations (approach 7). However, past decisions may have been made in different circumstances. The initiator of this approach, Robinson Gregory [1955], mentions open forage areas in forests ambiguously, in terms both of financical optimization, and as a means of judging broader multiple use. Even if an ethical conservation motive can confidently be attributed for past sacrifices, there are further possibilities of confusion: human judgement is influenced by human interest, and a quite unwarranted bias towards 'cuddly' and towards rare species may have prompted such decisions. Seen from the 'viewpoint' of wildlife itself, existence as an individual in a population of individuals, rather than existence as a member of a rare or attractive species, must be considered the principal source of value.

The relevance of such metaphysical considerations for agroforestry development is this: ecosystems whose structural diversity encourages a high biomass per hectare should be deemed to contribute a high intrinsic value, even if the represented species are neither rare nor of particular human interest.

Attributes of amenity

Hedonic pricing

While contingent valuation has become the 'lead methodology' for valuing wildlife, recent attention in valuing landscape has concentrated on the *hedonic pricing method.* This interprets expenditure, typically on house purchase or travel to countryside sites, as a valuation of attributes to which the expenditure gives access (approach 8): these attributes include environmental as well as social, locational and structural attributes of houses, or the environment and activities available on countryside visits.

One study [Garrod and Willis, 1992] found that the woodland attribute which best accounted for variation in house price was whether tree cover in proximity to the house exceeded 20% or not. Preference for 20–30% tree cover also appeared in a survey of designated community forestry areas [Whiteman, 1994]. These studies, however, make no reference to the *arrangement* of woodland and open ground – whether, for example, the components were in large blocks, or intimately mixed as in classical agroforestry systems. Moreover, they do not consider changing appearance through time. In any event, results given for conventional, closed canopy forestry are unlikely to translate into an agroforestry context.

Of more relevance are survey results revealing greater willingness to pay for access to forests which included open areas and diversity of height [Hanley and Ruffell, 1993]. However, in terms of statistical significance and agreement between techniques, the results of comparison of several statistical approaches to forest attributes were described as 'disappointing'. Given the theoretical objections to hedonic pricing [Price, 1991], this should cause no surprise.

Expert judgement

Approach 6 is also based on attributes. Here, however, money values are derived from subjective estimates of what is 'reasonable' and what commands consensus among other experts. Helliwell's system [Helliwell, 1990] requires a score on a scale 1–4 to be given for each of seven attributes, some relevant and readily quantified for an agroforestry system. For example, the scoring for a hectare of poplar, in a flood plain, spaced at 8 × 8 m, halfway through a rotation of 30 years would be

Size of tree	Medium	2
Useful life expectancy	10–20 years	1
Importance of position in landscape	Some	2
Presence of other trees	Many	1
Relation to the setting	Very suitable	3
Form	Good	3
Special factors	None	1

provided the trees were regarded as individuals. Scores are multiplied together, and further multiplied by £10 to give a value per tree – in this case £360. The value for the whole hectare would be £56,250. However, if the trees are considered collectively as part of a wood, the aggregate score would be derived as follows:

Size of woodland	Medium	2
Position in landscape	Average	2
Viewing population	Average	2
Presence of other trees and woodland	Area 5–25% wooded	2
Composition and structure	Even-aged crop	2
Compatibility in the landscape	Good	3
Special factors	None	1

The aggregate score, 96, is to be multiplied by a 'points' value of £25, giving a value of £2400.

Clearly, whether an agroforestry system is seen as individual trees or as an integral wood has an important bearing on its ascribed value. There are also problems in interpreting the descriptive categories. Notably, 'appropriateness to setting' depends on cultural background. The English Countryside Commission has been negative about poplar agroforestry in the traditional English landscape (D. Gear, pers. comm.).

Recreational agroforestry

The open structure and variety of agroforestry systems makes them in one sense amenable to the kind of recreational development which has been widely canvassed as a justification for conventional and community forestry. On the other hand, regularity of structure and intensive management of the agricultural component are less favourable. One proposal for introducing agroforestry to a recreation park [Broder and Odronic, 1990] saw it simply as a means of raising funds for recreation management.

The value of sustainability

Sustainability of land use systems may be seen not only from physical and economic viewpoints, but also in short- and long-term perspectives. In the short-term economic sense, sustainability implies an ability to survive fluctuations in the conditions of production and sale. Several authors [Blandon, 1985; Lilieholm and Reeves, 1991] have pointed out that agroforestry, by delivering several products, is less vulnerable to fluctuations in markets. In a subsistence setting, many reasons have been identified for expecting tree crops to enhance the stability of consumption [Falconer, 1990].

However, the systems which are most diversified may not have the highest value measured in DCF terms [Reeves and Lilieholm, 1993]. A means of trading off profitability and stability of income is offered by examining behaviour in the market for a similar product (approach 3), that is, a diversified portfolio of shares on the stock market. An elaborate body of theory (the capital assets pricing model) 'prices' the advantage of a stabilized cash flow in terms of the lower rate of return habitually accepted for products whose price varies little, or even inversely, with fluctuations in prices generally.

But many economists have argued that using such variation in rates of return as a basis for adjusting discount rates has no adequate theoretical basis, and gives misleading and perverse results when the production cycle is not an annual one [Price, 1993b]. In particular, it favours the forestry component over the agricultural component, when favour should be given to the *very jointness* of the products. Thus a price justified by the diversification that agroforestry offers may actually work to reduce the system's diversity. The value of stable income is more reliably interpreted as a percentage reduction in revenue accepted as the price of stability.

Note that the advantages of diversification in the face of market risks may be achieved by mixing enterprises on a coarse scale as well as by adopting a single enterprise which embodies intimately mixed components: once again, the case for agroforestry *per se* must hang on the benefits noted in the 'financial values' section above. Such *particular* benefits include robustness to biophysical threats, but not to market fluctuations [Price, 1989, Chapter 18].

In relation to long-term economic sustainability of land uses, economists have recently advanced some striking, counter-intuitive, and dangerous claims. Sustainability as most widely understood entails that resource use should avoid 'compromising the ability of future generations to meet their own needs' [World Commission on the Environment and Development, 1987]. Conventional DCF, on the other hand, automatically gives negligible weight to benefits and costs accruing beyond a couple of generations into the future, and so seems a very efficient way of compromising sustainability. Yet economists argue discounting is justified precisely to *protect* the interests of future generations: firstly because discounting reduces the profitability of investments, and thus the probability that environmentally damaging investments will achieve an acceptable economic performance [Pearce and Turner, 1990];

secondly, because it ensures allocation of resources to investments in human-made capital, which offer greater returns to future generations than keeping resources in low-yielding natural assets [Kay and Mirrlees, 1975].

These ideas can be traced back to development theory in the 1950s and 1960s [Westoby, 1962]. Yet the actual outcome in that era was that natural resource destruction financed, not real development, but immediate consumption and unproductive investment [Price, 1990]. There is little reason to suppose that the current round of forest destruction will be accompanied by a more benign use of the funds made available.

As for the role of discounting in barring environmentally destructive investments, it is more aptly achieved by focusing on the sustained and undiscounted environmental losses in themselves, in so far as these effects are deemed to have undiminishing importance in the future.

But, as already emphasized, it should not be supposed that low or zero discount rates would favour agroforestry rather than pure forestry. The justification of agroforestry may be seen as a compromise technology: human-made capital that, while offering higher returns than unmodified ecosystems, may nonetheless preserve some environmental advantages of the original multilayer system. Such a realignment of productive assets is physically accountable within the ambit of the land manager, by contrast with the remission of exploitation revenues into a nebulous programme of 'development investment'.

Conclusion

Judging by relevant insights and results from environmental and forestry economics, agroforestry does not offer, consistently and without need for quantification, extraordinary benefits of the financial or non-financial kind. There may, indeed, be particular costs. The argument for agroforestry development continues to rely largely on biophysical and economic complementarity effects, and, in the tropical situation, on the ecological advantage over monocultures. No mystique, no propagandizing attached to a newly-topical production system will impress hard-nosed economists: a systematic and quantitative investigation of the claimed benefits and admitted disadvantages is what they require, as a precursor to fuller economic evaluation.

Acknowledgements

I am grateful to colleagues in the BioEconomic Agroforestry Modelling group in Bangor for useful discussions, and especially to Rob Willis for development of the carbon modelling spreadsheet.

References

Barrow P, Hinsley, AP and Price C (1986) The effect of afforestation on hydroelectricity generation: a quantitative assessment. Land Use Policy 3: 141–151

Benson JF (1994) Values for forest landscapes using travel cost and tokens. Landscape Research 19: 23–25

Blamey R and Common M (1994) Symbolic Responses in Contingent Valuation Studies. Centre for Resource and Environmental Studies, Australian National University, Canberra, Australia

Blandon P (1985) Agroforestry and portfolio theory. Agroforestry Systems 3: 239–249

Broder JM and Odronic BH (1990) Economic potential of agroforestry for public recreational parks. Agroforestry Systems 10: 99–112

Calder IR and Newson MD (1979) Land use and upland water resources in Britain – a strategic look. Water Resources Bulletin 15: 1628–1639

Cline WR (1992) The Economics of Global Warming. Institute for International Economics, Washington, DC, USA

Collett MEW (1970) External costs arising from the effects of forests upon streamflow in Britain. Forestry 43: 181–193

Ehui SK, Kang BT and Spencer DSC (1990) Economic analysis of soil erosion effects in alley cropping, no-till and bush fallow systems in South Western Nigeria. Agricultural Systems 34: 349–368

Falconer J (1990) Agroforestry and household food security. In: Prinsley RT (ed) Agroforestry for Sustainable Production, pp 215–239. Commonwealth Science Council, London, UK

Garrod GD and Willis KG (1992) Valuing goods' characteristics: an application of the hedonic pricing method to environmental attributes. Journal of Environmental Management 34: 59–76

Gee AS and Stoner JH (1988) The effects of afforestation and acid deposition on the water quality and ecology of upland Wales. In: Usher MB and Thompson DBA (eds) Ecological Change in the Uplands, pp 273–287. Blackwell Scientific Publications, Oxford, UK

Gregory GR (1955) An economic approach to multiple use. Forest Science 1: 6–13

Hanley ND and Ruffell R (1993) The contingent valuation of forest characteristics. Journal of Agricultural Economics 44: 218–229

Hanley ND and Spash C (1993) Cost-Benefit Analysis and the Environment. Edward Elgar, Cheltenham, UK

Harding RJ, Hall RL, Neal C, Roberts JM and Rosier PTW (1992) Hydrological Impacts of Broadleaf Woodlands. National Rivers Authority, Bristol, UK

Helliwell DR (1990) Amenity Valuation of Trees and Woodlands. Arboricultural Association, Romsey, UK

Kahneman D and Knetsch JL (1992) Valuing public goods: the purchase of moral satisfaction. Journal of Environmental Economics and Management 22: 57–70.

Kay JA and Mirrlees JA (1975) The desirability of natural resource depletion. In: Pearce DW and Rose J (eds) The Economics of Natural Resource Depletion. Macmillan, London, UK

Leeks GJL and Roberts G (1987) The effect of forestry on upland streams with special reference to water quality and sediment transport. In: Good JE (ed) Environmental Effects of Plantation Forestry in Wales, pp 9–25. Institute of Terrestrial Ecology, Bangor, UK

Lilieholm RJ and Reeves LH (1991) Incorporating economic risk aversion in agroforestry planning. Agroforestry Systems 13: 63–71

Milner NJ and Varallo PV (1990) Effects of acidification on fish and fisheries in Wales. In: Edwards RW, Gee AS and Stoner JH (eds) Acid Waters in Wales, pp 121–143. Kluwer Academic Publishers, Dordrecht, The Netherlands

Minter R (1994) Sharing common values. Landscape Research 19: 2–4

Mitchell RC and Carson RT (1989) Using Surveys to Value Goods: the Contingent Valuation Method. Resources for the Future, Washington, DC, USA

Pearce DW and Turner RK (1990) The Economics of Natural Resources and the Environment. Harvester Wheatsheaf, Hemel Hempstead, UK

Price C (1987) Upland land use: towards the elusive balance. In: Bell M and Bunce RGH (eds) Agriculture and Conservation in the Hills and Uplands, pp 156–159. Institute of Terrestrial Ecology, Grange-over-Sands, UK

Price C (1989) The Theory and Application of Forest Economics. Blackwell, Oxford, UK

Price C (1990) Deforestation and economic criteria. Project Appraisal 5: 159–166

Price C (1991) Landscape Valuation and Public Decision Making. Countryside Commission, Cheltenham, UK

Price C (1993a) Discounting diversity: a somewhat philosophical speculation. Scandinavian Forest Economics 34: 37–57

Price C (1993b) Time, Discounting and Value. Blackwell, Oxford, UK

Price C (1995) Pros and cons of alternative evaluation methods. In: Willis KG and Corkindale J (eds) Environmental Valuation: New Directions, pp 160–177. CAB International, Wallingford, UK

Price C and Willis RW (1993) Time, discounting and the valuation of forestry's carbon fluxes. Commonwealth Forestry Review 72: 265–271

Randall A (1991) The value of biodiversity. Ambio 20(2): 64–68

Reeves LH and Lilieholm RJ (1993) Reducing financial risk in agroforestry planning: a case study in Costa Rica. Agroforestry Systems 21: 169–175

Saville NM (1993) Bumblebee ecology in woodlands and arable farmland. PhD thesis, University of Cambridge, Cambridge, UK

Stone SW, Kyle SC and Conrad JM (1993) Application of the Faustmann principle to a short-rotation tree species: an analytical tool for economists, with reference to Kenya and leucaena. Agroforestry Systems 21: 79–90

Teklehaimanot, Z, Jarvis PG and Ledger DC (1991) Rainfall interception and boundary layer conductance in relation to tree spacing. Journal of Hydrology 123: 261–278

Wannawong S, Belt GH and McKetta CW (1991) Benefit-cost analysis of selected agroforestry systems in northeastern Thailand. Agroforestry Systems 16: 83–94

Westoby JC (1962) The role of forestry in the attack on economic underdevelopment. Unasylva 16: 168–201

White S (1990) The technical costs and benefits of agroforestry to the catchment as a whole. In: Prinsley RT (ed) Agroforestry for Sustainable Production, pp 323–350. Commonwealth Science Council, London, UK

Whitehead PG, Musgrove TJ and Cosby BJ (1990) Hydrochemical modelling of acidification in Wales. In: Edwards RW, Gee AS and Stoner JH (eds) Acid Waters in Wales, pp 255–277. Kluwer Academic Publishers, Dordrecht, The Netherlands

Whiteman A (1994) The supply and demand for timber, recreation and community forest outputs from forests in Great Britain. PhD thesis, University of Edinburgh, Edinburgh, UK

Willis RW, Thomas TH and Van Slycken J (1993) Poplar agroforestry: a re-evaluation of its economic potential on arable land in the United Kingdom. Forest Ecology and Management 57: 85–97

World Commission on the Environment and Development (1987) Our Common Future. Oxford University Press, Oxford, UK

Agroforestry Systems **30**: 87–103, 1995.

Farmer costs and benefits* from agroforestry and farm forestry projects in Central America and the Caribbean: implications for policy

D. CURRENT[1] and S. J. SCHERR[2]
[1] *University of Minnesota/EPAT Project, St. Paul, MN, USA;* [2] *IFPRI, 1200 17th Street N.W., Washington, DC 20036, USA*

Key words: economic analysis, extension, financial analysis

Abstract. This paper presents results of an evaluation of the benefits provided by agroforestry and farm forestry projects in Central America and the Caribbean, and the policy implications. Tree planting provided financial benefits to farmers, as well as social, economic and environmental benefits. These justify policy interventions to explicitly recognize the potential of on-farm tree-planting efforts and provide a policy environment favorable to such efforts. New or revised policies are needed, particularly in a period of declining public budgets, to limit external incentives to in-kind inputs, rather than financial subsidies; promote market development for tree products; adopt low-cost, community-based extension strategies; focus research support; and reduce regulatory disincentives.

Introduction

The technical potential of agroforestry and farm forestry to address a range of development and environmental challenges is now widely recognized. Efforts to integrate more trees into farming systems on a large scale have generated new interest in creating a supportive policy environment for these types of land use. A set of international workshops held to identify priority policy research issues for forestry and agroforestry noted, among the five predominant themes, the policy challenges of use and management of open woodlands, woody fallows and the role of agroforestry in supplying tree products and improved environment [CIFOR-ICRAF-IFPRI, 1994]. Referring to Latin America, in particular, De Camino [in press] highlighted several areas where policy change was necessary: public support, institutional reform, land tenure, access to resources, markets, and incentives.

Central America and the Caribbean have, for the last 10–15 years, been the scene of development projects and programs which have included the promotion of tree-growing on small and medium-sized farms. These projects have been a response to increasing deforestation and the subsequent environmental degradation affecting soil productivity and the general welfare of the large rural population in the region. The rural population depends on these diminishing tree resources for fuel, construction material, posts and poles, and protection for crops and homesteads from winds and excessive heat. Many initiatives are underway to modify public policy to be more favorable to afforestation and improved forest resource management. However, the design and implementation of these policies is hindered by lack of empirical information on the response of farm and forest users to policy action. Much of existing policy research on agroforestry and farm forestry in Central America is informal and not necessarily linked to policy review and formulation [Current, 1994].

This paper summarizes the results of the first major comparative study of financial and economic benefits to farmers from agroforestry practices promoted by these development projects. It briefly describes forestry policy challenges in Central America and the Caribbean and the status of agroforestry and farm forestry development and policy in the region. It describes the field studies used to collect data from 21 projects in the region and summarizes the results of the evaluation of financial and economic benefits provided by agroforestry and farm forestry systems.

Findings suggest some policy changes which will lower the costs of promoting widespread agroforestry adoption, by better matching agroforestry systems to local economic conditions and encouraging the development of latent markets, reducing the role of subsidies and subsidized credit, improving institutional support to extension and research, and modifying tenure and land use regulations currently constraining adoption.

Background

Forestry situation in Central America, Haiti and the Dominican Republic

Central America has a population of about 30 million, with approximately 50% of that population living in rural areas. This contrasts with South America and the Caribbean, where about 28% of the population lives in rural areas. This rural population depends heavily on wood for fuel, as well as for construction material and a variety of other tree products. For energy use alone, vegetable fuels (primarily fuelwood and charcoal) account for 56% of the total energy consumption in the region, compared to 23% for the Caribbean, and less than 20% in South America [Trossero, 1994]. There is also a relatively high commercial demand for tree products in Central America (see Table 1).

At the same time, there has been tremendous pressure put on the forest resources of Central America to provide land for agriculture and livestock production and sustenance for the large rural population. This high demand for tree products and forested land has led to high rates of deforestation. While all of Latin America and the Caribbean lost approximately 7% of their forest cover between 1981 and 1990, Central America, Haiti and the Dominican Republic lost approximately 18%. In 1991, Haiti and El Salvador had about 1% and 6% forest cover respectively. In these countries, only about 2.3% of the area deforested annually is replaced by reforestation efforts [FAO, 1993a].

Table 1. Key indicators of forestry in Central America and the Caribbean. (Adapted from Current et al., in press).

Country	Population density[a] (hab km^{-2})	Net value wood imports[b] (000 US$)	Net value wood exports[b] (00 US$)	% Forest cover in 1990[c]	% Change in forest cover 1981–1990[c]	Rate of deforestation 1981–1990[c]	Reforestation as % of area deforested per year[c]
Costa Rica	62.7	80567	21895	28.8	–23.8	2.9	7.5
Dominican Republic	154.2	85036	0	22.5	–24.2	2.8	1.1
El Salvador	257.7	21800	2725	6.2	–16.8	2.2	16.1
Guatemala	89.5	70028	19111	39.3	–15.6	1.7	3.1
Haiti	243.4	10187	0	1.3	–10.3	4.8	73.3
Honduras	48.7	136393	30921	41.2	–19.4	2.1	0.4
Nicaragua	33.4	6508	2634	50.8	–16.8	1.9	1.5
Panama	32.6	56176	1010	41.1	–17.0	1.9	0.8
Total	77.3	476695	78296	36.9	–18.1	n.a.	2.3
Latin America/Caribbean	24.2	n.a.	n.a.	56.2	–7.1	0.8	5.0

Sources:
[a] Banco Interamericano de Desarrollo (1993) Progreso Económico y social en América Latina. Informe 1993. Temp Especial: Inversión en Recursos Humanso. BID, Washington, DC, USA, 334 pp.
[b] FAO (1994) AGROSTAT-PC: Forestry Production. (Note: values are for 1991.)
[c] FAO (1993) Forest Resources Assessment 1990: Tropical Countries. FAO forestry paper 112. FAO, Rome, Italy.
n.a. = not available.

The consequences of deforestation and inappropriate land use in these countries have been environmental degradation, decreased productivity of soils requiring increased use of costly fertilizers, losses of capacity for generating hydroelectric power and, in general, a decrease in the well-being of the large rural population [Leonard and Jeffrey, 1987; Sain et al., 1994].

History of forestry and agroforestry development in Central America and the Caribbean

Prior to the early 1970s, reforestation efforts in Central America and the Caribbean concentrated on fairly large-scale reforestation. Government programs provided subsidized seedlings and fiscal incentives to large landowners, and undertook large-scale reforestation programs in which the government produced seedlings in central nurseries and paid laborers to plant the trees on government or community lands. These early efforts were largely unsuccessful, often because of a lack of interest in the trees by the local communities who had no stake in their survival.

Beginning in the mid to late seventies and early eighties more emphasis was placed on working with small and medium-sized landholders to promote reforestation. This was often motivated by the highly publicized 'fuelwood crisis', so many of these efforts were aimed at providing fuelwood. At about the same time, interest in agroforestry was growing. Agroforestry provided a means to combine agricultural production with tree growing, instead of promoting trees as a replacement for agricultural crops. For smaller farmers, who need to devote most of their small land holdings to agricultural production to survive, replacing agricultural production with large blocks of trees was never a viable option.

These small farmer programs have met with mixed results, but have been more successful than earlier large-scale reforestation efforts [Heckadon-Moreno, 1990]. As they have evolved, there has been greater emphasis on participation of the beneficiaries in the whole project process, leading to a greater feeling of 'ownership' of projects and the trees being planted. The emphasis has changed from fuelwood, which is an important product but is not generally considered very valuable by the farmer, to tree planting and management for the production of a variety of subsistence and commercial products and services.[1] Unfortunately, the monitoring and evaluation of these efforts has been limited. It is not clear whether the benefits claimed for farm forestry and agroforestry are actually being provided.

Policy environment

The policy context for effective implementation of these programs has, in the past, been quite mixed. Complementing a greater awareness and concern for the environment, there is presently a great deal of activity in policy reform in natural resource management in Central America. Most countries have appointed National Commissions for the Environment (CONAMA) which also meet on a regional basis. The individual countries have established or are in the process of establishing incentives programs to promote reforestation and are instituting or considering policy changes which will directly or indirectly influence the promotion of agroforestry and farm forestry systems. These efforts at policy reform have been supported by both governmental and non-governmental international aid agencies. Table 2 presents a brief summary of some recent policy activity in the region.

However, recent changes in national and donor policies are having an impact on the success of agroforestry activities. Structural adjustment programs are restricting and diminishing funding for state agencies which provide technical, logistical and research support to tree planting efforts. USAID, an important donor in the region supporting both field programs and research, is cutting funding and support. FINNIDA, which has sponsored a major regional effort, will be substantially cutting that program in 1996. Non-governmental organizations have expanded their role, but many of them depend on national agencies and international donors for a portion of their financial and logistical support.

A re-evaluation of policy objectives and instruments is now needed to ensure cost-effective, strategic support to the emerging farm forestry sector. This should be based upon a solid empirical understanding of actual farmer costs and benefits for agroforestry investment and management, and the most

Table 2. Examples of some recent policy activity in the region.

Country	Areas of recent policy activity
Costa Rica	Agroforestry included in incentives previously provided for commercial block planting. New forest policy. Forestry law of 1990 being reviewed for possible modifications.
El Salvador	National forest policy prepared (1993). Agroforestry included as major component in national agricultural research and extension service.
Guatemala	New forestry law of 1992 under revision and incentives legislation prepared.
Honduras	Law for modernization and development of the agricultural sector (1992) altered management of natural resources, passing much of the responsibility to private sector. Incentives legislation approved in 1993 but not published in official gazette.
Nicaragua	National forestry law in process. Incentives being prepared for reforestation. Study of natural resources policies.
Panama	Tropical Forestry Action Plan integrated as part of national development plan. Regulations for reforestation incentives program being prepared.
REGIONAL	USAID Agricultural Policy Analysis Project which includes natural resource policy and legislation. Regional forestry agreement signed by presidents in 1993.

Sources: Revista Forestal Centroamericana, 1994: Current, 1994; MAG-El Salvador, 1993.

binding economic, institutional and regulatory constraints to further adoption. Unfortunately, such information on farmer costs and benefits of agroforestry and farm forestry in the region was largely unavailable when this study began in 1992.

In order to develop information which could be used to evaluate and formulate effective agroforestry and farm forestry policies and programs, the Tropical Agricultural Research and Training Center (CATIE[2]), the International Food Policy Research Institute (IFPRI) and the World Bank, with funding from the United Nations Development Program (UNDP), undertook a comparative study. The project evaluated farmer benefits from 21 donor, NGO and national government agency-sponsored projects promoting agroforestry and farm forestry in Central America, Haiti and the Dominican Republic.

The total value of these projects was around US$ 150 million, and together they reached over 50,000 participants. Various technologies were evaluated, including woodlots, taungya,[3] managed woody fallows, perennial intercrops, homegardens, alley-cropping, windbreaks, trees on contours and trees in pastures. The projects had various objectives: to halt deforestation in the forest margin, to rehabilitate degraded lands, to establish new supplies for commercial wood industries, and to improve rural welfare [Current et al., in press].

Methods used for gathering information

Overall process

The evaluation was carried out by consultants in each of the countries, using guidelines prepared to ensure cross-country comparisons [Scherr, in press]. The process used in preparing the country case studies was the following:

- Projects in each country were selected for study.
- Guidelines were prepared to ensure uniformity of data gathering among countries and presented and discussed with consultants in a workshop.
- Project areas were visited to gather information from project personnel, individual farmers and farmer groups.
- Country studies of financial and other costs and benefits of individual agroforestry systems were prepared. In total, 56 agroforestry technologies were evaluated.

The first drafts of reports were presented in national workshops in each country. These workshops provided an opportunity to discuss issues not covered in the project case studies and also specific policy issues, bringing in the experience of other projects and individuals. In each individual case study, policy issues were identified which affected project performance at the field level. In a final regional workshop, national policy issues were presented and common policy issues at the regional level were discussed with national and regional representatives of the Tropical Forestry Action Plan, representatives of regional projects, government organizations and NGOs. This made it possible to discuss policy issues with an audience that represented a relatively broad range of backgrounds and interests.

Data on adoption were either taken from large studies previously undertaken by the projects themselves (seven cases) or were drawn from informal or partial samples of project participants. Consultants used different approaches in measuring and evaluating adoption: for example in Panama the consultant measured the level of adoption based upon the degree to which farmers repeated, maintained or abandoned plantings. In El Salvador, the consultant elicited reasons for adopting different species and systems. Information on effectiveness of different extension approaches was drawn from subjective evaluations by the consultants and workshop participants. Information on effects of tenure security were based on the adoption of tree planting by farmers with varying degrees of tenure security. The data on adoption has been complemented by data from extension monitoring by projects and programs cooperating with the Madeleña Project at CATIE.[4]

Financial/economic analysis

The focus of the project was on the costs and benefits of agroforestry systems for the farmers adopting them. Economic or 'societal' benefits were identified but no formal analysis was carried out. Consultants gathered detailed cost benefit information on the agroforestry systems most commonly adopted in each project studied. A financial analysis was undertaken of a specific system as applied by a farmer or 'models' were prepared which represented the typical costs and benefits of the system. Because many systems were relatively new and had not produced benefits, production was estimated based on experiences from similar areas. In other cases information was available for a complete rotation of the tree component. The following conventions were used in performing the analysis:

- The net present value (NPV) was calculated valuing the costs and benefits of the system as a whole. Alternative uses of the land were not included as opportunity costs. Market prices were used to value inputs and production.
- For comparison purposes NPVs were also calculated for an alternative land use.
- Family labor was valued at the same cost as hired agricultural labor.
- All inputs even though subsidized were valued at their market price.
- Production was valued using farm gate prices.
- Constant prices were used assuming that prices of all inputs and production would increase at an equal rate.
- No cost was assigned to the land because alternatives compared were on the same land base.

- A sensitivity analysis was performed varying input prices, product prices or production, and the discount rate.
- A real discount rate of 20% was used in presenting results unless otherwise noted.

To standardize results across countries, data gathered by consultants was entered into a standard spreadsheet for analysis.

Limitations of the study

Prior to this study, little financial and economic data was available at the field level in most projects promoting farm forestry. Due to the scale of the study and the relatively short time period in which it was implemented (an average of about 10 weeks was spent in the field collecting data), there are limitations which must be considered in interpreting the results.

Most of the financial data gathered was obtained from farmers through recall. It was possible to compare and sometimes complement this with data which had been collected through the socioeconomic research of CATIE's Madeleña Project which has been systematically gathering data on the costs and benefits of tree planting since 1988.

In most countries consultants were not able to select a random sample of participating farmers but selected farmers with systems they considered to be representative. Consultants also used different approaches in measuring and evaluating adoption.

General findings

The CATIE-IFPRI-World Bank study confirmed the ability of on-farm tree planting to: i) provide rural communities with the tree products they need for household use and sale; ii) protect and improve soil quality; iii) improve environmental conditions; and iv) create employment and income generating activities. These effects were reported by farmers and farmer organizations. Results varied, depending upon local conditions of scarcity, resource availability and markets. Local conditions played a significant role in determining the relative success of different systems.

The agroforestry systems most commonly accepted by farmers were woodlots and taungya arrangements, followed by line plantings (border plantings, contour plantings and live fences). As intensity of land use increased, systems which required more intensive management were more common (e.g., contour hedgerows in the Guatemalan Highlands, where land holdings are very small). Labor input varied from 0.5 man-days ha^{-1} yr^{-1} for one boundary planting of 33 trees to 211 man days ha^{-1} yr^{-1} in one intercropping system. Labor input for the agricultural intercropping systems was highly variable, ranging between 24 and 211 man-days ha^{-1} yr^{-1}; there is thus obviously considerable scope for farmers to modify spacing, management and intensity of agricultural crop production. *Eucalyptus camaldulensis*, *Gliricidia sepium* and *Leucaena leucocephala* were the most commonly used species, while *Tectona grandis*, *Acacia mangium* and *Cassia* spp. were mentioned in at least four projects.

Major policy findings and recommendations

One of the objectives of the study was to provide insights into policy issues and recommendations which could be used in formulating policy. The financial analysis was used to determine if systems were profitable and to determine the investment in time and other resources required to adopt agroforestry. Farmer interviews were used to determine if adopting farmers considered the agroforestry systems viable and worthwhile from their perspective. The national and regional workshops were designed to take advantage of the experience of project managers, technicians and extension agents to define policy issues. The combination of financial analysis, farmer interviews and workshop results were used to identify areas where policy interventions could help promote the adoption of agroforestry in a manner which would maximize the benefits for farmers and society. A brief discussion of specific issues and recommendations follows.

Agroforestry systems provide financial and economic benefits to farmers and society

The study showed most of the 56 agroforestry systems to be financially profitable for the farmers. In 75% of the agroforestry systems analyzed, there were positive net present values (NPV), at a 20% real discount rate. In two thirds of the cases, the NPV and the return to labor were superior to the alternative use of the land which was used for comparison. Agroforestry systems were compared to agricultural crops or pure stands of trees. Payback periods for woodlots were long (5–20 years), but all of the other systems were managed to fully pay back the initial investment within one to six years. Returns were sensitive to tree and agricultural product price. Even high labor input systems were profitable, so long as output prices or productivity were high. Technical compatibility between tree and crop components was important. Unfortunately, many high return systems also had high marketing risks or faced limited market demand. Table 3 presents a summary of the financial analysis of the systems.

Profitability was not necessarily a good predictor of adoption. Farmers adopted agroforestry systems for a number of reasons. Initially, most were interested in producing tree products for household use (construction material, posts and poles and fuelwood, fuelwood being of secondary importance) and in other cases for soil and crop protection and soil improvement. In Costa

Table 3. Summary results of the financial analyses.

AF system		Indicators								
		NPV 10	NPV 20	CB ratio	Return to labor	Ag.[a] wage	Payback period	Avg. $MDyr^{-1}$	Compared to altern.[b]	No. of obs.
Agricultural intercrop	Avg.	2863	1300	1.79			3.4	165	+	6
	Min.	–169	–366	0.89	1.92	2.41	1.0	61		
	Max.	3733	1600	2.59	7.09	3.70	6.0	302		
Alley cropping	Avg.	1335	847	2.10			1.9	56	++	9
	Min.	127	79	1.38	3.73	2.50	1.0	24		
	Max.	4667	2585	3.87	17.63	4.00	3.0	96		
Contour planting	Avg.	1426	761	1.63			2.0	116	++	4
	Min.	1013	433	1.25	1.87	1.29	1.0	51		
	Max.	2026	996	2.01	7.79	2.30	4.0	153		
Perennial intercrop	Avg.	2867	1405	1.75			4.0	139	++	3
	Min.	2324	963	1.71	3.27	1.43	3.0	104		
	Max.	3461	1932	1.80	4.87	2.26	6.0	188		
Taungya	Avg.	6797	2868	2.50			4.9	53	+	8
	Min.	261	–168	0.75	2.04	3.58	2.0	9		
	Max.	22892	9756	5.84	55.55	4.00	10.0	124		
Woodlot	Avg.	764	–33	0.97			9.2	12	–	10
	Min.	–566	–536	0.16	0	2.50	5	5		
	Max.	2433	222	1.43	7.95	3.70	20	29		

Unless otherwise noted, all indicators were calculated using a 20% discount rate and all monetary values are expressed in US dollars at a 1992 exchange rate.

[a] Agricultural wage reported in the project where the minimum and maximum values for return to labor occurred.

[b] Comparison to alternative: ++ = greater by 25% or more, + = greater by 1–25%, – = lower by 10–25%.

Rica and the Pacific coast of Guatemala, planting was motivated by markets and planting incentives. Once household needs were met, farmers became interested in marketing tree products to generate income. This finding is consistent with evidence from Africa and Asia [Warner, 1993; FAO, 1993b], although the Central American economy is more highly commercialized. Although financial benefits tend to be important in motivating tree planting, environmental improvement and aesthetics were also mentioned as reasons for planting.

Although not the major focus of the study, social and economic benefits were reported in several of the projects. In Costa Rica the effect of windbreaks as biological corridors was documented by a Cornell study. In Honduras farmers who previously had gathered wood products from the natural forest obtained them on-farm. In Guatemala tree plantings were used to provide posts in areas where the likely source of that material would have been mangrove forests, an important habitat supporting off-shore fisheries. In Costa Rica, reforestation efforts created employment opportunities and generated forestry based community development in a community which had lost over 50% of its population to out migration following a downturn in the local economy.

Policy recommendations:

- Policy makers should more aggressively use agroforestry as a tool for rural development and poverty alleviation, as well as environmental goals.

Incentives and subsidies, when properly designed, can be effective mechanisms for promoting tree planting

Some type of planting incentive or subsidy was provided in most of the projects studied. The most successful incentives were those providing minimal in-kind material inputs, combined with technical assistance for seedling production and planting. These kinds of incentives help eliminate 'out of pocket' expenditures for farm families for whom cash is a scarce resource, and reduce the perceived risk of new practices. The evidence of this came principally from farmer interviews and discussions with technicians and project managers during the national workshops. Food-for-work type incentives have caused problems of dependency, lack of continuity of tree planting and management once the food assistance is withdrawn, and reluctance to participate in subsequent tree planting efforts without some kind of compensation.

In many of the projects community nurseries and family nurseries allowed farmers to substitute family labor (often that of women and children) for cash. This largely eliminates the need for a cash investment to establish a planting. Seedlings are obtained from the nurseries and family labor is used for planting. Cash expenditures are limited to the purchase of fertilizer and pesticides.

Self-financing or minimal in-kind incentives produced a slower, but possibly more sustainable, pace of tree planting. Extension registers show that farmers often start with a few trees but gradually expand their plantings. Once farmers are trained and see benefits, they have been able to continue without subsidies. The most important incentives for new and repeat plantings, once projects have started, are the benefits provided by the plantings. Once farmers see concrete results, incentives can be gradually eliminated.

Policy recommendations:

- Policy makers should favor the use of minimal, in-kind incentives and family and community nurseries rather than credit and financial incentives.
- Policy guidelines should consider incentives and subsidies as temporary measures, eliminating them as the benefits from tree planting become evident.

Markets were important to project success

Markets provided important incentives to tree planting. In Guatemala, the market for posts for drying tobacco was an important incentive for the expansion of tree plantings on the Pacific coast. Where markets did not exist, management of existing trees and the diffusion of agroforestry systems were discouraged. In Costa Rica, farmers who had expected to sell trees from thinning woodlots did not perform the required thinnings when the markets did not materialize. This lack of thinning will ultimately lower the productivity and benefits of their plantings.

Projects have demonstrated that farmers will plant trees for household use. Beyond that point market incentives become important to reach a level of planting which can: i) help meet the demand for tree products avoiding the destruction of natural forests; ii) provide raw materials for local industries which can generate employment and income for rural communities; and iii) in many areas substitute an unsustainable land use system with a more appropriate one. There are also possibilities for exporting wood.

Marketing problems often arise when farmers start looking to markets as an outlet for surplus production. Marketing channels for wood and tree products are generally poorly developed but, with present scarcities, are needed to link producers with consumers. Marketing questions to be addressed include the identification of markets, organizing small landowners to be able to produce the quality and quantity demanded by local, national and international markets and processing small diameter logs in local industries to generate employment and income.

Policy recommendations:

- Demand at the household and market level for tree products or services should be important considerations when locating on-farm tree planting promotion efforts.
- Assistance in the identification and development of markets should be provided as an effective incentive to tree planting.

- Guaranteed markets and mechanisms to ameliorate risk in the long term are further measures which can be used to promote more tree planting.

Tenure and regulatory concerns have inhibited project participation

In most countries, harvesting laws and regulations created to protect forests have been a major disincentive to tree-planting activities. Farmers fear possible expropriation of the land or future inability to harvest trees without permits. Most projects have developed mechanisms to overcome this, such as contracts and written agreements giving the farmer explicit rights to harvest trees. Formal land titles were not found to be associated with greater tree-planting efforts. However, farmers with insecure land rights (for example tenants and squatters) faced particular constraints for agroforestry investment.

Policy recommendation:

- Mechanisms should be promoted which exempt on-farm plantings from costly permitting procedures perhaps through some kind of registration system. Simple systems to register tree planters could serve this purpose. Registration systems would maintain protection for natural forest areas while removing constraints to tree planting.

Research support has been an important factor in project success

Research support provided to the projects evaluated was important to their success. Exotic species for which there was little or no local knowledge have been important and attractive species to farmers in most countries. In the Dominican Republic, the Enda Caribe Projects' early research trials identified productive exotic and native species. In Central America, species elimination and provenance testing trials and the development of nursery techniques at CATIE and national research agencies provided valuable information to projects. *Eucalyptus camaldulensis* is a good example. Introduced through research efforts, it was the species most frequently used in the projects reviewed.

Most research to date has been related to nursery practices and on-farm tree establishment. There has been little research on tree management to provide the desired end-products and optimize economic and financial returns. Socio-economic research has received little support. Few projects have the personnel or assigned funding to do socioeconomic research to indicate optimal planting strategies from an economic, not simply technical, point of view.

Many farmers reported improvements in the farm environment associated with agroforestry and farm forestry. Yet poor tree establishment and management practices have also led to accelerated erosion. At this time, the environmental contributions of these systems are still assumed, rather than scientifically confirmed, because of a near-total lack of research.

International donor funding has been important in the past in integrating and complementing national research efforts, but it is diminishing. Unfortunately, national research agencies do not receive the funding necessary to carry on research which is not considered a priority by the national governments. With the recent austerity measures adopted in many countries, research capacity has been further eroded.

Policy recommendations:

- Policy support should be provided for research on tree management to produce the products (fuel, posts, sawn timber) and services (soil conservation and enrichment, protection) desired by the farm family and market, and to reduce production costs.
- More socio-economic research is needed to develop systems which provide the greatest economic and social benefits from on-farm tree planting.
- International donors should take a leading role in financing research efforts.

A variety of extension strategies were used with varying degrees of success

The projects analyzed used a variety of extension strategies. In Costa Rica incentives were channelled through local farmer organizations, who were responsible for hiring foresters to prepare management plans for participating farmers. In community and family nurseries in El Salvador, foresters and extension agents were used. In about half of the projects, persons from local communities were hired and trained as para-technicians to promote tree planting and to provide technical assistance to their neighbors. Some projects provided free or subsidized seedlings while others relied on family and community nurseries for production of planting stock.

Three mechanisms worked especially well in the projects studied. The first was training and using members of local communities as liaisons between project extension agents and the community. This provides a direct link with the community, and a person locally available for consultation when the extension agent cannot be present. It also provides for continuity beyond the termination of a project, by leaving a trained person who can provide technical assistance to the community and access technical assistance from government agencies.

Second, training of farmers through participation in community and family nurseries reduces the need for continued technical assistance, making local communities more self sufficient and lowering the need for and costs of government assistance in the future. Third, seeing the successful results of past on-farm tree planting efforts managed by farmers themselves has been an effective way to convince farmers to engage in tree planting, especially when they can discuss the results with the farmer who implemented the planting. Ten to fifteen years ago there were few demonstrations of the benefits of tree planting outside of traditional systems. At present there are numerous demonstrations in most countries of the region which can serve to promote further diffusion of tree planting in communities without experience.

All of these community-centered approaches still call for strong (though not necessarily large), publicly-supported organizations to provide information exchange, specialist technical consultation, training and community support.

Policy recommendations:

- Involving local communities in the promotion of tree planting activities through the preparation of local para-professionals should be promoted.
- Training local populations in the production of planting stock and planting and management of their agroforestry plantings should be emphasized in agroforestry efforts.
- Demonstrations of successful implementation of agroforestry on farms should be a key component of promotion efforts.

Agroforestry efforts in the region have suffered from lack of coordination

Common themes from the national and regional workshops were the lack of coordination, and even outright competition, between projects operating in the same geographical regions, and the lack of well-defined national development strategies which explicitly recognize the potential of agroforestry and establish it within an organizational structure. Agroforestry efforts are often lost between agricultural and forestry agencies. This lack of coordination has led to duplication of efforts, poor location of development efforts and general inefficiency in the use of relatively scarce financial and human resources.

Policy recommendation:

- Government policy should promote coordination of agroforestry efforts by including agroforestry in development plans; incorporating the presently dispersed and uncoordinated efforts into those plans, and directing new resources to areas of the country where they can be most effectively used. Efforts in the region to unite field projects should be supported by public agencies, donors, and projects.

There is little monitoring of agroforestry promotion efforts

The ability to evaluate the projects included in this study was hampered by the lack of monitoring data. There is currently little monitoring and evaluation of many agroforestry efforts, even listings of farmer participants. This is a point which the individual projects recognized as a shortcoming. Individual projects and extension agencies need to develop capabilities to evaluate and improve their efforts. National governments could use monitoring to gauge their effectiveness in meeting development goals and to better direct national efforts.

Policy recommendations:

- Policy making bodies should ensure that monitoring activities are put in place to monitor the impacts of policies and determine whether they are having the impact expected.
- More monitoring is needed at the project level and should be instituted by donors and project developers.

Summary

Agroforestry systems have been shown to produce private and social benefits which justify their promotion and which are recognized by farmers. They allow farm families the opportunity to integrate trees into their farms enhancing diversity and increasing overall productivity without taking agricultural land out of production. Because tree planting for the production of wood products is a relatively new activity for farmers and many of the successful species are new, diffusion has not yet become widespread. Thus, extension activities to promote tree planting are needed.

The decline in external support for agroforestry and farm forestry investment in Central America and the Caribbean need not lead to reduced farmer adoption, so long as supportive national and local policies are put in place, and donor support is strategically directed. Less expensive approaches to agroforestry promotion have been found to work, although the pace may be slower.

Public resources should be focused on applied research linked to development efforts, training support for community-based extension, minimal in-kind subsidies, and removal of regulatory constraints. Public agencies, farmer organizations and NGOs can work directly with the private sector to expand markets and marketing services for new products or new sources of supply. Extension should continue to make available high-quality germplasm of tree species preferred by farmers for household use and by commercial markets. Improved inter-agency coordination, particularly at a local level, can support regular exchange of information and innovative practices. Public support for agroforestry and farm forestry can be integrated into a wide range of on-going development and environmental efforts.

Notes

* Benefits as used here are defined broadly and refer to outputs of agroforestry plantings which enhance well being and include financial, social, aesthetic and environmental benefits as perceived by the farmers themselves.

1. Services include the protection of agricultural crops and livestock, soil protection and improvement, and using trees as living fences.
2. Centro Agronomico de Investigación y Enseñanza. CATIE is a regional scientific and educational institution located in Turrialba, Costa Rica.

3. Taungya here refers to the practice by landowners of growing trees and agricultural crops together during the establishment phase of the trees planted. Once the trees are established and shade out the crops, cropping is discontinued. It should not be confused with the more commonly described practice of allowing landless farmers access to public lands to raise crops during the establishment phase of government tree plantations.
4. The data gathered by these monitoring efforts includes: i) farmers attending extension activities (meetings, field days, etc.); ii) farmers deciding to try tree planting; iii) species, agroforestry systems, and size of plantings; and iv) follow up data on care and maintenance of plantings in some cases.

References

De Camino R (in press) Interventions for the success of social agroforestry and forestry in Latin America: Guidelines for policy research and institutions. In: Workshop on Forestry and agroforestry policy research priorities in Latin America, sponsored by IICA, CIFOR, and IFPRI, San José, Costa Rica

Centre for International Forestry Research (CIFOR)/International Centre for Research in Agroforestry (ICRAF)/International Food Policy Research Institute (IFPRI) (1994) A review of tropical forestry and agroforestry problem areas and policy research needs and the planned response of the CGIAR system. Draft

Current DA, Lutz E and Scherr S (eds) (in press) Farmer Adoption and Economic Benefits of Agroforestry: Project Experience in Central America and the Caribbean. CATIE-IFPRI-World Bank, Washington, DC, USA

Current DA (1994) La investigacion de politicas con impactos sobre el sector forestal y los sistemas agroforestales en America Central: estado actual. In: Alfaro M, De Camino R, Mora MI and Oram P (eds) Taller Regional: Necesidades y Prioridades de Investigacion en Politicas Forestales y Agroforestales para Latinoamerica, 19–23 Julio de 1993, pp 193–204. San José, Costa Rica

Food and Agricultural Organization of the United Nations (FAO) (1993a) Forest resources assessment 1990: tropical countries. FAO forestry paper 112. FAO, Rome, Italy

Food and Agricultural Organization of the United Nations (FAO) (1993b) Forestry policies of selected countries in Asia and the Pacific. FAO forestry paper 115. FAO, Rome, Italy

Heckadon-Moreno S (1990) Madera y lena de las milpas: los viveros comunales una alternativa para el desarrollo forestal en El Salvador. CATIE-ROCAP. Turrialba, Costa Rica, 106 pp

Leonard HJ (1987) Natural Resources and Economic Development in Central America: A Regional Environmental Profile. IIED, Washington, DC, USA, 279 pp

Ministerio de Agricultura y Ganaderia (MAG) de El Salvador (1993) Politica Forestal. Unidad de Analisis de Politicas, February. CENTA, San Andres, El Salvador

Revista Forestal Centroamericana (1994) El sector forestal en America Central: el punto de vista oficial. Revista Forestal Centroamericana 7: 29–36

Sain G, Tripp R and Brenes ER (1994) Desafios Presentes y Futuros del Medio Ambiente y la Productividad en la Agroempresa Centroamericana. INCAE, San José, Costa Rica, 96 pp

Scherr SJ (in press) Economic analysis of agroforestry systems from the farmers' perspective. In: Current D, Lutz E and Scherr S (eds) Farmer Adoption and Economic Benefits of Agroforestry: Project Experience in Central America and the Caribbean

Trossero M (1994) Los sistemas dendroenergeticos optimizados. In FAO Memoria: Seminario Regional sobre los Sistemas Dendroenergeticos Optimizados para el Desarrollo Rural y la Proteccion Ambiental. Tegucigalpa, Honduras, 18–22 de Octubre de 1993, pp 67–74

Warner K (1993) Patterns of farmer tree growing in Eastern Africa: a socioeconomic analysis. Tropical forestry papers 27. International Centre for Research in Agroforestry and Oxford Forestry Institute, Oxford, UK

Agroforestry Systems **30**: 105–124, 1995.

Contribution of agroforestry trees to nutrient requirements of intercropped plants

C. A. PALM
Tropical Soil Biology and Fertility Programme, P.O. Box 30592, Nairobi, Kenya

Key words: lignin, nutrient content, nutrient release, polyphenols, prunings, synchrony

Abstract. A major tenet of agroforestry, that trees maintain soil fertility, is based primarily on observations of higher crop yields near trees or where trees were previously grown. Recently objective analyses and controlled experiments have addressed this topic. This paper examines the issues of tree prunings containing sufficient nutrients to meet crop demands, the timing of nutrient transfer from decomposition to intercrops, the percent of nutrients released that are taken up by the crop, and the fate of nutrients not taken up by the crop.

The amount of nutrients provided by prunings are determined by the production rate and nutrient concentrations, both depending on climate, soil type, tree species, plant part, tree density and tree pruning regime. A large number of screening and alley cropping trials in different climate-soil environments indicate that prunings of several tree species contain sufficient nutrients to meet crop demand, with the notable exception of phosphorus. Specific recommendations for the appropriate trees in a given environment await synthesis of existing data, currently only general guidelines can be provided.

Tree biomass containing sufficient nutrients to meet crop demand is not enough, the nutrients must be supplied in synchrony to crop needs. Nutrient release patterns from organic materials are, in part, determined by their chemical composition, or quality. Leguminous materials release nitrogen immediately, unless they contain high levels of lignin or polyphenols. Nonlegumes and litter of both legumes and nonlegumes generally immobilize N initially. There is little data on release patterns of other nutrients. Indices that predict nutrient release patterns will assist in the selection of species for synchronizing with crop demand and improve nutrient use-efficiency.

Field trials with agroforestry species ranging in quality show that as much as 80% of the nutrients are released during the course of annual crop growth but less than 20% is captured by the crop, a low nutrient-use efficiency. There are insufficient data to determine how much of the N not captured by the crop is captured by the trees or is in the soil organic matter, the availability of that N to subsequent crops, or how much of that N is lost through leaching, volatilization or denitrification. Longer term trials are needed.

Introduction

One of the main tenets of agroforestry is that trees maintain soil fertility. This hypothesis is based partially on studies of the efficient transfer of nutrients from litter to trees in natural ecosystems [Vitousek and Sanford, 1986]. The assumption is that trees in agroforestry systems will likewise transfer nutrients to intercropped plants. Other support is based on observations of higher crop yields near trees, such as the famous case of *Faidherbia albida* [Vandenbelt, 1992], or where trees have been recently removed, as the case of bush and tree fallows [Nye and Greenland, 1960]. The effect of trees on crop growth is actually due to a combination of factors such as water and temperature relations, soil structure, soil organic matter, in addition to recycling nutrients. Experiments with comparable treatments and proper controls have only recently been designed to determine the relative importance of the various contributions of trees in enhancing crop production [Coe, 1994].

Intercropping of agroforestry trees with crop plants includes sequential systems, where the trees and crops occupy the same piece of land at different times, and simultaneous systems, where the trees and crops are grown on the same piece of land at the same time. Simultaneous systems can vary greatly in the relative proportions of trees and crops and in their spatial arrangement [Young, 1989]. Spatially mixed systems such as coffee and cacao plantations include upper storey trees that provide nutrients, from litterfall and periodic prunings, and shade to the lower storey tree crops [Heuvelop et al., 1988]. Spatially zoned systems include alley cropping where annual crops are grown between hedges of trees that are pruned periodically to provide nutrients to the intercropped plants [Kang et al., 1981]. In both the spatially mixed and zoned systems the intercrops, whether they are annuals or perennials, receive nutrients from the agroforestry trees. This review will focus primarily on alley cropping and biomass transfer systems where trees are planted expressly to serve as a source of nutrients for the intercrops. The concepts and processes discussed, however, equally apply to other agroforestry systems.

Several issues related to the transfer of nutrients from agroforestry trees to intercropped plants are addressed. The first issue is a simple question: Do tree prunings and litter contain enough nutrients to meet crop demand? If yes, then one can ask several more questions. What proportion of the nutrients are released from the decomposing plant materials during the course of crop growth? What proportion of those nutrients are taken up by the crop and what proportion by the trees? What is the fate of the nutrients that were released but were not taken up by the crops? These questions are concerned with the efficiency of nutrient transfer from the tree prunings to the crops. This concept of supplying nutrients when demanded by the crop is now commonly referred to as synchrony [Swift, 1987]. In addition to the more direct, short term supply of nutrients from decomposing plant materials, nutrients can also be supplied indirectly from the mineralization of soil organic matter that is formed from the continued application of organic inputs. Another important issue to address is the relative importance of synchrony or soil organic matter for supplying nutrients to crops from the addition of tree prunings.

For most low-input systems that depend even partially on organic inputs as a source of nutrients, answers to such questions are crucial for making appropriate recommendations for the selection and management of organic inputs, including those from agroforestry trees, for improving crop production.

Amount of nutrients

To answer the first question, do tree prunings contain enough nutrients to meet crop demand, it is necessary to know the amount of prunings produced and their nutrient concentration, as well as the nutrient requirements of specific crops. A large number of biomass production screening trials and alley cropping trials have been conducted in a range of climate and soil environments with a variety of leguminous trees, and even a few non-leguminous trees. Reviews indicate that leguminous trees in alley cropping systems produced up to 20 t ha^{-1} yr^{-1} dry matter of prunings, containing as much as 358 kg N, 28 kg P, 232 kg K, 144 kg Ca, and 60 kg Mg [Young, 1989; Szott et al., 1991], more than enough to meet most crop requirements. In situations where trees are interplanted with crops, fine roots can also supply nutrients to crops through root turnover and root dieback, caused by pruning aboveground biomass. There is little data on the amount of nutrients supplied through roots in agroforestry systems so this study will focus on aboveground inputs. It is, however, important to note that fine root and mycorrhizal turnover in forest systems can contribute two to four times more nitrogen and six to ten times more phosphorus than aboveground litterfall [Bowen, 1984], so tree roots in agroforestry systems are likely to contribute a considerable amount of nutrients to intercropped plants.

Despite the large number of biomass screening trials, it is difficult to make recommendations for a given environment as to which trees produce sufficient pruning biomass. This difficulty results from poor documentation of site characteristics and research methodologies. Details of the climate and soils of the study area are frequently not provided. Trials are compared that have been conducted using different tree densities and pruning regimes. Prunings in one study may consist of leafy materials but a combination of woody and leafy components in another study. It is therefore important to assess data reported as tonnes of biomass per hectare per year taking these factors into account. In addition, often only pruning biomass is given and not the nutrient concentration or content of the prunings.

The nutrient content of the prunings depends on many factors, including tree species and the relative proportions of leaves and stems in the prunings and their respective nutrient concentrations. Even among leguminous trees the N concentration of the leaves varies from 1.5 to 3.4% [Young, 1989], or more. Within a species nutrient concentrations can vary by a factor of two or more as shown in Budelman's [1989] review of the information on nutrient content of leaves of *Leucaena leucocephala* and *Gliricidia sepium*. Despite the higher nutrient concentrations in *G. sepium*, *L. leucocephala* produced more pruning biomass and therefore served as a better source of nutrients for crops [Budelman, 1989], showing that the combination of factors is important.

Differences in nutrient concentrations within a species can be due to a number of factors, including differences in provenances, soil fertility, climate, season, age of leaves or plant, frequency of pruning, or even differences in the laboratories or methodologies used to analyze nutrients. As with biomass production studies, many studies do not provide sufficient information to allow a cause and effect analysis and synthesis, although Budelman [1989] suggests that soil nutrient status is the most important factor influencing nutrient concentrations within a species. The magnitude of the effect, however, varies among species.

Thus, despite the considerable data on biomass and nutrient content of prunings which exist for several agroforestry species, data are often not comparable because of different methodologies or incomplete documentation. There is currently no way to predict the biomass and nutrient production for a particular species for a given soil, climate, and management practice, and there is not likely to be in the near future, except for a few well-documented species such as *L. leucocephala* and *G. sepium*. Fernandes et al. [1994] and Young [1989] have recommended several species with high biomass production and nutrient content for use in different environments. Databases such as ICRAF's multipurpose tree database [Von Carlowitz et al., 1991] and reference compilations such as TREE-CD [CABI, 1991] can be useful in selecting species but care must be taken in extrapolating information from one site to another when selecting tree species.

Given the constraints of the existing data for making recommendations, examples are given to illustrate the amount of nutrients provided in the leaves from tree prunings and other organic inputs compared with the nutrients required by a maize crop (Table 1). The nitrogen (N) content of 4 t ha^{-1} of leaf material from a variety of agroforestry trees, except for the two non-leguminous species, is sufficient to meet the demands of 2 t of maize (plus 3 t stover). Whereas N was the focus of early studies in agroforestry and alley cropping in particular, in general tree prunings can meet crop N requirements, even though as little as 14% and often less than 50% of the N in prunings is from N fixation [Giller and Wilson, 1991].

Calcium demands were met by all the species and magnesium demands were close to being met. Potassium was not supplied in sufficient quantities by many of the species but if crop residues are recycled the nutrient balance is positive.

Phosphorus (P) was not provided in sufficient quantities to meet crop demand by any of the species. Even if crop residues are recycled there is still a negative P balance. Nutrient budgets accounting for the nutrients added in prunings and the nutrients removed in crop harvest from alley cropping systems on both fertile and infertile soils also show an insufficient amount of P in prunings [Palm et al., 1991; Salazari et al., 1993]. In order to meet P requirements, either more pruning biomass must be applied or the concentration of P in the prunings must be higher – both of these options are somewhat limited. Many of the soils in the tropics are P limited [Sanchez, 1976] so there is an obvious need to channel some research efforts to increasing P cycling and availability in intercropping systems.

Table 1. Nutrients required by a crop of maize compared to the nutrients contained in 4 t of organic inputs.

A. Nutrients required by a crop of maize

	Nutrients kg ha^{-1}				
	N	P	K	Ca	Mg
Maize[a]					
Grain (2 t)	50	12	30	6	4
Stover (3 t)	30	6	36	9	6
Total	80	18	66	15	10

B. Nutrients added in 4 t of leaves of various organic inputs

Species	Nutrients kg ha^{-1}				
	N	P	K	Ca	Mg
Leucaena leucocephala[b]	154	8	84	52	13
Erythrina poeppigina[c]	132	7	46	61	–
Inga edulis (fertile soils)[d]	142	11	40	45	6
Inga edulis (infertile soils)[e]	127	9	50	30	7
Senna siamea[f]	105	6	44	110	7
Dactyladenia barteri[f]	60	4	31	40	8
Grevillea robusta[g]	52	2	24	60	7
Maize stover[a]	40	8	48	13	8

[a] Adapted from Sanchez (1976).
[b] Budelman (1989).
[c] Russo and Budowski (1986).
[d] Salazari et al. (1993).
[e] Palm and Sanchez (1990).
[f] Tian et al. (1992a).
[g] C.A. Palm (unpublished).

When discussing nutrient balances and nutrients applied via prunings of agroforestry trees, it is necessary to make an important distinction between nutrients recycled within a system and nutrients added to the system. Prunings added to intercropped plants are recycling nutrients within the soil-plant system, they are not a nutrient addition to the system. Some of the N may be added to the system by N fixation but the amount of N fixed varies greatly depending on the trees used and the environmental conditions [Giller and Wilson, 1991]. External inputs to the system from precipitation are small. Some nutrients, otherwise considered unavailable to crops because they are below the rooting zone of the crops, might be brought into the system from deeper layers in the soil by trees with deeper roots but the magnitude of this 'input' is not known. If crop products are harvested, then there is a net loss of nutrients from the system. Recycling of nutrients through prunings does not offset those losses. Eventually a decline in productivity of the systems would be expected, unless external nutrient inputs are supplied in the form of inorganic fertilizers or organic amendments.

Release of nutrients

Once it is established that prunings contain sufficient nutrients to meet crop requirements, those nutrients should be utilized efficiently and losses minimized by attempting to synchronize nutrient release with crop demands. Synchrony might be achieved by manipulating plant demands, through time of planting and crop selection, or by manipulating the nutrient release pattern of the inputs, through timing and placement of application or the resource quality of the organic input [Myers et al., 1994]. In this paper the emphasis will be on manipulating nutrient release patterns by the selection of agroforestry species based on their organic resource quality.

Resource quality generally refers to the relative rates of decomposition and nutrient release of organic materials, which is determined in large part by their chemical composition. Rapid decomposition and nutrient release are associated with high quality and conversely, immobilization or slow release refer to low quality [Swift et al., 1979]. Past decomposition and nutrient release studies focused on crop residues and cover crops in agricultural systems or leaf litter in natural ecosystems. These studies were concerned primarily with nitrogen dynamics. For agricultural crops, plant materials with %N greater than 1.73 or C:N ratio less than 20 mineralize N and other materials immobilize N [Iritani and Arnold, 1960; Frankenberger and Abdelmagid, 1985]. For natural ecosystems the lignin to N ratio, LIG/N, improved prediction of the percentage of initial N released from litter: the higher the LIG/N the slower the decomposition and N release [Melillo et al., 1982].

Recent studies focusing on the leaves of agroforestry trees indicate a wider range in the percent of initial N released and in the patterns of immobilization and mineralization than those described for crops and forest litters. The %N or LIG/N ratio of the materials also do not always serve as indices of release patterns for agroforestry materials. To facilitate the screening of agroforestry trees for their nutrient supplying capacity, better indices are needed that predict release or immobilization patterns. Until such indices, analogous to chemical formulations provided for inorganic fertilizers, are available for plant materials it will be difficult to recommend particular species for their fertilizer value.

Work by Constantinides and Fownes [1994] showed that it is first of all important to categorize materials into leguminous and nonleguminous plants and also to distinguish between nonsenesced leaves and litter, or senesced, leaves. The N release patterns of the categories are different; litters of legumes and nonlegumes immobilize N; fresh, nonlegume materials do not mineralize or immobilize N to a great extent; and fresh legumes mineralize N immediately. Percent N in the leaves serves as the best index of initial N released if all the materials are grouped together [Tian et al., 1992a; Constantinides and Fownes, 1994]. If legumes are considered as a separate category, %N does not serve as a good index of N release. There are enough cases where there is immobilization or N release is less than expected to merit attention.

Several laboratory incubation studies have followed the N release patterns from leaves of agroforestry legumes [Fox et al., 1990; Palm and Sanchez, 1991; Oglesby and Fownes, 1992; Tian et al., 1992a; Kachaka et al., 1993; Constantinides and Fownes, 1994; Handayanto et al., 1994]. Although the methods are not directly comparable (the length of the incubation varied and some studies included ground leaves while others used whole leaves), some interesting patterns still emerge (Table 2). Nonsenesced leaves, as opposed to litter, are compared unless otherwise specified. All comparisons here are based on cumulative N released at eight weeks, except for the study by Tian et al. [1992a] in which seven weeks was used.

Gliricidia sepium consistently released a higher percentage, 30–70%, of its N compared to other materials. *Leucaena leucocephala* leaves, with %N greater than 3.5 in all but one study, generally released less than 25% of its initial N. *Senna siamea*, a non N-fixing legume, released an intermediate percentage, 10–40%, of the N added. In all cases *S. siamea* showed initial immobilization that lasted up to six weeks before net mineralization occurred. Some of the studies have also shown temporary immobilization from other legumes. *Calliandra calothrysus* and *Inga edulis* released less than 20% of the applied N. In contrast, *Dactyladenia barteri*, a nonlegume, showed net immobilization throughout the incubation.

The combined results from these studies provide preliminary indices for predicting relative amounts of initial N released or immobilized from the legume materials. Correlation analyses show that both polyphenol and lignin content are inversely related to the percent of N released. Both compounds lower the quality of the plant materials. Multiple regression analyses indicate that %N is the primary factor and %polyphenol the secondary factor in determining the percent of initial N released [Tian et al., 1992a; Constantinides and Fownes, 1994]. The polyphenol to N, PP/N, ratio may serve as an index for short term immobilization patterns observed for legumes with relatively high polyphenol content [Palm and Sanchez, 1991; Oglesby and Fownes, 1992]. The lignin plus polyphenol to nitrogen ratio (LIG + PP)/N may serve as an index for longer term release patterns [Fox et al., 1990; Handayanto et al., 1994; Constantinides and Fownes, 1994]. The legume materials with high polyphenol and lignin contents release more N than the nonlegume materials and as such would not necessarily be considered of low quality when compared to most crop residues.

Exact critical values at which N release is inhibited cannot yet be given because different methodologies have been used for measuring polyphenol content in the plant materials. Also, different types of polyphenolic are extracted together, some of which are quite reactive with N while others are

Table 2. Percent of added N released from leaves of agroforestry trees during eight weeks of incubation from various studies.

Material added	Oglesby and Fownes [1992][a]	Constantinides and Fownes [1994]	Tian et al. [1992a]	Palm and Sanchez [1991]	Handayanto et al. [1994]	Kachaka et al. [1993]
Gliricidia sepium	60	70	42	46	31	
Leucaena leucocephala	35	25	23	2[b]	16	58
Calliandra calothrysus	11	28			18	
Senna siamea	30[b]	38[b]	10[b]			40[b]
Inga edulis	11	28		–4		
Dactyladenia barteri			–15			–54

[a] Used leaves and stems.
[b] Denotes initial immobilization.

not [Swain, 1979]. The amount of N immobilized and the time before net mineralization occurs have not been thoroughly investigated but the model developed by Aber and Melillo [1982] could be applied to these plant materials.

Until more precise indices are available for selection of agroforestry materials, the following general guidelines will help in selecting organic inputs for their N release patterns. If %N in the organic material is less than 1.74, as is the case for most litters and many nonlegumes, net N immobilization will occur and will be relatively long term, perhaps beyond the length of a cropping season. If %N is greater than 1.74, then net mineralization will occur but will decrease as a function of the (LIG + PP)/N ratio. It is not yet possible to give details on the critical value at which immobilization will occur in plant materials with greater than 1.74% N and high polyphenol content, although the immobilization does appear to be short term in all cases. As rough estimates, if lignin concentrations are higher than 15% or soluble polyphenol concentrations are higher than 3%, there is likely to be reduced or delayed N release.

Most nutrient release work has focused on nitrogen. Once nitrogen requirements are met, phosphorus becomes limiting in many tropical soils and it may often be the primary limiting nutrient. Although release patterns of P generally follow that of biomass there can be periods of immobilization [Palm and Sanchez, 1990; Tian et al., 1992b; Thomas and Asakawa, 1993] that could exacerbate soil solution P deficiencies relative to crop demand. There is little understanding of controls on P release and immobilization patterns but combined with the net negative P balance reported for alley cropping systems, there is an obvious need for research similar to that conducted on N release patterns. In addition organic inputs can indirectly affect soil P availability by reducing the P fixation capacity of the soil [Singh and Jones, 1976]. The reduced P fixation capacity is temporary and probably depends on the resource quality of the inputs. The potential of plant materials to increase P availability either by release from decomposing materials or by reducing the P fixation capacity of the soil could be an important criterion for selecting agroforestry species.

Plant materials, once characterized for their nutrient release patterns, could be selected to match the nutrient demand pattern of a specified crop. Various scenarios have been proposed that illustrate attempts at synchronizing release and demand (Fig. 1). It is hypothesized that high quality materials will release nutrients too quickly, similar to the use of inorganic fertilizers, resulting in large losses and low nutrient use efficiency whereas low quality materials will release them too slowly or do not contain enough nutrients to meet crop demand. Based on this hypothesis materials of intermediate quality would be expected to show nutrient release patterns in closer synchrony with plant nutrient demand. This might be achieved with the use of specific plant materials or by a mixture of high and low quality materials with release of nutrients slowly at first, when crop demand is low, and providing an increasing

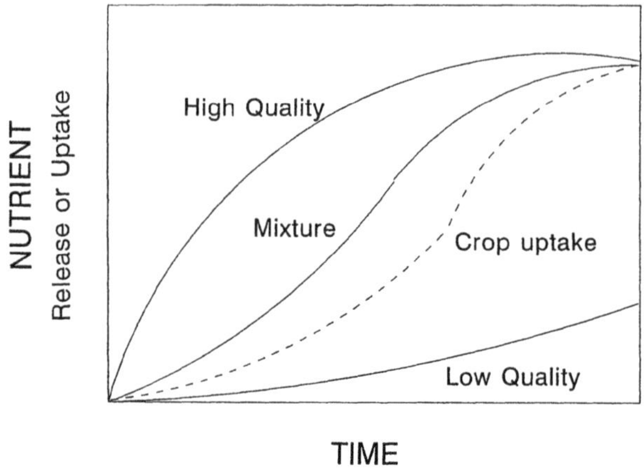

Fig. 1. Hypothetical nutrient release patterns of different quality plant materials compared with nutrient uptake patterns (adapted from Swift [1987]).

rate of release with time, as the crop grows and demands more [Swift, 1987].

There is no clear evidence that mixtures of high and low quality materials alter the patterns of nutrient release. In general, mixtures show release patterns that are parallel to those of the high quality materials rather than altered release patterns with an increase in release rates at later stages [Constantinides and Fownes, 1994; Handayanto et al., 1994]. A few studies cited by Myers et al. [1994] do show changes in release patterns with mixtures. More experimentation is needed in varying the proportions of high and low quality materials, to understand the relative importance of nutrient release versus immobilization in determining the combined pattern. Agroforestry species with high N content but with delayed release patterns or initial immobilization, caused by lignin and polyphenol contents, may provide release patterns more similar to crop demand and also provide sufficient N. Release patterns of *S. siamea*, that contains high N and high polyphenol contents, exhibit temporary immobilization followed by rapid net mineralization of N [Tian et al., 1992a; Constantinides and Fownes, 1994] and may come close to approximating crop demand.

Fate of nutrients

Since the primary purpose of alley cropping and biomass transfer systems is to provide nutrients to the crop from tree prunings, it is important to ask: How much of the nutrients from the applied prunings are actually taken up by the crop? The amount of N added from the leaves or prunings of agroforestry species taken up by the crop is quite low. Recovery values are generally less than 20% and more frequently closer to 10% (Table 3). These recovery values

Table 3. The % N mineralized, % recovery of total N added, % recovery of the N mineralized, % remaining in undecomposed plant material, and the % of that is unaccounted for following the additions of the leaves of several agroforestry trees to crops. Values within a study are comparable but not among studies.

Source of N	% Mineralized	% Recovery of total N	% Recovery of mineralized N	% Remaining in mulch	% Lost or in SOM	References
L. leucocephala	76	4	5	24[b]	71	1
Sesbania sesban	89	31	35	11[c]	58	2
L. leucocepahla	88	10	12	12[c]	78	
Calliandra calothyrsus	40	−3	–	60[c]	> 40	
Acacia cunninghamii	66	2	3	34[c]	64	
L. leucocephala	47	11	23	53[d]	36	3
G. sepium	48	13	23	52[d]	35	
Peltophorum dasyrachis	24	8	32	76[d]	16	
C. calothyrsus	32	7	22	68[d]	25	
G. sepium	100	21[a]	21[a]	0[e]	79	4, 5
L. leucocephala	90	26[a]	28[a]	10[e]	64	
Inga edulis	68	12	18	32[f]	56	6
Erythrina sp.	94	21	22	6	73	
Cajanus cajan	73	20	27	27	53	

[a] Compared to fertilized control.
[b] 84 days.
[c] 70 days.
[d] 64 days.
[e] 100 days.
[f] 4 rice crops.
References: 1. Mulongoy and Van der Meersch (1988); 2. Gutteridge (1992); 3. Handayanto et al. (1994); 4. Tian et al. (1992b); 5. Tian et al. (1993); 6. Palm (1988).

are similar to those for crop legumes but less than those for inorganic fertilizers. The materials that released a higher percentage of initial N, such as *G. sepium* and *Sesbania sesban*, had higher percent recoveries than materials that released a smaller percent of N, such as *C. calothrysus*. It is not clear what the implication of these results is with respect to the relative degree of synchrony attained by the different quality materials. Even though the materials that released a higher percentage of initial N also had higher percent recoveries, they may also have a greater percent loss of nutrients. So to assess synchrony it is important to know the potential loss of nutrients in addition to percent recovery of nutrients.

Why do the intercrops receive so little of the applied nutrients? What is the fate of the remaining 80–90% of the added N? Unfortunately, few agroforestry experiments have been designed to answer these questions or the data are not presented in sufficient detail to allow such an analysis. Using data presented in several key papers, an attempt is made here to try to answer some of the questions (Table 3). It is important to point out that some of the experiments are pot studies and others are field experiments, the time spans of the experiments are also different; nevertheless, some of the issues can be addressed. Most of these experiments do not include the competitive effect of intercropped trees, an important fact to keep in mind. Where possible results from trials with crop legumes are used to reinforce results or provide an indication if there are no data from tree legumes.

The amount of N released from the prunings during the course of crop growth is not always 100% of that added, and in many cases as much as 50% of the N may still be held in the undecomposed plant material (Table 3). Percent N recovery values improve to greater than 20% if they are calculated based on the N mineralized from the plant material, rather than the total N added in the plant material. The recovery values for the lower quality materials are more similar to those of the higher quality materials when calculated in this manner because a smaller percent of the N was mineralized. These modified recovery values are also more comparable to those obtained with inorganic fertilizer.

If the amounts of N taken up by the crop and remaining in the undecomposed plant material are summed, the amount of N unaccounted for ranges from 16 to 80%, the larger values corresponding to the higher quality materials. The N that is unaccounted for could be in the soil organic matter; in the soil as inorganic N, although much of it could have leached below the crop rooting zone; or have been lost by volatilization or denitrification. Nitrogen can also be taken up by the trees in intercropping situations, although trees were not included in these particular studies. Evidence for the relative importance of the different pools or loss pathways of N are presented below.

Haggar et al. [1993], based on a detailed labelling study of *Erythrina poeppigiana* and *G. sepium*, concluded that the majority of the N ends up in some readily mineralizable fraction of the soil organic matter. They also showed that this fraction is not the microbial biomass because the biomass contained only 3–5% of the added N and there was no difference in microbial N among treatments despite different quantities of organic input. They speculated that it is a fraction with a longer turnover time. Ladd et al. [1981] found for a legume-wheat rotation that the first crop recovered 11–17% of the N added as legume and that 72–78% was found in the soil organic matter (SOM). They did not attempt to separate different SOM fractions. Both studies concluded that the benefit of legume inputs to crop production is through the long term buildup of soil organic N rather than the direct release of N from the decomposing legume.

Controlled experiments are needed that compare the effects of different quality materials on fractions of soil organic matter, and the relative N supplying capacity of those fractions. For example, plant material containing high N but also high polyphenol and lignin contents may result in the formation of different soil organic matter fractions, with differing N availability, compared with that formed from organic inputs with lower lignin and polyphenol contents. Size and density fractions of SOM, such as the Ludox fractions [Hassink, 1995] or particulate organic matter, POM [Cambardella and Elliott, 1992], would be good candidate measurements.

The N remaining in the soil organic matter and the undecomposed plant material should have a residual effect on subsequent crops. There are no agroforestry experiments yet that have looked specifically at the residual nitrogen effect, although some are currently underway (J. Henrot and M. van Noordwijk, pers. comm.). Studies with crop legumes provide an indication of the magnitude of the residual effect. Twenty-two percent of the N from labelled alfalfa was recovered by a first crop of wheat and an additional 4% by the second crop [Ladd et al., 1983]. Sisworo et al. [1990] followed the N added from a cowpea crop and found that more than 70% was recovered in six consecutive crops. There is no reason to believe the same would not hold for agroforestry systems with leguminous trees. The amounts of lignin and polyphenols in the prunings may influence the residual effect but it is not known if the residual effect would be enhanced or reduced.

Losses of N from leaching or gas emissions would depend on many soil and climatic factors and again few measurements are available. Volatilization only occurs if the prunings are surface applied, so the importance of this loss pathway will vary depending on management practice. Loss of N via volatilization from green manures measured in laboratory incubations ranges from 5 to 50% but it is usually less than 20% of the added N [Cornforth and Davis, 1968; Costa et al., 1990; Glasener, 1991; Janzen and McGinn, 1991]. Higher quality materials that release N rapidly tend to lose more N via volatilization [Glasener, 1991].

Leaching of nitrate is possible when there is a large pool of inorganic N in the soil, as could happen if there was asynchrony between release and demand. Leaching is only of concern, however, in more humid environments where the depth of wetting in the soil is deeper than the crop rooting zone.

This also implies that asynchrony is not an important issue in drier climates as proposed by Myers et al. [1994]. Even if leaching is a factor in humid environments, deep roots of trees might serve as a safety net for capturing N that leaches below the crop rooting zone [Van Noordwijk at al., 1995]. Again there are few field measurements to give an indication of the relative importance of N leaching in systems with or without agroforestry trees.

The biggest and perhaps the most important unknown is the amount of nutrients released from added plant material that is taken up by the intercropped trees, rather than the crop. Several studies have shown reduction in crop yield in the rows nearest to the trees [Rosecrance et al., 1992; Salazari et al., 1993; Fernandes et al., 1993]. The reduction may be due to shading, competition for nutrients or water, or a combination of these factors. Competition for nutrients, rather than water, between trees and crops would be likely on infertile soils of the humid tropics where water is usually not limiting to plant production. Although agroforestry trees have deeper rooting systems than crops, the majority of the roots are found in the topsoil and this proportion increases when the trees are pruned [Hairiah et al., 1992], increasing the likelihood of competition between trees and crops. Root pruning along the tree hedges has partially reduced the loss in crop yield near trees in some cases [Fernandes et al., 1994], suggesting there was in fact competition but root pruning has not helped in other cases [Sitompul et al., 1992].

The results above on the fate of N added from prunings indicate that as much as 80% is found in the soil, in SOM or inorganic form. The nutrient benefit to crops from tree prunings is more from longer term buildup of SOM rather than the short term release from decomposing prunings. Potential losses by volatilization and leaching are larger for higher quality materials. Scenarios for the fate of N from the prunings of two legumes, one with high N but low lignin and polyphenols concentrations and the other with high N and high lignin and polyphenol concentrations, are presented in Fig. 2. These scenarios are based on data from Table 3 and from estimates of volatilization cited above.

Crop response to different quality inputs

This section is not intended to review crop production in agroforestry intercropping systems but to see if there are obvious effects of different quality inputs on crop yields. To make such comparisons it is best to have the inputs applied at equal N rates or the results expressed on an added N basis. In most agroforestry trials this is not the case so interpretations must be made with caution. One type of input may result in higher crop yields simply because it was applied at a higher rate, not because of increased synchrony or nutrient use efficiency.

Crop growth in pot experiments reflects the different qualities of the inputs. In an experiment by Fox et al. [1990], several legumes were added to sorghum-

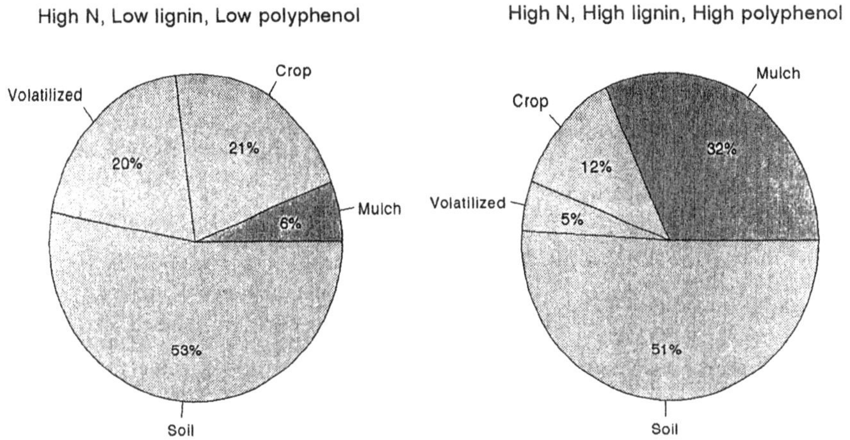

Fig. 2. The fate of N of prunings of different quality after a cropping season; in undecomposed mulch; in the crop; in the soil as SOM or inorganic N, or lost via volatilization (adapted from Palm [1988]). Dark shading represents the N not yet mineralized and light shading represents N mineralized during the cropping season.

sudan at equal N rates. Alfalfa released 47% of its N in 12 weeks and resulted in higher crop growth compared with *L. leucocephala* and *Cassia rotundifolia*, which released 29% and 11.2%, respectively, of their N in the same period (Fig. 3). Gutteridge [1992] applied a variety of tree legumes at equal biomass rates and compared the crop response at the rate of N applied. If compared at a rate equivalent to 100 N kg ha^{-1}, *S. sesban* released 86% of its N and produced about 60 g of maize stover per pot compared with *L. leucocephala*, which also released 86%, but only produced 40 g of stover. *Calliandra calothyrsus* released 34% of its N and produced only 15 g of stover per pot. It is interesting to note that even though *S. sesban* and *L. leucocephala* released similar proportions of their N, the crops grown with *S. sesban* produced more mass. This difference could be a result of different N release patterns and different degrees of synchrony or that the N from *L. leucocephala* is somehow not available. Both these studies suggest that the higher quality material produced higher yields and higher %N recovery but the relative degree of synchrony is not known.

Many field experiments do not show differential response to the quality of inputs. Yamoah et al. [1986a,b] reported that maize yields were similar although 58% of *Senna siamea* prunings compared with 96% of *G. sepium* prunings decomposed during the cropping period. This lack of response to such different release patterns can be explained by two observations. More prunings of *S. siamea* were applied, resulting in twice the amount of N compared to that added in the *G. sepium* prunings. Even if only 58% of the N was released from *S. siamea*, it would be equal to the amount released by *G. sepium*. Other information suggests that the soil was not N limited, so a maize response to additional N would not be expected. Rosecrance et al.

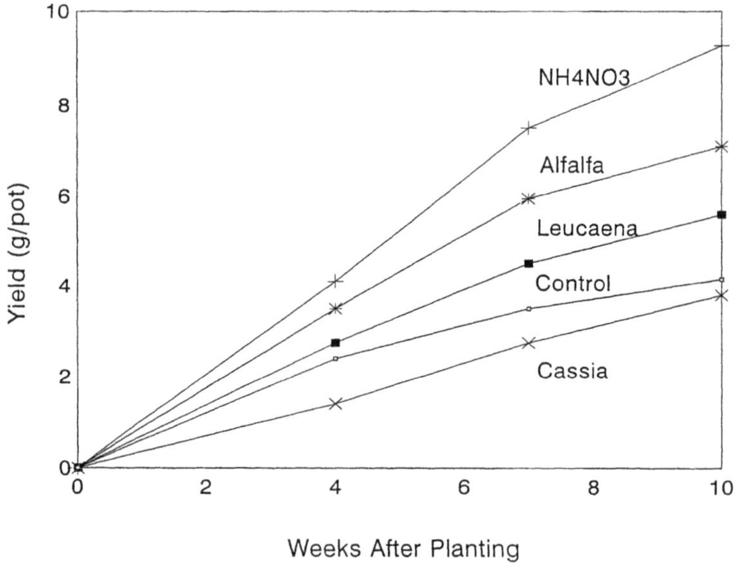

Fig. 3. Growth of sorghum-sudan in response to legume mulch inputs of different quality in a pot study [Fox et al., 1990].

[1992] found that maize yields were related to the amount of N added from the prunings of nine leguminous trees even though materials of strikingly different qualities, such as *Sesbania sesban* and *C. calothyrsus*, were used. Mulongoy et al. [1993] found no difference in maize yields when grown with 10 t ha^{-1} prunings of *G. sepium*, *L. leucocephala*, and *Senna siamea*, materials with differing N release patterns. Only 70% of the N was added in *S. siamea* as compared with the other species. This soil was N limiting, so the results suggest that *S. siamea* has a higher N use efficiency, a possible indication of greater synchrony as speculated above. They also found that date of application of prunings relative to crop planting date affected maize growth: yields were higher when prunings were applied close to planting date, also an indication of better synchrony.

Tian et al. [1993] found interesting results when comparing maize response to different quality inputs. Although there were no significant differences in maize yields using the same quantity of different quality tree prunings, different trends were observed over two seasons. Yields in the first season, with abundant rainfall, were directly related to the amount of N released from the materials. Higher quality *L. leucocephala* and *G. sepium* produced higher yields than the nonlegume *Dactyladenia barteri*. In the second season, with insufficient rainfall, *L. leucocephala* produced the highest yields but yields in *D. barteri* plots were higher than those of *G. sepium*. The authors contend that high quality materials enhance crop production through the direct contribution of nutrients from decomposition and that lower quality materials enhance production through mulching effects on microclimate. In the dry season, the temperature was lower and moisture higher in plots with *D. barteri* compared with plots where the mulches had already decomposed; these conditions are more favorable to maize growth and possibly enhance SOM mineralization.

The results of the pot and field studies give contradictory evidence on the effect of leguminous prunings of different quality on crop yields. Why? If only 10% of the N released from the decomposing material is recovered by the crop, that would be 20 kg of N from 100% decomposition of a material containing 200 kg N, or 10 kg N from a material that decomposed by only 50%. Significant responses to these small differences in inputs are difficult to detect in field studies but may be detectable in pot studies were conditions are more controlled and uniform. In addition, other factors, such as micro-climate, can be affected by input quality and confound the interpretation of results.

The need for well-designed experiments that address these issues systematically should now be obvious. First of all, these types of experiments must be done in soils known to be N limiting, or limiting in the nutrient in question for each particular study. The amounts of N added should be within the responsive part of the curve, in order to detect differences. Including N response curves as part of the study would help in the interpretation of results. Equal rates of N should be applied, or the results expressed in terms of N added so as not to confound the amount of mulch added with the amount of nutrient added. Other factors that may also be affected by input quality and affect crop yields, independently of nutrients, must be monitored. These factors include soil temperature and moisture, pests, disease, and allelopathy.

Conclusions

Many leguminous agroforestry trees produce sufficient pruning biomass and contain enough nutrients, except for P, to meet crop demand in intercropping systems. The nitrogen release patterns, or quality, of the prunings differs greatly, from 100% mineralization to net immobilization during the course of crop growth. These patterns are reflected in differential crop growth in pot studies but not in field studies. In general, only 10 to 20% of the N released is taken up by the first crop and a large portion is in the soil organic matter, indicating that the N benefit of the pruning additions is in the long term rather than immediate. The effect of different quality inputs on the various soil organic fractions and their N supplying capacity is not known.

Acknowledgements

The author would like to thank Nancy Wangari for assistance in the review and synthesis of literature. Comments on an earlier draft of this paper by E.

Barrios, M. van Noordwijk, M. Swift, G. Tian, and two anonymous reviewers were very helpful.

References

Aber J and Melillo J (1982) Nitrogen mineralization in decaying hardwood leaf litter as a function of initial nitrogen and lignin content. Canadian Journal of Botany 60: 2263–2269

Bowen GD (1984) Tree roots and the use of soil nutrients. In: Bowen GD and Nambiar EKS (eds) Nutrition of Plantation Forests, pp 147–179. Academic Press, London, UK

Budelman A (1989) Nutrient composition of the leaf biomass of three selected woody leguminous species. Agroforestry Systems 8: 39–51

CABI (1991) TREE-CD. CAB International, Wallingford, UK

Cambardella CA and Elliott ET (1992) Particulate soil organic-matter changes across a grassland cultivation sequence. Soil Science Society of America Journal 56: 777–783

Coe R (1994) Through the looking glass: 10 common problems in alley-cropping research. Agroforestry Today 6: 9–11

Constantinides M and Fownes JH (1994) Nitrogen mineralization from leaves and litter of tropical plants: relationship to nitrogen, lignin and soluble polyphenol concentrations. Soil Biology and Biochemistry 26: 49–55

Cornforth IS and Davis JB (1968) Nitrogen transformations in tropical soils. I. Mineralization of nitrogen-rich organic materials added to soil. Tropical Agriculture 45: 211–221

Costa FJSA, Bouldin DR and Suhet AR (1990) Evaluation of N recovery from mucuna placed on the surface or incorporated in a Brazilian oxisol. Plant and Soil 124: 91–96

Fernandes ECM, Davey CB and Nelson LA (1993) Alley cropping on an acid soil in the upper Amazon: mulch, fertilizer, and hedgerow root pruning effects. In: Ragland J and Lal R (eds) Technologies for Sustainable Agriculture in the Tropics. ASA Special Publication 56, pp 77–96. ASA, Madison, WI, USA

Fernandes ECM, Garrity DP, Szott LT and Palm CA (1994) Use and potential of domesticated trees for soil improvement. In: Leakey RRB and Newton AC (eds) Tropical Trees: The Potential for Domestication and the Rebuilding of Forest Resources, pp 137–147. HMSO, London, UK

Fox RH, Myers RJK and Vallis I (1990) The nitrogen mineralization rate of legume residues in soil as influenced by their polyphenol, lignin and nitrogen contents. Plant and Soil 129: 251–259

Frankenberger WT and Abdelmagid HM (1985) Kinetic parameters of nitrogen mineralization rates of leguminous crops incorporated into soil. Plant and Soil 87: 257–271

Giller KE and Wilson KJ (1991) Nitrogen Fixation in Tropical Cropping Systems. CAB International, Wallingford, UK

Glasener KM (1991) Ammonia volatilization losses from tropical legume mulches. MSc thesis, North Carolina State University, Raleigh, NC, USA

Gutteridge RC (1992) Evaluation of the leaf of a range of tree legumes as a source of nitrogen for crop growth. Experimental Agriculture 26: 195–202

Haggar JP, Tanner EVJ, Beer JW and Kass DCL (1993) Nitrogen dynamics of tropical agroforestry and annual cropping systems. Soil Biology and Biochemistry 25: 1363–1378

Hairiah K, Van Noordwijk M, Santoso B and Syekhfani MS (1992) Biomass production and root distribution of eight trees and their potential for hedgerow intercropping on an ultisol in southern Sumatra. Agrivita 15: 54–68

Handayanto E, Cadisch G and Giller KE (1994) Nitrogen release from prunings of legume hedgerow trees in relation to quality of the prunings and incubation method. Plant and Soil 160: 237–248

Hassink, J (1995) Density fraction of macro-organic matter and microbial biomass as predictors of C and N mineralization. Soil Biology and Biochemistry (in press)

Heuvelop J, Fassbender HW, Alpizar L, Enriquez G and Folster H (1988) Modelling agroforestry systems of cacao (*Theobroma cacao*) in Costa Rica. II. Cacao and wood production, litter production and decomposition. Agroforestry Systems 6: 37–48

Iritani WM and Arnold CY (1960) Nitrogen release of vegetable crop residues during incubation as related to their chemical composition. Soil Science 89: 74–82

Janzen HH and McGinn SM (1991) Volatile loss of nitrogen during decomposition of legume green manure. Soil Biology and Biochemistry 23: 291–297

Kachaka S, Vanlauwe B and Merckx R (1993) Decomposition and nitrogen mineralization of prunings of different quality. In: Mulongoy K and Merckx R (eds) Soil Organic Matter Dynamics and Sustainability of Tropical Agriculture, pp 199–208. John Wiley and Sons Ltd., West Sussex, UK

Kang BT, Wilson GF, and Sipkens L (1981) Alley cropping maize (*Zea mays* L.) and leucaena (*Leucaena leucocephala* LAM) in Southern Nigeria. Plant and Soil 63: 165–179

Ladd JN, Oades JM and Amato M (1981) Distribution and recovery of nitrogen from legume residue decomposing in soils sown to wheat in the field. Soil Biology and Biochemistry 13: 251–256

Ladd JN, Amato M, Jackson RB and Butler JHA (1983) Utilization by wheat crops of nitrogen from legume residues decomposing in soils in the field. Soil Biology and Biochemistry 15: 231–238

Melillo JM, Aber JD and Muratore JF (1982) Nitrogen and lignin control of hardwood leaf litter decomposition dynamics. Ecology 63: 621–626

Mulongoy K and Van der Meersch MK (1988) Nitrogen contribution by leucaena (*Leucaena leucocephala*) prunings to maize in an alley cropping system. Biology and Fertility of Soils 6: 282–285

Mulongoy K, Ibewiro EB, Oseni N, Kilumba AO, Opara-Nadi and Osunubi O (1993) Effect of management practices on alley-cropped maize utilization of nitrogen derived from prunings on a degraded alfisol in south western Nigeria. In: Mulongoy K and Merckx R (eds) Soil Organic Matter Dynamics and Sustainability of Tropical Agriculture, pp 223–230. John Wiley and Sons Ltd, West Sussex, UK

Myers RJK, Palm CA, Cuevas E, Gunatilleke IUN and Brossard M (1994) The synchronisation of nutrient mineralization and plant nutrient demand. In: Woomer PL and Swift MJ (eds) The Biological Management of Tropical Soil Fertility, pp 81–116. John Wiley and Sons, West Sussex, UK

Nye PH and Greenland DJ (1960) The soil under shifting cultivation. Technical communication 51. Commonwealth Bureau of Soils, Harpenden, UK

Oglesby KA and Fownes JH (1992) Effects of chemical composition on nitrogen mineralization from green manures of seven tropical leguminous trees. Plant and Soil 143: 127–132

Palm CA (1988) Mulch quality and nitrogen dynamics in alley cropping system in the peruvian Amazon. PhD dissertation, North Carolina State University, Raleigh, NC, USA

Palm CA and Sanchez PA (1990) Decomposition and nutrient release patterns of the leaves of three tropical legumes. Biotropica 22: 330–338

Palm CA and Sanchez PA (1991) Nitrogen release from the leaves of some tropical legumes as affected by their lignin and polyphenolic contents. Soil Biology and Biochemistry 23: 83–88

Palm CA, McKerrow AJ, Glasener KM and Szott LT (1991) Agroforestry systems in lowland tropics: is phosphorus important? In: Tiessen H, Lopez-Hernandez D and Salcedo IH (eds) Phosphorus Cycles in Terrestrial and Aquatic Ecosystems. Regional Workshop 3: South and Central America, pp 134–141. Saskatchewan Institute of Pedology, Saskatoon, Canada

Rosecrance RC, Brewbaker JL and Fownes JH (1992) Alley cropping of maize with nine leguminous trees. Agroforestry Systems 17: 159–168

Russo RO and Budowski G (1986) Effect of pollarding frequency on biomass of *Erythrina poeppigiana* as a coffee shade tree. Agroforestry Systems 4: 145–162

Salazari A, Szott LT and Palm CA (1993) Crop-tree interactions in alley cropping systems on alluvial soils of the Upper Amazon Basin. Agroforestry Systems 22: 67–82

Sanchez, PA (1976) Properties and Management of Soils in the Tropics. John Wiley and Sons, New York, NY, USA

Singh BB and Jones JP (1976) Phosphorus sorption and desorption characteristics of soil as affected by organic residues. Soil Science Society of America Journal 40: 389–394

Sisworo WH, Mitrosuhardjo MM, Rasjid H and Myers RJK (1990) The relative roles of N fixation, fertilizer, crop residues and soil in supplying N in multiple cropping systems in a humid, tropical upland cropping system. Plant and Soil 121: 73–82

Sitompul SM, Syekhfani MS and Van der Heide J (1992) Yield of maize and soybean in a hedgerow intercropping system. Agrivita 15: 69–75

Swain T (1979) Tannins and lignins. In: Rosenthal GA and Janzen DH (eds) Herbivores; Their Interactions with Secondary Plant Metabolites. Academic Press, New York, NY, USA

Swift MJ, Heal OW and Anderson JM (1979) Decomposition in Terrestrial Ecosystems. Studies in Ecology. University of California Press, Berkeley, CA, USA

Swift MJ (1987) Tropical Soil Biology and Fertility: Interregional Research Planning Workshop. Biology International Special issue 13. IUBS, Paris, France

Szott LT, Fernandes ECM and Sanchez PA (1991) Soil-plant interactions in agroforestry systems. Forest Ecology and Management 45: 127–152

Thomas RJ and Asakawa NM (1993) Decomposition of leaf litter from tropical forage grasses and legumes. Soil Biology and Biochemistry 25: 1351–1361

Tian G, Kang BT and Brussaard L (1992a) Effects of chemical composition on N, Ca, and Mg release during incubation of leaves from selected agroforestry and fallow plant species. Biogeochemistry 16: 103–119

Tian G, Kang BT and Brussaard L (1992b) Biological effects of plant residues with contrasting chemical compositions under humid tropical conditions – decomposition and nutrient release. Soil Biology and Biochemistry 24: 1051–1060

Tian G, Kang BT and Brussaard L (1993) Mulching effect of plant residues with chemically contrasting compositions on maize growth and nutrients accumulation. Plant and Soil 153: 179–187

Van Noordwijk M, Sitompul SM, Hairiah K, Listyarini E and Syekhfani MS (1995) Nitrogen supply from rotational or spatially zoned inclusion of Leguminosae for sustainable maize production on an acid soil in Indonesia. Plant and Soil (in press)

Vandenbelt RJ (ed) (1992) *Faidherbia albida* in the West African Semi-Arid Tropics. ICRISAT, Patancheru, India

Vitousek PM and Sanford Jr RL (1986) Nutrient cycling in moist tropical forest. Annual Review of Ecology and Systematics 17: 137–167

Von Carlowitz PG, Wolf GV and Kemperman REM (1991) Multipurpose Tree Database: An Information and Decision Support System. ICRAF, Nairobi, Kenya

Yamoah CF, Agboola AA and Mulongoy K (1986a) Decomposition, nitrogen release and weed control by prunings of selected alley cropping shrubs. Agroforestry Systems 4: 229–246

Yamoah CF, Agboola AA and Wilson GF (1986b) Nutrient contribution and maize performance in alley cropping systems. Agroforestry Systems 4: 257–254

Young, A (1989) Agroforestry for Soil Conservation. International Council for Research and Agroforestry and CAB International, Wallingford, UK

Agroforestry Systems **30**: 125–143, 1995.

Tree root characteristics as criteria for species selection and systems design in agroforestry

G. SCHROTH
Institute of Soil Science and Soil Geography, University of Bayreuth, D-95440 Bayreuth, Germany

Key words: nutrient leaching, nutrient recycling, root competition, root distribution, root management, tree-crop interface

Abstract. This literature review presents information about the role of tree root systems for the functioning of agroforestry associations and rotations and attempts to identify root-related criteria for the selection of agroforestry tree species and the design of agroforestry systems. Tree roots are expected to enrich soil with organic matter, feed soil biomass, reduce nutrient leaching, recycle nutrients from the subsoil below the crop rooting zone and improve soil physical properties, among other functions. On the other hand, they can depress crop yields in tree-crop associations through root competition. After a brief review of favourable tree root effects in agroforestry, four strategies are discussed as potential solutions to the dilemma of the simultaneous occurrence of desirable and undesirable tree root functions: 1) the selection of tree species with low root competitiveness, eventually supplemented by shoot pruning; 2) the identification of trees with a root distribution complementary to that of the crops; 3) the reduction of tree root length density by trenching or tillage; and 4) the use of agroforestry rotations instead of tree-crop associations. The potential and limitations of these strategies are discussed, and deficits in current understanding of tree root ecology in agroforestry are identified. In addition to the selection of tree species and provenances according to root-related criteria, the development of management techniques that allow the manipulation of tree root systems to maximize benefit and minimize competition are proposed as important tasks for future agroforestry research.

Introduction

A principal aim of agroforestry is to create interactions between woody perennials, herbaceous crops or pastures, and their biotic and abiotic environments which increase the overall productivity of the land use system and/or its sustainability. To design new agroforestry systems or to optimize existing ones, these component interactions need to be understood. The prediction of the behaviour of agroforestry systems when components are changed or interactions modified by management options, such as pruning, tillage and fertilization, is a major aim of agroforestry research and modelling.

Interactions in agroforestry systems can be divided into above-ground and below-ground processes, and it is evident that much more is currently known about the former than the latter [Anderson and Sinclair, 1993]. The aim of this paper is to review some aspects of below-ground interactions in agroforestry and to give at least tentative answers to the following questions: 1) Are root characteristics important for the agronomic success and sustainability of agroforestry systems? 2) What root system attributes are relevant in the selection of agroforestry tree species? 3) In what ways should root characteristics be taken into consideration when designing agroforestry systems?

In this paper, the term agroforestry includes both spatial and sequential combinations of trees and crops, which will be named agroforestry associations and agroforestry rotations, respectively. Although the discussion should be relevant to most types of agroforestry system, emphasis is mainly on combinations of annual crops with service trees. Priority has been given to the presentation of concepts, rather than to a complete coverage of the available literature.

Benefits expected from the presence of tree roots in agroforestry

A common feature of agroforestry is that several benefits for soil fertility and nutrient cycling which partly or entirely depend on the activity of root systems are expected from the integration of trees into land-use systems. These processes are only briefly reviewed here; additional information is provided by Young [1989] and Anderson and Sinclair [1993].

Soil enrichment with organic matter, maintenance of the soil biomass and enhanced nutrient cycling through root production and turnover

Although numerous studies have been conducted on root mass and its distribution in the soil in tropical agroforestry systems (see below), very little is known about the turnover of the below-ground biomass and thus the input of C into the soil. According to different authors cited by Young [1989], root production in natural and plantation forests is 30 to 70% of total biomass production. In a coffee plantation shaded by *Erythrina* spp. and *Inga* spp. on acid soils in the Venezuelan highlands, root production was estimated to be 6600 kg ha^{-1} yr^{-1} in the top 7.5 cm of soil [Cuenca et al., 1983], whereas values of only about 1100 kg ha^{-1} yr^{-1} in 0–50 cm soil depth were calculated for an alley cropping system with *Gliricidia sepium* on a Plinthic Lixisol[1] in the West African rainforest zone (compared with 13,600 kg ha^{-1} yr^{-1} of above-ground biomass production) [Schroth and Zech, 1995b]. Nye and Greenland [1960] suggested that root litter may be more efficiently converted into soil humus than above-ground litter. However, living roots may also increase organic matter mineralization by supplying easily degradable C sources [Helal and Sauerbeck, 1986], and the net effect of tree roots on the soil C balance may thus be reduced. Decomposing roots are a nutrient source for other plants, especially when there is little retranslocation of nutrients from the roots during senescence [Ferrier and Alexander, 1991; Nambiar, 1987], or when non-senescent roots are cut off during tillage. Bowen [1984] summarized data from forest stands outside the tropics suggesting that the amounts of nutrients which are cycled through tree roots and their mycorrhizas can be far higher than corresponding values for litterfall, but similar information from tropical agroforestry systems is lacking. The role of roots in carbon and nutrient cycling in agroforestry is particularly important when the above-ground biomass is removed from the system, for example as fodder. Also, because of their considerable lateral extension (see below), tree roots may influence the soil at distances from trees far beyond the area which is affected by natural litterfall.

Uptake of water and mobile nutrients from the soil, thus reducing leaching losses

Nutrient-conserving processes are particularly relevant when crops are absent or insufficiently developed but mineralization rates of soil organic matter are high, as at the onset of the rains after a dry season. Trees have been shown to dry out the soil of adjacent fields, thus not only competing with associated crops [Malik and Sharma, 1990; Schroth et al., 1995b], but presumably also reducing percolation rates. Kühne [1993] found considerably lower leaching of N, Mg and K in alley cropping with *Leucaena leucocephala* and maize compared to a maize monocrop on a Nitosol[1] in subhumid southern Benin. Seyfried and Rao [1991] measured much lower nutrient concentrations in the soil solution under a perennial cropping system than under an annual monocrop on a Typic Dystropept[2] in Costa Rica and calculated higher leaching losses from the latter system, but the systems also differed in fertilization rates and presumably in soil tillage.

Pumping up nutrients from subsoil layers below the main rooting zone of the crops

The recycling of nutrients which have been leached from topsoil or released by rock weathering in subsoil is considered an important mechanism of soil regeneration during forest fallows in shifting cultivation. The extension of tree roots into deep subsoil layers and the ability of these roots to take up nutrients which are incorporated into the above-ground biomass and eventually released by litterfall have been demonstrated [Comerford et al., 1984; Stone and Kalisz, 1991; Van Rees and Comerford, 1986]. However, the actual contribution of this process to nutrient cycling in agroforestry systems and nutrient enrichment of topsoils is not known. Woody perennials usually have their root maximum in the topsoil (see below) where they also take up most of their nutrients [IAEA, 1975], although the contribution of the subsoil to total nutrient and water uptake increases when the topsoil dries out [Comerford et al., 1984; Eastham et al., 1990]. It would thus be expected that nutrient pumping by tree roots is least efficient in perhumid climates, where soils are usually most weathered and leaching rates highest so that the recycling of subsoil nutrients would be most needed.

Improvement of soil penetrability for crop roots and of other soil physical properties

Woody perennials can improve the root development of associated or subsequent annual crops [Hulugalle and Lal, 1986; Schroth and Zech, 1995b; Yamoah et al., 1986]. Possible mechanisms include the creation of root channels [Schroth and Zech, 1995b; Stone and Kalisz, 1991; Van Noordwijk et al., 1991b], an improved soil structure through increased earthworm activity [Hulugalle and Lal, 1986] and organic matter enrichment, or a general improvement of crop growth through favourable effects on soil fertility and microclimate. Tree roots may also increase the rooting volume of subsequent crops by penetrating hardened subsoil layers. On a shallow Ferric Acrisol[1] in the savanna of central Togo, roots of three to four years old *Senna siamea* and *Acacia auriculiformis* trees penetrated a subsoil horizon which could only be dug with hammer and chisel, whereas the weaker roots of *Cajanus cajan* in the first season were seldom observed in this horizon (Schroth et al., unpub.). The organic debris in old tree root channels may protect the roots of sensitive crop species from toxic aluminium in acid subsoil horizons [Van Noordwijk et al., 1991b]. The 'bypass flow' of surplus water which rapidly infiltrates through such channels may reduce nutrient leaching from the soil matrix, although the nutrient loss from fertilizers or mulch applied to the soil surface may be increased by reducing their contact with the soil and the plant roots [Cameron and Haynes, 1986]. The overall effect of enhanced macroporosity thus depends on the site concerned and its management, and the difficulty of quantifying macropore flow on a per-area basis is a major problem in the study of tree root effects on nutrient leaching. Improved water infiltration and soil structure also reduce erosion hazards [Young, 1989], although the risk of soil shear may be increased.

Fixation of atmospheric N

With the exception of stem-nodulated tree species such as *Sesbania rostrata*, symbiotic N-fixation in trees is related to tree root systems and the conditions in which they develop. Symbiotic N-fixation reduces the dependence of trees on soil N reserves and may increase N-availability in an ecosystem if the fixed N is returned to the soil via litterfall, root turnover or prunings [Young, 1989]. A high N-fixation potential is thus an important criterion for tree species selection. This aspect of tree root ecology has recently been reviewed [Danso et al., 1992; Giller and Wilson, 1991] and will not be further considered here.

The afore-mentioned root-related benefits are supplemented by above-ground effects, such as improvement of microclimate (although shading may be detrimental), covering of soil by leaf litter and, of course, tree products. The listed processes seem to suggest that the presence of tree root systems with high biomass, rapid turnover and efficient nutrient uptake in the soil is necessary for soil fertility maintenance in agroforestry. Unfortunately, this situation is likely to provoke problems of root competition between trees and crops when grown together. Root competition occurs when one root system reduces the access of another root system to soil water and nutrients, either by taking them up itself or by impeding the development of the associated root system, for example through allelopathic effects [May and Ash, 1990] or the promotion of root diseases [Mchowa and Ngugi, 1994]. Avoiding the inherent conflict between fertility maintenance through tree roots and competition from tree roots is a key problem in agroforestry. Four possible strategies for alleviating this dilemma are discussed in the following section.

1) Selection of tree species with low root competitiveness

Root competition can be avoided if trees and crops are planted at sufficient distance so that the root systems do not intermingle, but this would necessarily also negate the favourable effects of tree roots. In addition, the roots of tropical tree species, such as *Adansonia digitata*, *Bombax costatum*, *Eucalyptus camaldulensis* and *Senna siamea*, may extend several tens of meters from the trunk [Kessler and Breman, 1991; Stone and Kalisz, 1991; Schroth et al., 1995b], and even in agroforestry systems where the trees are widely spaced, much of the area between them may be influenced by tree roots to some extent. An extensive root system is particularly useful where soils are infertile and often dry, especially when nutrient distribution is irregular [Stone and Kalisz, 1991] as in many savanna areas. In alley cropping, where hedgerow spacing is normally only a few meters, the whole alleys are usually permeated by tree roots [Hauser, 1993; Rao et al., 1993; Schroth and Lehmann, 1995; Schroth and Zech, 1995a]. However, the presence of tree roots in adjacent fields does not necessarily indicate root competition, nor does it indicate favourable root effects on soil fertility. On shallow Acrisols[1] in central Togo, root mass and length density decreased logarithmically with increasing distance from *S. siamea* trees [Schroth et al., 1995b], and in much of the area within which roots were found their quantity may have been too low to affect soil conditions significantly. On a Haplic Lixisol[1] in semiarid Kenya, Huxley et al. [1994] observed a patchy root distribution of *Grevillea robusta* trees and associated maize within the shared volume of soil and suggested that this may have reduced interspecific root competition.

Several authors have compared different tree species according to their competitiveness. On sandy soils in the Sudan/Sahel zone of Senegal, Cazet [1989] found that line plantings of *Acacia tortilis* subsp. *raddiana*, *Acacia nilotica* subsp. *adstringens*, and *Acacia senegal* were more competitive with associated crops than *Prosopis juliflora* and *Faidherbia albida*, but this could only partly be explained by the particularly dense root systems of the *Acacia* species in the topsoil because of their limited lateral root extension. In humid to subhumid India, *Eucalyptus* sp. and *L. leucocephala* were found to be more competitive for soil moisture than banana (*Musa × paradisiaca*) and coconut (*Cocos nucifera*) when intercropped with cassava (*Manihot esculenta*) [Ghosh et al., 1987]. Budelman [1990] explained higher yam (*Dioscorea alata*) yields in association with *Gliricidia sepium* compared with *L. leucocephala* on a sandy ferralitic soil in the rainforest area of the Côte d'Ivoire partly by the higher root mass in the topsoil of the latter species. The denser root system of *L. leucocephala* compared with *G. sepium* was confirmed on a Lixisol[1] in semiarid Kenya, with *S. siamea* being intermediate [Rao et al., 1993]. Accordingly, experiments with root barriers on Alfisols[2] in semiarid India demonstrated the intensive root competition between *L. leucocephala* and associated crops in alley cropping [Ong et al., 1991; Rao et al., 1991; Singh et al., 1989], whereas no yield increases of the crops were found with barriers in a similar experiment with *G. sepium* on a Ferralic Cambisol/Plinthic Lixisol[1] in the subhumid forest area of the Côte d'Ivoire despite water-limited conditions [Schroth and Zech, 1995a]. On a shallow Plinthic/Ferric Acrisol[1] in central Togo, maize yield gradients as a function of the distance from two year old *S. siamea*, *G. sepium* and *Calliandra calothyrsus* hedgerows were measured with and without root barriers [Schroth and Lehmann, 1995]. Only the maize in the *C. calothyrsus* plots showed a yield depression at the tree-crop interface, which was significantly reduced by the installation of root barriers, indicating that this tree species had a more competitive root system than *G. sepium* and *S. siamea*. Unfortunately, the root characteristics of the tree species could not be related to their differing competitiveness, because the roots of *C. calothyrsus* and *G. sepium* could not be quantitatively distinguished from weed and crop roots. Rapid root growth, leading to a high root length density, is considered a particularly important parameter for nutrient uptake [Caldwell and Richards, 1986; Smethurst and Comerford, 1993], but efficient physiological uptake mechanisms, mycorrhizal infection or allelopathic effects may also contribute to high competitiveness.

The selection of tree species with low root competitiveness is a reasonable strategy when 'above-ground' functions of the trees, such as the production of wood, fodder or mulch, are to be achieved with minimum competition. However, where major benefits for the cropping system are expected from favourable root functions, it becomes critical to know if root systems of low competitiveness are also less efficient in the reduction of leaching losses and in soil enrichment with organic matter. There seems to be no direct evidence which shows this, but a comparison of the competitive *L. leucocephala* with the less competitive *G. sepium* may be instructive. As mentioned above, *L. leucocephala* was able to reduce nutrient leaching in an experiment in southern Benin, but reduction in soil water content and nitrate concentration in the soil solution was held responsible for negative effects on associated crop yields [Horst, 1991; Kühne, 1993]. In the aforementioned alley cropping experiment with *G. sepium* in the Côte d'Ivoire, on the other hand, a tendency for increased root length densities in the subsoil (30–

50 cm) of the agroforestry plots compared with sole cropping was observed during phases of high leaching risk; in general, however, the root length density of the trees was much less than that of the crops they replaced, and it decreased during the cropping season, presumably as an effect of frequent shoot pruning [Schroth and Zech, 1995a]. It also appeared as though total root production and thus C inputs into the soil through root turnover were not increased by tree integration in this system [Schroth and Zech, 1995b]. It was concluded that a tree species with a more intensive root system would be preferable in associations where the creation of a closed nutrient cycle and soil enrichment with organic matter are important objectives.

The latter example also shows that root competitiveness is not a fixed characteristic of a tree species, but may be influenced by tree management. Tree roots in forests and plantations follow soil moisture conditions in their seasonal growth rates [Cuenca et al., 1983; Kummerow et al., 1990; Srivastava et al., 1986], and their competitiveness would thus be expected to be highest during the cropping season. Intensive shoot pruning of the trees during the rainy season, on the other hand, can induce an 'anti-cyclic' tree root development with decreasing root mass and length density during crop development and a regeneration of the root systems during the dry season, thereby leading to a certain complementarity in the time of tree and crop root development [Schroth and Zech, 1995a]. The reason is probably the depletion of energy and possibly nutrient reserves of the trees by frequent removal of the aboveground biomass [Kandiah et al., 1984], although Erdmann et al. [1993] found that stem starch reserves were sufficient for initial regrowth of pruned *G. sepium*. Mycorrhizal infection may also be negatively affected by defoliation [Gehring and Whitham, 1994]. Increased root mortality and reduced formation of new roots after tree pruning should temporarily reduce root competition with associated crops, but increased nutrient losses after crop harvest may be a consequence of low tree root length densities during the rainy season when the risk of nutrient leaching is highest [Schroth and Zech, 1995a].

For maximizing C inputs into the soil through tree root turnover at minimum competition cost, tree species that take up low amounts of nutrients per unit C investment in roots should be advantageous. Also, the trees should have a lower flexibility in terms of root formation to react to local nutrient enrichment, for example from fertilizer application or mulch decomposition, than the crops. Efficient infection of tree roots with mycorrhiza can increase both nutrient uptake per unit root mass and flexibility of a root system [Atayese et al., 1993; Bowen, 1985] and would thus be a disadvantage in this case. There seems to be a tendency that plants with thick roots have less 'opportunistic' root systems than those with fine roots; that is, they reallocate their roots less readily in favourable volumes of soil, shed them less readily in unfavourable soil and have a lower capacity for water and nutrient uptake [Eissenstat, 1992]. If this tendency can be substantiated, tree species with thick roots would be advantageous in agroforestry systems where the C input through root turnover is a primary aim of the association and root competition needs to be reduced. However, there is also a tendency for thicker roots to have lower turnover rates [Caldwell and Richards, 1986], and this would reduce their value for soil C enrichment. Trees with dense, flexible and mycorrhiza-infected root systems, on the other hand, would be most efficient in the interception of leached nutrients, but would be more competitive in tree-crop associations. Thus, the criteria for species selection depend on the priorities given to different tree root functions and the relative importance of tree and crop productivity.

2) Selection of tree species with a root distribution complementary to that of the crops

An ideal tree from a nutrient-cycling perspective should possess a deep root system with limited lateral extension in the topsoil, thus recycling nutrients from the subsoil and forming a 'safety net' against nutrient leaching below the crop rooting zone without interfering much with the crop root systems in the topsoil [Van Noordwijk et al., 1991a]. This theoretical tree root distribution is in contradiction to the observation that most trees, like crops, normally have the maximum root length density in the topsoil [Dhyani et al., 1990; Jonsson et al., 1988; Kessler and Breman, 1991; Ruhigwa et al., 1992; Schroth et al., 1995a; Toky and Bisht, 1992], although some species with a relatively uniform root distribution in the soil profile have been identified, such as *Eucalyptus* spp. [Jonsson et al., 1988] and *Dactyladenia (Acioa) barteri* [Ruhigwa et al., 1992]. Many tropical soils are too shallow or too acidic and nutrient-deficient in the subsoil to allow a spatial separation of the rooting zones of crops and trees [Szott et al., 1991]. Furthermore, several factors are likely to reduce rooting depth of trees in agroforestry associations: fertilization and biomass application make topsoil more attractive for tree roots and may favour the formation of shallow root systems. Accordingly, the percentage of roots near the soil surface increased with increasing age of forest successions in Costa Rica, corresponding to the increasing importance of litterfall for nutrient cycling [Berish, 1982]. Mulching has been shown to reduce the rooting depth of fruit trees in temperate climates, and to increase their lateral root extension [Kutschera, 1983; Yocum, 1937]. Shoot pruning of trees, a common procedure in many associations, seems to increase root branching in the topsoil and to restrict tree roots to shallower soil depths compared with roots of unpruned trees [Rao et al., 1991; Van Noordwijk et al., 1991a]. It should also be stressed that, although the 'inverse' tree root distribution may be ideal from a nutrient cycling perspective, it is not desirable for C cycling as it would negate C inputs from tree root turnover in the topsoil.

Although the presence of tree roots in the topsoil is thus desirable, agroforestry trees should also be deep-rooted and take up a maximum of water and nutrients from subsoil horizons beneath the main rooting zone of the crops. The presence of competing root systems in the topsoil, either from neighboring trees planted at narrow spacing or from associated pastures or spontaneous undergrowth, has been shown to increase the rooting depth of trees and the use of subsoil resources [Comerford et al., 1984; Eastham et al., 1990]. This may improve resource sharing in tree-pasture associations [Anderson and Sinclair, 1993] and the recycling of subsoil nutrients. In associations of trees with annual crops, tree spacing has to be wide to leave room for the crops, and competing grasses are normally removed by weeding. Although the root systems of annual crops are less dense than those of perennial grasses and are not present permanently, competitive crops such as maize are able to restrict the lateral spread of tree roots and to force them into the subsoil. Yocum [1937] described how apple tree (*Malus sylvestris*) roots turned downward when they came into competition with the root systems of associated maize and continued to spread laterally below the crop rooting depth on a deep loess soil in Nebraska. Huxley et al. [1994] observed on a Haplic Lixisol[1] in semiarid Kenya that the lateral root development of *Grevillea robusta* hedgerows was less (and the trees were also smaller) when maize was grown adjacent to the hedgerows than when the adjacent area was uncropped and weeded, and the tree root density under the hedgerows seemed to be higher in the association than it was without competing root systems (Fig. 9 in Huxley et al., 1994). These observations suggest that strips of crops or grasses with dense root systems may be used to protect less competitive crop species from tree root competition and may help to promote desirable tree root systems that take up a major proportion of their water and nutrient requirements from the subsoil. Evidently, significant nutrient uptake from the subsoil is largely limited to sites where at least some weatherable minerals have been retained within the reach of tree root systems during soil formation.

3) Reduction of tree root density by root trenching or tillage

In arid regions, where trees are planted as shelterbelts at relatively wide distances, the digging of open trenches between the trees and the fields reduces root competition and increases crop yields near the tree-crop interface [Onyewotu et al., 1994; Singh, 1987]. The effort required to dig trenches, however, is not commensurate with unmechanized farming, and the exclusion of tree roots from fields would also negate any benefits of roots in C and nutrient cycling. An interesting aspect of root trenching might be that it forces tree roots into deeper soil [Kang, 1993], where they may spread laterally and take up nutrients below the main rooting zone of the crops. However, the efficiency of this strategy depends on the ability of the tree root systems to colonize the topsoil of neighbouring fields by roots growing upward [Singh et al., 1989] and may only be of short duration for some species.

Soil tillage may be used to destroy tree roots in the topsoil before sowing crops, thereby protecting the crops for some time from tree root competition, although the tree roots remain active in the subsoil and may serve as a 'safety net' against nutrient leaching. Zero-tillage provoked yield-reductions of maize and cowpea (*Vigna unguiculata*) in different alley cropping experiments with *L. leucocephala* which were explained by root competition [Ssekabembe, 1985], whereas deep hoeing to destroy the superficial tree roots reduced bean (*Phaseolus vulgaris*) yield depressions around *G. robusta* trees on an Ultisol[2] in the highlands of Rwanda [Neumann, 1984]. On a Typic Chromustert[2] in semiarid India, 30 cm deep ploughing ('root pruning') along *L. leucocephala* hedgerows before sowing increased grain and stover yields of alley cropped sorghum, and higher soil water contents in the alleys with tree roots pruned indicated that root competition had been reduced [Korwar and Radder, 1994]. In addition, nutrients may have been released from the decomposing tree roots and taken up by the crop. Interestingly, a significant interaction of root pruning and the frequency of shoot pruning of the trees was only exceptionally observed in this study, despite the above-mentioned effect of shoot pruning on root mass and distribution. In some agroforestry associations, like boundary planting, it may be possible to repeat deep ploughing between trees and crops during the cropping season, thus further reducing the competition from regrowing tree roots and increasing the input of dead roots into the cropped soil. No effect of two different tillage methods, traditional hoe-ridging and level ploughing with an ox-plough, was found on root competition from *S. siamea* trees bordering agricultural fields on a Ferric Acrisol[1] in central Togo, although ploughing led to generally higher crop yields and accelerated soil degradation [Schroth et al., 1995b].

Besides the reduction of tree root competition, a desirable aspect of soil tillage in agroforestry is that it probably increases root turnover, and thus C inputs into the soil, by the periodic destruction of the tree roots. A considerably higher fine root mass of *S. siamea* was found in ploughed than in untilled soil [Schroth et al., 1995b]; however, as live and dead roots were not distinguished, part of the increase may have been caused by a stimulation of root growth by increased mineralization of soil organic matter. By increasing both root turnover and mineralization rates of soil organic matter, tillage is likely to have contrasting effects on the soil C balance even when a possible increase of erosion losses is not considered, and the net effect requires investigation.

A further desirable effect of soil tillage is that nutrients are released from sloughed-off tree roots and may become available for the crops. In a study with nine leguminous tree species on a Ferralic Cambisol[1] in the Côte d'Ivoire, N-fixing species had higher N-concentrations particularly in the coarser roots (1–5 mm) than non-N-fixing species, but the latter had the higher total root mass in most cases [Schroth et al., 1995a]. According to the 'synchrony' hypothesis [Swift, 1984], nutrients released from decomposing organic matter should be used most efficiently by the crops if the release is matched with crop demand; hence the need to predict nutrient release dynamics. The leaf biomass of tropical legumes is not always a source of readily available N [Palm and Sanchez, 1991], and the decomposition of legume roots and even root nodules does also not necessarily lead to immediate net N mineralization [Nnadi and Balasubramanian, 1978; Sandhu et al., 1990; Wardle and Greenfield, 1991]. C/N ratios are often unreliable as predictors for N release from legume biomass, and more sensitive predictors, such as the polyphenol/N ratio [Palm and Sanchez, 1991] or the (polyphenol + lignin)/N ratio [Fox et al., 1990] are currently being developed. Lehmann et al. [1995] compared the N release from decomposing leaves, twigs and two diameter classes of roots of three agroforestry species in a field experiment in central Togo and found a curvilinear relationship between initial N release and the (polyphenol + lignin)/N ratios of these substrates (Fig. 1). At ratios greater than about 10, the N release from the substrates, if any, was quite low and was not further reduced by increasing ratios. Nutrient release from *Gliricidia sepium* fine roots was considerably faster than from *C. calothyrsus* and *S. siamea* roots, but was slower than from leaves. The slow nutrient release from roots is a disadvantage when these are immediately required by annual crops, but may be an advantage in tree-production systems with more continuous nutrient demand [Palm and Sanchez, 1990]. The temporary immobilization of mineral N in decomposing roots with low N or high lignin and polyphenol contents may be desirable under conditions of high leaching risk. A possibly interesting aspect is that excised roots from temperate-zone tree species, including their mycorrhiza, may remain alive for prolonged time intervals [Ferrier and

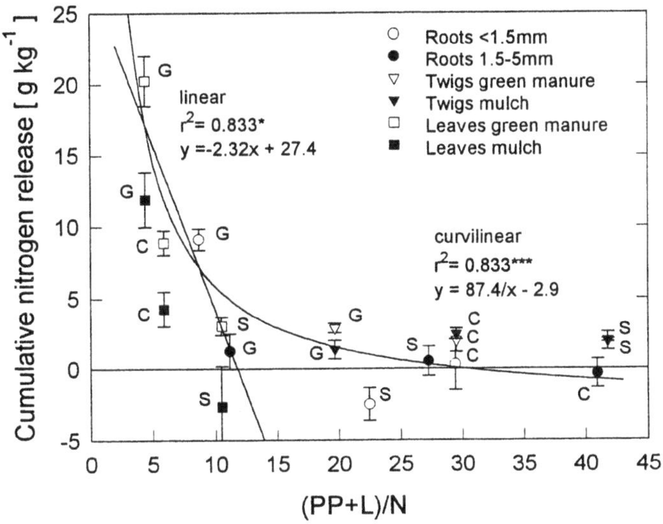

Fig. 1. Relationship between initial polyphenol+lignin/N ratios and cumulative N release during the first eight days of incubation of different materials in an alley cropping field experiment in central Togo. Roots and 'green manure' samples were incubated in litterbags at 10 cm soil depth, and 'mulch' samples were incubated in similar litterbags on the soil surface. The linear regression only includes data with a (PP + L)/N ratio < 15. G = *Gliricidia sepium*; S = *Senna siamea*; C = *Calliandra calothyrsus* (redrawn from Lehmann et al. [1995]).

Alexander, 1985; Publicover and Vogt, 1993], so that nutrient release from roots that were cut off during tillage could be retarded.

Despite these uncertainties, deep tillage could be an ideal management tool as it reduces root competition and increases at the same time the input of C and nutrients from the tree roots into the topsoil while retaining the 'safety net' of tree roots in the subsoil. Ways should be sought by which the efficiency and frequency of tree root destruction in the fields by tillage can be increased, but the simultaneous increase of organic matter mineralization and eventually erosion losses deserves careful attention. Also, information about the speed with which roots of different tree species recolonize the topsoil after tillage is lacking.

4) Agroforestry rotations

A complete solution to the dilemma of the simultaneous occurrence of desirable and undesirable tree root functions is probably not possible in agroforestry associations, but requires the separation of trees and crops in time. In rotational systems, the problem of root competition between trees and crops does not exist, and the tree species may be selected exclusively according to maximum fertility regeneration and economic output. However, improved tree fallows are not always an alternative to permanent tree-crop associations, but rather a supplement. In high rainfall areas, a disadvantage of agroforestry rotations is that during the cropping phase leached nutrients may infiltrate into the soil too deeply to be recycled by fallow trees, so that permanent asso-ciations would be necessary for creating closed nutrient cycles. Fallow rotations would only be suitable in subhumid areas where infiltration rates are lower [Van Noordwijk, 1989], although the efficiency of the recycling of leached nutrients during the fallow phase needs to be quantified experimentally. As in humid climates, the restriction of trees to fallows may also not be possible in semi-arid areas because of low growth rates of trees which require long fallow periods for significant wood production. On sloping or wind-exposed terrain, the permanent presence of trees in the fields may be necessary for reasons of soil and microclimatic protection.

In view of the regeneration of soil fertility, one would expect deep-rooted trees with, in addition, a high root mass and rapid root turnover in the topsoil to be particularly efficient fallow species, but the relationships between tree root characteristics and soil regeneration in fallows have rarely been investigated experimentally. In the afore-mentioned study in the central Côte d'Ivoire, the soil regenerating effect of nine leguminous tree species, of which six were N-fixing, and a spontaneous control were compared [Schroth et al., 1995a]. The trees were grown in replicated plots for five years on a Ferralic Cambisol[1], where the original topsoil had been partly removed during earlier land preparation. Litterfall and root mass in 0–60 cm soil depth were measured and were related to chemical and biological properties of the soil in the plots in order to identify characteristics of efficient fallow species. Despite considerable differences between tree species in above- and below-ground biomass and litterfall, total C and N in the soil differed little between species. However, N-mineralization in topsoils (0–10 cm) which were incubated in the laboratory showed a significant species effect. For the N-fixing species, tree root mass was significantly correlated with N-mineralization of the soil (Fig. 2). N-mineralization was also correlated with litterfall for seven of the tree species. High N-mineralization rates resulted in better N-nutrition as well as higher biomass and grain yield of rice (*Oryza sativa*), which was grown in 15 l bags filled with topsoil from all plots as a bioassay of soil fertility. Apparently, tree species with high root mass and litterfall were most efficient in increasing the available N-pool in the soil, in agreement with a suggestion by Haggar et al. [1993], that agroforestry maintained soil N availability at a higher level than sole cropping mainly by providing high organic matter inputs from the trees. Root intensive tree species also suppressed weeds efficiently. A potential problem was however that the fallow species that promoted high N-mineralization rates tended to accumulate less N in the soil and it could, therefore, be expected that there would be a more rapid depletion of labile soil N-pools after the fallow phase when compared with species that promoted lower N-mineralization rates. There may thus be a trade-off between short-term yield improvement and sustainability in the selection of

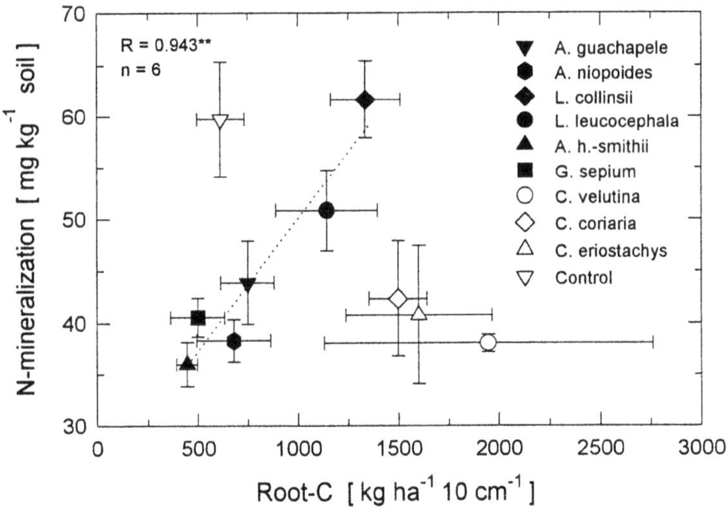

Fig. 2. Relationship between root-C in the topsoil (0–10 cm) under nine different tree species and spontaneous fallow vegetation (*Chromolaena odorata*) in a field experiment in the central Côte d'Ivoire, and N-mineralization after the initial mineralization flush in incubated topsoil from the same plots (means and S.E.). The Spearman correlation was calculated for the N-fixing species (black symbols), that is without the *Caesalpinia* species and the spontaneous control (redrawn from Schroth et al., 1995a).

fallow species, and long-term studies are needed to investigate this [Szott et al., 1991].

Fallow trees should be deep-rooted for an efficient recycling of subsoil nutrients, and in addition to an appropriate selection of tree species and provenances [Vandenbeldt, 1991], the design and management of the fallow should be planned with the aim of expressing this characteristic as much as possible. Close spacing of trees leading to early intraspecific root competition may be an efficient way of achieving this [Atkinson et al., 1976; Comerford et al., 1984; Eastham et al., 1990]. In addition, a root-intensive understorey vegetation that dries out the topsoil and forces the tree roots into the subsoil may be desirable for nutrient recycling [Comerford et al., 1984], but this advantage has to be balanced against the danger of increased weed infestation during the subsequent cropping phase.

Conclusions

Root characteristics and root processes are clearly important factors for the success of agroforestry systems. Tree roots contribute to the maintenance and regeneration of soil fertility, but they are also important in competition with crops. Tree species and even provenances differ markedly in their root characteristics, with respect to both their competitiveness and their effects on the soil. For this reason, root characteristics must become criteria in the selection of tree species for agroforestry, but more research is necessary before these can be defined on a sound basis for different agroforestry systems and pedoclimatic situations in the tropics. The undisputable deficits in our knowledge about tree root ecology should not be taken as a justification for selecting agroforestry trees merely according to easily measurable, above-ground production criteria, as this could implicate the partial sacrifice of desirable tree root functions [Sinclair et al., 1994].

The temporal sequence of trees and crops would be an ideal solution to the conflict between desirable and undesirable tree root effects in agroforestry. However, in many situations fallow improvement techniques cannot replace tree-crop associations, although they may supplement them. In fallow improvement, root intensive tree species are probably preferable to species with smaller root systems, although some evidence suggests that this might not necessarily be so. In tree-crop associations, the situation is even more complicated, as a balance needs to be found between the contrasting aims of low competitiveness, high carbon inputs into the topsoil through root turnover and efficient nutrient recycling from the subsoil. Depending on the relative weight given to these aims, desirable tree root characteristics vary from one system to another: limited lateral root extension and low root length density in the topsoil for low competitiveness; high root mass and turnover rates in the topsoil for high carbon inputs; and high root length density in the subsoil and deep root systems for efficient nutrient recycling.

Tree root systems are dynamic and flexible and react to environmental and management conditions. Thus, the genetic potential of a tree, for example for deep rooting, may not be fully expressed under the conditions of a given agroforestry system unless this is facilitated (or at least not impeded) by the growth conditions of the tree in this system. For this reason, the development of 'root management' [Bowen, 1984] packages, such as suitable combinations of tree spacing, tillage methods, fertilization, pruning regime and association with competitive herbaceous plants, which allow the manipulation of tree root systems according to the requirements of a specific agroforestry situation, should be given similar priority in agroforestry research to the screening of species and provenances for root characteristics.

A consequence of the contradictory nature of tree roots in agroforestry for research projects is that whenever possible, different aspects of tree roots should be studied simultaneously in order to avoid the selection of tree species or management options according to a single objective, because this may be unsuitable for meeting the requirements of another objective. For example, when management options for reducing the root length density of trees in neighboring fields are investigated, the effects on carbon cycling and nutrient leaching in the fields should also be quantified. Only when this is done can the optimum balance between the contrasting aims of reducing root competition and maximizing the soil improving effects of tree roots be found. Tree selection for agroforestry according to multiple objectives has been discussed in more detail by Sinclair et al. [1994].

This review has concentrated on the direct effects of tree roots and how these can be manipulated. However, trees may also have important indirect effects on the root ecology of a cropping system. Not only may the crop root development be affected by associated trees, but the composition of the weed flora may also be changed [Siaw et al., 1991], and this may have significant consequences for the quantity and distribution of the weed root systems [Schroth and Zech, 1995a]. These indirect effects increase further the considerable complexity of the root ecology of agroforestry systems.

Acknowledgements

Part of the work cited in this paper has been funded by the European Community Commission (DG XII STD 2/1145), the Deutsche Forschungsgemeinschaft (Ze 154/12–1), the Deutsche Gesellschaft für Technische Zusammenarbeit (GTZ) and a PhD fellowship from the University of Bayreuth. The author thanks L. Kutschera and R.J. Buresh for discussions and suggestions on root processes in agroforestry and possibilities of their management, and F.L. Sinclair, P. Drechsel, J. Lehmann and anonymous reviewers for valuable comments on earlier versions of the manuscript.

Financial support from the Universitätsverein of the University of Bayreuth for travel to present this paper is gratefully acknowledged.

Notes

1. Classified according to FAO/Unesco [1990, and earlier versions].
2. Classified according to US Soil Taxonomy [Soil Survey Staff, 1992, and earlier versions].

References

Anderson LS and Sinclair FL (1993) Ecological interactions in agroforestry systems. Agroforestry Abstracts 6: 57–91

Atayese MO, Awotoye OO, Osonubi O and Mulongoy K (1993) Comparisons of the influence of vesicular-arbuscular mycorrhiza on the productivity of hedgerow woody legumes and cassava at the top and the base of a hillslope in alley cropping systems. Biol Fertil Soils 16: 198–204

Atkinson D, Naylor D and Coldrick GA (1976) The effect of tree spacing on the apple root system. Hortic Res 16: 89–105

Berish CW (1982) Root biomass and surface area in three successional tropical forests. Can J For Res 12: 699–704

Bowen GD (1984) Tree roots and the use of soil nutrients. In: Bowen GD and Nambiar EKS (eds) Nutrition of Plantation Forests, pp 147–179. Academic Press, London, UK

Bowen GD (1985) Roots as a component of tree productivity. In: Cannell MGR and Jackson JE (eds) Attributes of Trees as Crop Plants, pp 303–315. Institute of Terrestrial Ecology, Huntingdon, TN, USA

Budelman A (1990) Woody legumes as live support systems in yam cultivation I. The tree-crop interface. Agroforestry Systems 10: 47–59

Caldwell MM and Richards JH (1986) Competing root systems: morphology and models of absorption. In: Givnish TJ (ed) On the Economy of Plant Form and Function, pp 251–273. Cambridge University Press, Cambridge, UK

Cameron KC and Haynes RJ (1986) Retention and movement of nitrogen in soils. In: Haynes RJ (ed) Mineral Nitrogen in the Plant-Soil System, pp 166–241. Academic Press, Orlando, FL, USA

Cazet M (1989) Les plantations linéaires denses sur les sols sableux dégradés de la zone centre-nord du Sénégal. Bois Forêts Trop 222: 27–37

Comerford NB, Kidder G and Mollitor AV (1984) Importance of subsoil fertility to forest and non-forest plant nutrition. In: Stone EL (ed) Forest Soils and Treatment Impacts. Proceedings of the Sixth North American Forest Soils Conference, Knoxville, June 1983, pp 381–404. University of Tennessee, Knoxville, TN, USA

Cuenca G, Aranguren J and Herrera R (1983) Root growth and litter decomposition in a coffee plantation under shade trees. Plant Soil 71: 477–486

Danso SKA, Bowen GD and Sanginga N (1992) Biological nitrogen fixation in trees in agro-ecosystems. Plant Soil 141: 177–196

Dhyani SK, Narain P and Singh RK (1990) Studies on root distribution of five multipurpose tree species in Doon Valley, India. Agroforestry Systems 12: 149–161

Eastham J, Rose CW, Cameron DM, Rance SJ, Talsma T and Charles Edwards DA (1990) Tree/pasture interactions at a range of tree densities in an agroforestry experiment. II. Water uptake in relation to rooting patterns. Aust J Agric Res 41: 697–707

Eissenstat DM (1992) Costs and benefits of constructing roots of small diameter. J Plant Nutr 15: 763–782

Erdmann TK, Nair PKR and Kang BT (1993) Effects of cutting frequency and cutting height on reserve carbohydrates in *Gliricidia sepium* (Jacq.) Walp. For Ecol Manage 57: 45–60

FAO/Unesco (1990) Soil Map of the World, revised legend. Rome, 119 pp

Ferrier RC and Alexander IJ (1985) Persistence under field conditions of excised fine roots and mycorrhizas of spruce. In: Fitter AH, Atkinson D, Read DJ and Usher MB (eds) Ecological Interactions in Soil, pp 175–179. Blackwell Scientific Publications, Oxford, UK

Ferrier RC and Alexander IJ (1991) Internal redistribution of N in Sitka spruce seedlings with partly droughted root systems. For Sci 37: 860–870

Fox RH, Myers RJK and Vallis I (1990) The nitrogen mineralization rate of legume residues in soil as influenced by their polyphenol, lignin, and nitrogen contents. Plant Soil 129: 251–259

Gehring CA and Whitham TG (1994) Interactions between aboveground herbivores and the mycorrhizal mutualists of plants. Trends Ecol Evol 9: 251–255

Ghosh SP, Nair GM, Pillai NG, Ramanujam T, Mohankumar B and Lakshmi KR (1987) Growth, productivity and nutrient uptake by cassava in association with four perennial species. Trop Agric (Trinidad) 64: 233–236

Giller KE and Wilson KJ (1991) Nitrogen Fixation in Tropical Cropping Systems. CAB International, Wallingford, UK, 313 pp

Haggar JP, Tanner EVJ, Beer JW and Kass DCL (1993) Nitrogen dynamics of tropical agroforestry and annual cropping systems. Soil Biol Biochem 25: 1363–1378

Hauser S (1993) Root distribution of *Dactyladenia (Acioa) barteri* and *Senna (Cassia) siamea* in alley cropping on Ultisol. I. Implication for field experimentation. Agroforestry Systems 24: 111–121

Helal HM and Sauerbeck D (1986) Effect of plant roots on carbon metabolism of soil microbial biomass. Z Pflanzenernähr Bodenk 149: 181–188

Horst WJ (1991) Cropping systems and their effect on sustained soil productivity in the tropics. Göttinger Beitr Land Forstwirtsch Trop Subtrop 71: 243–268

Hulugalle NR and Lal R (1986) Root growth of maize in a compacted gravelly tropical alfisol as affected by rotation with a woody perennial. Field Crops Res 13: 33–44

Huxley PA, Pinney A, Akunda E and Muraya P (1994) A tree/crop interface orientation experiment with a *Grevillea robusta* hedgerow and maize. Agroforestry Systems 26: 23–45

IAEA (1975) Root Activity Patterns of Some Tree crops. Technical Report Series 170. International Atomic Energy Agency, Vienna, Austria, 154 pp

Jonsson K, Fidjeland L, Maghembe JA and Högberg P (1988) The vertical distribution of fine roots of five tree species and maize in Morogoro, Tanzania. Agroforestry Systems 6: 63–69

Kandiah S, Wettasinghe DT and Wadasinghe G (1984) Root influence on shoot development in tea (*Camellia sinensis* (L.) O.Kuntze) following shoot pruning. J Hortic Sci 59: 581–587

Kang BT (1993) Alley cropping: past achievements and future directions. Agroforestry Systems 23: 141–155

Kessler JJ and Breman H (1991) The potential of agroforestry to increase primary production in the Sahelian and Sudanian zones of West Africa. Agroforestry Systems 13: 41–62

Korwar GR and Radder GD (1994) Influence of root pruning and cutting interval of *Leucaena* hedgerows on performance of alley cropped *rabi* sorghum. Agroforestry Systems 25: 95–109

Kummerow J, Castillanos J, Maas M and Larigauderie A (1990) Production of fine roots and the seasonality of their growth in a Mexican deciduous dry forest. Vegetatio 90: 73–80

Kutschera L (1983) Wurzel-Tropismen als Funktionen der Wasserabgabe und -aufnahme, die 'Wassertheorie'. In: Böhm W, Kutschera L and Lichtenegger E (eds) Root Ecology and its Practical Application. International Symposium, Gumpenstein, 1982, pp 301–322. Bundesanstalt Gumpenstein, Irdning, Austria

Kühne RF (1993) Wasser- und Nährstoffhaushalt in Mais-Maniok-Anbausystemen mit und ohne Integration von Alleekulturen ('Alley cropping') in Süd-Benin. Hohenheimer Bodenkundl. Hefte 13. Universität Hohenheim, Germany, 244 pp

Lehmann J, Schroth G and Zech W (1995) Decomposition and nutrient release from leaves, twigs and roots of three alley-cropped tree legumes in central Togo. Agroforestry Systems 29: 21–36

Malik RS and Sharma SK (1990) Moisture extraction and crop yield as a function of distance from a row of *Eucalyptus tereticornis*. Agroforestry Systems 12: 187–195
May FE and Ash JE (1990) An assessment of the allelopathic potential of *Eucalyptus*. Aust J Bot 38: 245–254
Mchowa JW and Ngugi DN (1994) Pest complex in agroforestry systems: the Malawi experience. For Ecol Manage 64: 277–284
Nambiar EKS (1987) Do nutrients retranslocate from fine roots? Can J For Res 17: 913–918
Neumann I (1984) Vom Nutzen der Bäume in der kleinbäuerlichen Landwirtschaft tropischer Bergländer. In: Rottach P (ed) Ökologischer Landbau in den Tropen, pp 250–262. C.F. Müller, Karlsruhe, Germany
Nnadi LA and Balasubramanian V (1978) Root nitrogen content and transformation in selected grain legumes. Trop Agric (Trinidad) 55: 23–32
Nye PH and Greenland DJ (1960) The Soil under Shifting Cultivation. Commonwealth Bureau of Soils, Harpenden, UK, 144 pp
Ong CK, Corlett JE, Singh RP and Black CR (1991) Above and below ground interactions in agroforestry systems. For Ecol Manage 45: 45–57
Onyewotu LOZ, Ogigirigi MA and Stigter CJ (1994) A study of competitive effects between a *Eucalyptus camaldulensis* shelterbelt and an adjacent millet (*Pennisetum typhoides*) crop. Agric Ecosyst Environ 51: 281–286
Palm CA and Sanchez PA (1990) Decomposition and nutrient release patterns of the leaves of three tropical legumes. Biotropica 22: 330–338
Palm CA and Sanchez PA (1991) Nitrogen release from the leaves of some tropical legumes as affected by their lignin and polyphenolic contents. Soil Biol Biochem 23: 83–88
Publicover DA and Vogt KA (1993) A comparison of methods for estimating forest fine root production with respect to sources of error. Can J For Res 23: 1179–1186
Rao MR, Ong CK, Pathak P and Sharma MM (1991) Productivity of annual cropping and agroforestry systems on a shallow Alfisol in semi-arid India. Agroforestry Systems 15: 51–63
Rao MR, Muraya P and Huxley PA (1993) Observations of some tree root systems in agroforestry intercrop situations, and their graphical representation. Expl Agric 29: 183–194
Ruhigwa BA, Gichuru MP, Mambani B and Tariah NM (1992) Root distribution of *Acioa barteri*, *Alchornea cordifolia*, *Cassia siamea* and *Gmelina arborea* in an acid ultisol. Agroforestry Systems 19: 67–78
Sandhu J, Sinha M and Ambasht RS (1990) Nitrogen release from decomposing litter of *Leucaena leucocephala* in the dry tropics. Soil Biol Biochem 22: 859–863
Schroth G, Kolbe D, Balle P and Zech W (1995a) Searching for criteria for the selection of efficient tree species for fallow improvement, with special reference to carbon and nitrogen. Fertilizer Res (in press)
Schroth G, Poidy N, Morshäuser T and Zech W (1995b) Effects of different methods of soil tillage and biomass application on crop yields and soil properties in agroforestry with high tree competition. Agric Ecosyst Environ 52: 129–140
Schroth G and Lehmann J (1995) Contrasting effects of roots and mulch from three agroforestry tree species on yields of alley cropped maize. Agric Ecosyst Environ (in press)
Schroth G and Zech W (1995a) Root length dynamics in agroforestry with *Gliricidia sepium* as compared to sole cropping in the semi-deciduous rainforest zone of West Africa. Plant Soil 170: 297–306
Schroth G and Zech W (1995b) Above- and below-ground biomass dynamics in a sole cropping and an alley cropping system with *Gliricidia sepium* in the semi-deciduous rainforest zone of West Africa. Agroforestry Systems (in press)
Seyfried MS and Rao PSC (1991) Nutrient leaching loss from two contrasting cropping systems in the humid tropics. Trop Agric (Trinidad) 68: 9–18
Siaw DEKA, Kang BT and Okali DUU (1991) Alley cropping with *Leucaena leucocephala* (Lam.) de Wit and *Acioa barteri* (Hook.f.) Engl. Agroforestry Systems 14: 219–231
Sinclair FL, Verinumbe I and Hall JB (1994) The role of tree domestication in agroforestry. In:

Leakey RRB and Newton A (eds) Tropical Trees: Potential for Domestication and the Rebuilding of Forest Resources, pp 124–136. HMSO, London, UK

Singh GB (1987) Agroforestry in the Indian subcontinent: past, present and future. In: Steppler HA and Nair PKR (eds) Agroforestry: A Decade of Development, pp 117–138. International Council for Research in Agroforestry, Nairobi, Kenya

Singh RP, Ong CK and Saharan N (1989) Above and below ground interactions in alley-cropping in semi-arid India. Agroforestry Systems 9: 259–274

Smethurst PJ and Comerford NB (1993) Simulating nutrient uptake by single or competing and contrasting root systems. Soil Sci Soc Am J 57: 1361–1367

Soil Survey Staff (1992) Keys to Soil Taxonomy. SSMS technical monograph no. 19. Pocahontas Press, Inc, Blacksburg, VA, USA, 556 pp

Srivastava SK, Singh KP and Upadhyay RS (1986) Fine root growth dynamics in teak (*Tectona grandis* Linn. F.). Can J For Res 16: 1360–1364

Ssekabembe CK (1985) Perspectives on hedgerow intercropping. Agroforestry Systems 3: 339–356

Stone EL and Kalisz PJ (1991) On the maximum extent of tree roots. For Ecol Manage 46: 59–102

Swift MJ (1984) Soil biological processes and tropical soil fertility. Biol Intern Spec Issue 5: 1–38

Szott LT, Fernandes ECM and Sanchez PA (1991) Soil-plant interactions in agroforestry systems. For Ecol Manage 45: 127–152

Toky OP and Bisht RP (1992) Observations on the rooting patterns of some agroforestry trees in an arid region of north-west India. Agroforestry Systems 18: 245–263

Vandenbeldt RJ (1991) Rooting systems of western and southern African *Faidherbia albida* (Del.) A. Chev: (syn. *Acacia albida* Del.) – a comparative analysis with biogeographical implications. Agroforestry Systems 14: 233–244

Van Noordwijk M (1989) Rooting depth of cropping systems in the humid tropics in relation to nutrient use efficiency. In: Van der Heide J (ed) Nutrient Management for Food Crop Production in Tropical Farming Systems, pp 129–144. Institute for Soil Fertility, Haren, The Netherlands

Van Noordwijk M, Hairiah K, Ms S and Flach E N (1991a) *Peltophorum pterocarpa* (DC.) Back (Caesalpiniaceae), a tree with a root distribution suitable for alley cropping on acid soils in the humid tropics. In: McMichael BL and Persson H (eds) Plant Roots and Their Environment, pp 526–532. Elsevier, Amsterdam, The Netherlands

Van Noordwijk M, Widianto, Heinen M and Hairiah K (1991b) Old tree root channels in acid soils in the humid tropics: Important for crop root penetration, water infiltration and nitrogen management. Plant Soil 134: 37–44

Van Rees KCJ and Comerford NB (1986) Vertical root distribution and strontium uptake of a slash pine stand on a Florida spodosol. Soil Sci Soc Am J 50: 1042–1046

Wardle DA and Greenfield LG (1991) Release of mineral nitrogen from plant root nodules. Soil Biol Biochem 9: 827–832

Yamoah CF, Agboola AA and Wilson GF (1986) Nutrient contribution and maize performance in alley cropping systems. Agroforestry Systems 4: 247–254

Yocum WW (1937) Root development of young delicious apple trees as affected by soils and by cultural treatments. Univ Nebrasca Agric Exp Stat Res Bull 95: 1–55

Young A (1989) Agroforestry for Soil Conservation. CAB International, Wallingford, UK, 276 pp

Agroforestry Systems **30**: 145–159, 1995.

Soil amelioration and root symbioses of *Parkia biglobosa* (Jacq.) Benth. in West Africa

H. TOMLINSON[1], Z. TEKLEHAIMANOT[1], A. TRAORÉ[2] and E. OLAPADE[3]
[1] *School of Agricultural and Forest Sciences, University of Wales, Bangor, Gwynedd, LL57 2UW, UK;* [2] *Centre National de Semences Forestières, B.P. 2682, Ouagadougou, Burkina Faso;* [3] *Forest Research Institute of Nigeria, Jericho, Ibadan, Nigeria*

Key words: endomycorrhiza, multipurpose tree, rhizobium, soil fertility

Abstract. *Parkia biglobosa* is an important multipurpose tree from the savanna zone of West Africa. It has been reported to increase soil fertility and crop yields beneath its crown. However, no work has been conducted to determine the role of root symbioses in soil amelioration by this species. The existing reports of nodulation in *Parkia biglobosa* are contradictory and the presence of mycorrhizae is not documented in the literature.

Fieldwork was conducted at six sites in Burkina Faso and Nigeria, to investigate the amelioratory effect of *Parkia biglobosa* on soil fertility and to ascertain the occurence of root symbioses. Soil samples were collected from points close to the trunk, the mid-crown, crown edge and the open areas away from individual trees. The samples were analysed chemically to assess pH, organic matter, nitrogen, phosphorus and potassium content. Root samples were also collected from the field and examined for nodules and for evidence of mycorrhizal infection.

The results of the soil analysis showed a significant amelioration for total nitrogen and available potassium with proximity to the tree, but organic matter, available phosphorus and soil pH showed no significant trend. This pattern was similar for all sites. The results also indicated that amelioration of nitrogen and potassium with proximity to the tree increased with tree size.

Root nodules were absent from the roots of trees examined in the field, both in Burkina Faso and Nigeria. A pot experiment was set up using the soil collected from beneath the trees and a laboratory grown cowpea miscellany culture as rhizobial inoculum for *Parkia biglobosa* seedlings. The roots of these seedlings also showed no evidence of nodulation. Ectomycorrhizae were not present in roots collected from the field, but endomycorrhizal infection was prolific.

A study of root distribution in Burkina Faso suggested that one mechanism for increased soil fertility beneath *Parkia biglobosa* is the widespread lateral root system, which redistributes nutrients from an extensive radius to a concentrated area beneath the crown. The root system is aided in the efficient uptake of nutrients by endomycorrhizae. Other possible mechanisms include interception of dust by the crown and deposition by throughfall and stemflow, defacation by animals beneath the tree and decomposition of leaf litter.

Introduction

Parkia biglobosa is a leguminous tree (sub-family Mimosoideae) indigenous to the savanna zones of West Africa, with a natural range extending from 5° N to 15° N and 16° W to 32° E [Hopkins and White, 1984]. The tree is normally protected around homesteads [Okigbo, 1983], but is rarely actively regenerated and maintained [von Maydell, 1987]. It is highly valued by the rural populations, and is one of the commonest tree species found in farmed parklands in West Africa [Gijsbers et al., 1994]. The seeds are fermented to produce a condiment used extensively in the preparation of local dishes. In Nigeria, the daily consumption of fermented seeds has been estimated at 1–17 g per head [Campbell-Platt, 1980]. It is sometimes described as a meat or cheese substitute because of its savoury taste and high protein content [Booth and Wickens, 1988]. In addition to human nutrition, the leaves and pods are used for fodder; the bark, leaves and roots for medicine and the timber for fuelwood and light construction. The crude protein content of the leaves is low compared with *Leucaena leucocephala* [le Houérou, 1980], but the leaf biomass yield is similar to that of *Gliricidia sepium* at a site in humid Nigeria [Sabiitti and Cobbina, 1992a].

The multipurpose nature of this tree has attracted attention from agroforesters. However, little scientific information exists regarding the possibilities for its incorporation into agroforestry systems other than the traditional farmed parklands, and in particular its potential for soil amelioration. Published reports are conflicting regarding the tree's suitability for intercropping. Some researchers in Nigeria [Sabiitti and Cobbina, 1992b; Young, 1989] have observed that crops grow better under *Parkia biglobosa* compared with the open field. In contrast, in Burkina Faso it has been reported that sorghum yields were only 30% of those in the open field, as light intensity was reduced under the large spreading crown by as much as 80% [Kessler, 1992]. In a study in south Mali, yields of cotton, sorghum and millet under *P. biglobosa* trees decreased by as much as 60–65% compared with the open field. This was attributed to shading, competition from the trees' root systems and increased humidity under the tree resulting in susceptibility of the crop to fungal attack [Kater et al., 1992]. However, the potential negative effect on cereal yields can be more than compensated by the income received from the trees' seeds. A study in Burkina Faso in 1992 showed that the average benefit accrued from *P. biglobosa* seeds was estimated to be approximately 5,000 FCFA (US$20) per hectare, based on an average density of two trees per hectare [Kessler, 1994]. In addition, pruning has been shown to reduce the negative impacts on crop production and is commonly practised in many parts of Nigeria and Burkina Faso. Where trees have recently been cut or have died, crop yields are often higher than in the open field, apparently as a result of the combination of relatively high soil fertility and an absence of shading [Kessler, 1992].

The aim of this research was to investigate the potential for soil amelioration under isolated *P. biglobosa* trees growing on farmland in Burkina Faso and Nigeria in three savanna ecozones; Derived (Forest-Savanna Mosaic), Guinea and Sudan savanna [Keay, 1959]. The methodology chosen was a transect, whereby a series of soil samples were taken from the base of a tree to open land beyond the crown [Young, 1986]. This technique has been used successfully in other savanna ecosystems [Belsky et al., 1989; Bernhardt-Reversat, 1982; Kater et al., 1992; Kellman, 1979]. In addition to soil sampling, the root systems of the trees were examined for the presence of root symbioses to determine their contribution to soil amelioration. The genus *Parkia* has been listed as having potential for nitrogen fixation [Brewbaker et al., 1984], and early workers [Allen and Allen, 1936] reported nodulation in *Parkia africana*, previously synonymous with *Parkia biglobosa* [Hopkins, 1983]. However, more recent investigations have found no evidence of nodulation in nursery-grown *P. biglobosa* seedlings [Workman, 1986] or mature trees (J. Sprent, pers. comm.). There are no reports in the existing literature regarding the mycorrhizal status of the tree.

Materials and methods

Soil and root sampling

In July to October 1993, fifty-four *Parkia biglobosa* trees were selected; nine in each of the six locations listed in Table 1. The trees at each location were selected to represent three size classes; 10–25 cm diameter at breast height (d.b.h.), 35–70 cm d.b.h. and > 70 cm d.b.h., with three replicates of each size class at each location. The trees were well isolated from other trees and located mainly in cultivated and fallow areas. However, it was not always possible to find trees of each size class growing under similar conditions at each location. Table 2 characterises the conditions at each location for each of the three size classes.

Soil samples were taken at the four cardinal points (north, east, south and west) from the base of the trunk, mid-crown, the crown edge and from the open area outside the crown. The sample from the open was taken at 5 m from the crown edge for the two larger size classes, and at 2 m from the crown edge for the smallest size class. This was because the smaller trees were less well isolated from surrounding vegetation as they were generally found in the fallow fields as a result of natural regeneration.

Each soil sample was taken to a depth of 30 cm and mixed thoroughly. The four samples for each direction were bulked according to distance from the tree; making a total of 216 samples. The bulked samples were combined

Table 1. Locations of trees sampled.

Location	Lat. Long.	Rainfall (mm)	Savanna type	Soil type
Saki, Nigeria	8° 30′ N 3° 25′ E	1200–1400	Derived	Loam/sand
Afaka, Nigeria	9° 30′ N 7° 25′ E	1000–1200	Guinea	Loam/clay
Toussiana, Burkina Faso	10° 50′ N 4° 36′ W	900–1000	Guinea	Sand/loam
Pama, Burkina Faso	11° 14′ N 0° 42′ E	700–800	Sudan	Sand/clay
Kano, Nigeria	12° 00′ N 8° 30′ E	700–800	Sudan	Sand
Gonse, Burkina Faso	12° 15′ N 1° 40′ W	600–700	Sudan	Sand/loam

Table 2. Distribution of trees between vegetation types at each location.

Location	Tree size	Site characteristics
Saki	10–35 cm	Cultivated fields
	35–70 cm	Cultivated fields
	> 70 cm	Cultivated fields/fallow fields
Afaka	10–35 cm	Fallow fields
	35–70 cm	Cultivated fields
	> 70 cm	Cultivated fields
Toussiana	10–35 cm	Fallow fields
	35–70 cm	Cultivated fields
	> 70 cm	Cultivated fields
Pama	10–35 cm	Fallow fields
	35–70 cm	Cultivated fields
	> 70 cm	Cultivated fields
Kano	10–35 cm	Cultivated fields
	35–70 cm	Cultivated fields
	> 70 cm	Cultivated fields
Gonse	10–35 cm	Plantation
	35–70 cm	Cultivated fields
	> 70 cm	Secondary forest

and a sub-sample of approximately 300 g of soil taken. 200 g of this sub-sample was air dried for soil chemical analyses, assessments of mycorrhizal spores and for use as mycorrhizal inoculum in greenhouse experiments. The remaining 100 g was kept moist to be used as rhizobial inoculum in greenhouse experiments.

A main lateral root was selected for each tree and traced up to a distance of 4 m. The ability to trace a root depended on the size of the tree, the soil type and depth of the root. The fine roots (diameter < 2 mm) were examined for any evidence of nodulation. Samples of fine roots were taken along the length of the exposed root, washed and stored in acidic glycerol, prior to staining for mycorrhizal infection.

Soil chemical analysis

The air dried soil was analysed for soil organic matter by loss on ignition (LOI) at 450 °C for 16 h. Soil pH was determined using a 1:2.5 (w/v) suspension of soil in distilled water. Total nitrogen was analysed using a Kjeldahl digestion; available phosphorus and potassium were extracted using a 1 M ammonium acetate solution and determined by the Olsen method and flame photometry respectively.

Rhizobial inoculation

Each 100 g sample of moist soil was mixed with sufficient perlite to fill a 7 cm diameter plastic plant pot. The perlite had been sterilised in an autoclave at 121 °C for 15 min at 1 atm. Seeds collected from three provenances in Nigeria were pre-germinated on blotting paper in petri dishes. The provenances were selected to represent the three ecozones under investigation; Derived, Guinea and Sudan savanna, to account for any possibility of specificity between rhizobial strains and the host plant. Once the radicle had emerged, the seeds were transferred into the pots containing the relevant soil type. Emergence of the cotyledons occurred within a week, and any seedlings which became susceptible to fungal disease were replaced within the first week. Ten seedlings from each provenance were also grown in perlite inoculated with a cowpea miscellany culture and a further ten were placed in uninoculated perlite as a control. The seedlings were grown in a heated greenhouse with supplementary lighting. The day length was 16 hours, with a day time temperature of 25 °C and a night time temperature of 20 °C. After four weeks, when all cotyledon reserves had been utilised, a modified Long Ashton solution, without nitrogen, was applied to the plants twice weekly. After twelve weeks the seedlings were harvested and the roots gently washed and examined for nodules.

Assessment of mycorrhizal infection

The roots collected from trees in the field were examined for both ectomycorrhizal and endomycorrhizal infection. The root tips were stained with 0.1% toludine blue and examined for any evidence of a mantle or fungal sheath. Further samples were stained with 0.05% Trypan blue [Phillips and Hayman, 1970] with some recent modifications to the original methodology [Koske and Gemma, 1989]. The percentage of mycorrhizal infection was assessed by counting fields of observation of infected root length under a light microscope at a magnification of ×100.

Statistical analysis

The data from the chemical analyses was analysed by a three-way analysis of variance using 'location', 'tree size' and 'distance from tree' as factors. The percentage mycorrhizal infection was analysed using a two-way analysis of variance with 'location' and 'tree size' as factors.

Results and discussion

Soil analyses

Figure 1 shows soil properties at different distances from the tree at each location. There were significant differences between locations for all soil characteristics ($p < 0.001$), probably as a consequence of the underlying parent rock, and climatic and vegetation effects, but no significant interactions were observed between location and distance for any of the sites surveyed. Soil pH was lowest at Toussiana, possibly because of the combined effect of high leaching rates and low soil fertility, and highest in Saki and Pama. The organic matter content at each site can be directly correlated with percentage vegetation cover and biomass production at each of the locations [Ahn, 1970], with the highest percentage LOI at Saki in the loamy Derived savanna soil and lowest at Kano in the sandy Sudan savanna soil. The high organic matter in the soils at Gonse can be attributed to the samples taken from the secondary forest. There were no significant differences between pH and organic matter between samples taken at different distances from the trees within a site. At all sites except Kano there was an amelioration of total nitrogen with proximity to the tree ($p < 0.001$). The lack of amelioration under the trees at Kano may be a consequence of high leaching rates in the sandy soil. Available potassium showed an significant amelioration with proximity to the tree for all sites ($p < 0.001$).

The results for phosphorus were extremely variable between sites. A correlation between the percentage LOI and phosphorus concentrations for each sample gave an r value of 0.05, suggesting that some of the available phosphorus in the soils was derived from sources other than organic matter. At Saki, Afaka, Pama and Gonse there were no significant differences in phosphorous concentrations with distance from the tree. However, at Toussiana, the phosphorus concentrations were approximately double beyond the canopy compared with close to the trunk. Toussiana had the most impoverished soils of the locations analysed and it is known that fertiliser is applied to the cultivated fields beyond the tree crowns at this site. At Pama the reverse was observed, with almost twice as much phosphorus found close to the tree trunk compared with the open field. Cattle are known to shelter around the trees at this location and the elevated phosphorus concentrations may be attributable to deposition of faeces and urine. However, similarly high concentrations of nitrogen and potassium were not found at either Toussiana or Pama. As the ions of these elements are more mobile in the soil than phosphate, it is possible they may have been leached, or in the case of ammonium, volatilised.

The concentrations of available phosphorus found in this survey were extremely low, on average 0.05–0.25 $\mu g\ g^{-1}$ soil. However, a previous soil survey at Gonse gave results of a similar order of magnitude [Anon, 1990]. These low concentrations can partly be attributed to the traditional methods

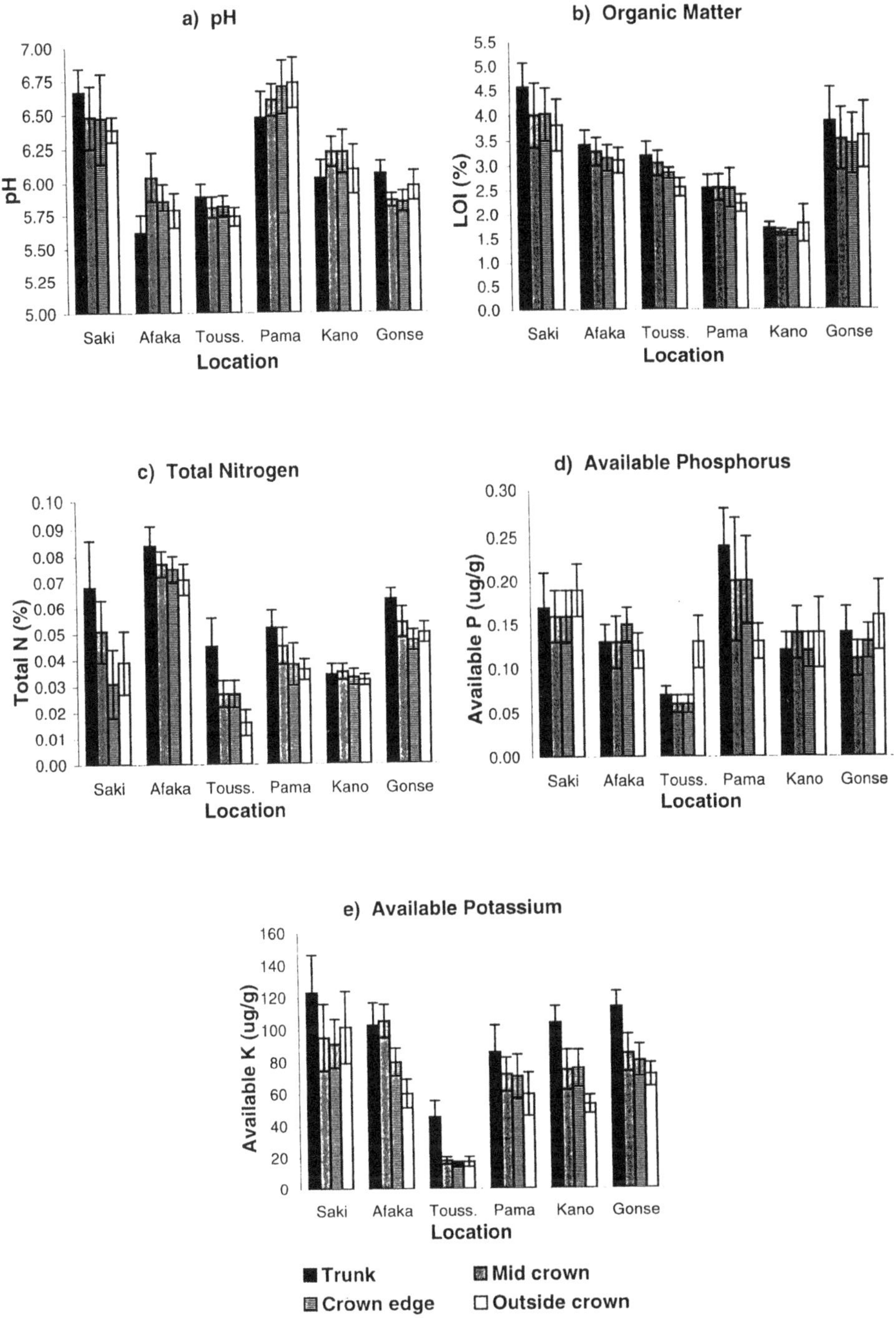

Fig. 1. Soil properties according to location and distance from *Parkia biglobosa* trees (n = 9) Touss. = Toussiana.

of cultivation used in West Africa, whereby the topsoil is overturned, introducing the less fertile subsoil into the upper soil horizons. Loss of fertile topsoil during the Harmattan, and the lateritic nature of the parent rock at some of the sites such as Toussiana, are other possible explanations. When concentrations of nutrients are limiting, small changes in actual nutrient concentrations, such as those observed for phosphorus in this survey, can have significant effects on plant growth.

The results of the soil analyses according to tree size class and distance from the tree are shown in Fig. 2. There were significant differences for all soil characteristics between size classes, with more organic matter, total nitrogen, available phosphorus and available potassium beneath the larger trees ($p = 0.019$). Soil pH, organic matter and phosphorus showed no significant differences with increasing distance from the tree bole for any of the size classes. Only nitrogen and potassium were significantly higher with increased proximity to the tree ($p = 0.001$ and $p < 0.001$, respectively) and this amelioration was more pronounced for the larger trees. The phosphorus concentrations were significantly higher for all soil samples taken from the larger trees. This can be explained by fertilisation at Toussiana and defecation by animals at Pama, as elevated concentrations of phosphorus were found mostly around the larger trees at these locations. The interaction between location and size class was significant for all soil characteristics. For nitrogen ($p = 0.009$), phosphorus ($p = 0.001$) and potassium ($p = 0.012$) there were higher nutrient concentrations under the bigger trees at all locations, but the magnitude of the amelioration varied between sites. For pH ($p < 0.001$) and organic matter ($p = 0.006$), no consistent pattern was observed and the variation appeared to be site specific.

Soil amelioration by savanna trees has been reported by many researchers [Belsky et al., 1989; Bernhardt-Reversat, 1982; Dancette and Poulain, 1969; Kater et al., 1992; Keay, 1959; Mazzarino et al., 1991]. In each case organic matter, nitrogen, potassium and magnesium decreased with radial distance from the tree bole or centre of crown, although in one case [Kater et al., 1992] available phosphorus increased away from the influence of the tree. Phosphorus uptake may decrease in the area outside the crown because of the lower tree root densities, resulting in higher phosphorus availability in the soil. Yet in this study, a genuine amelioration of soil fertility beneath *Parkia biglobosa* was observed only for nitrogen and potassium.

The results also suggest that soil amelioration is augmented with increasing tree size, particularly in the case of nitrogen and potassium. A study of soil amelioration under individual savanna trees in Senegal [Bernhardt-Reversat, 1982] showed that there was a positive correlation between tree size and nitrogen concentration beneath the crown, implying that soil enrichment by a tree continues throughout its life. Larger trees also have the ability to establish a larger plant-litter-soil nutrient cycle by prolonged persistence at one site [Isichei and Muoghalu, 1992]. However, the form of soil survey used in this study is limited because the vegetation varied between tree size classes

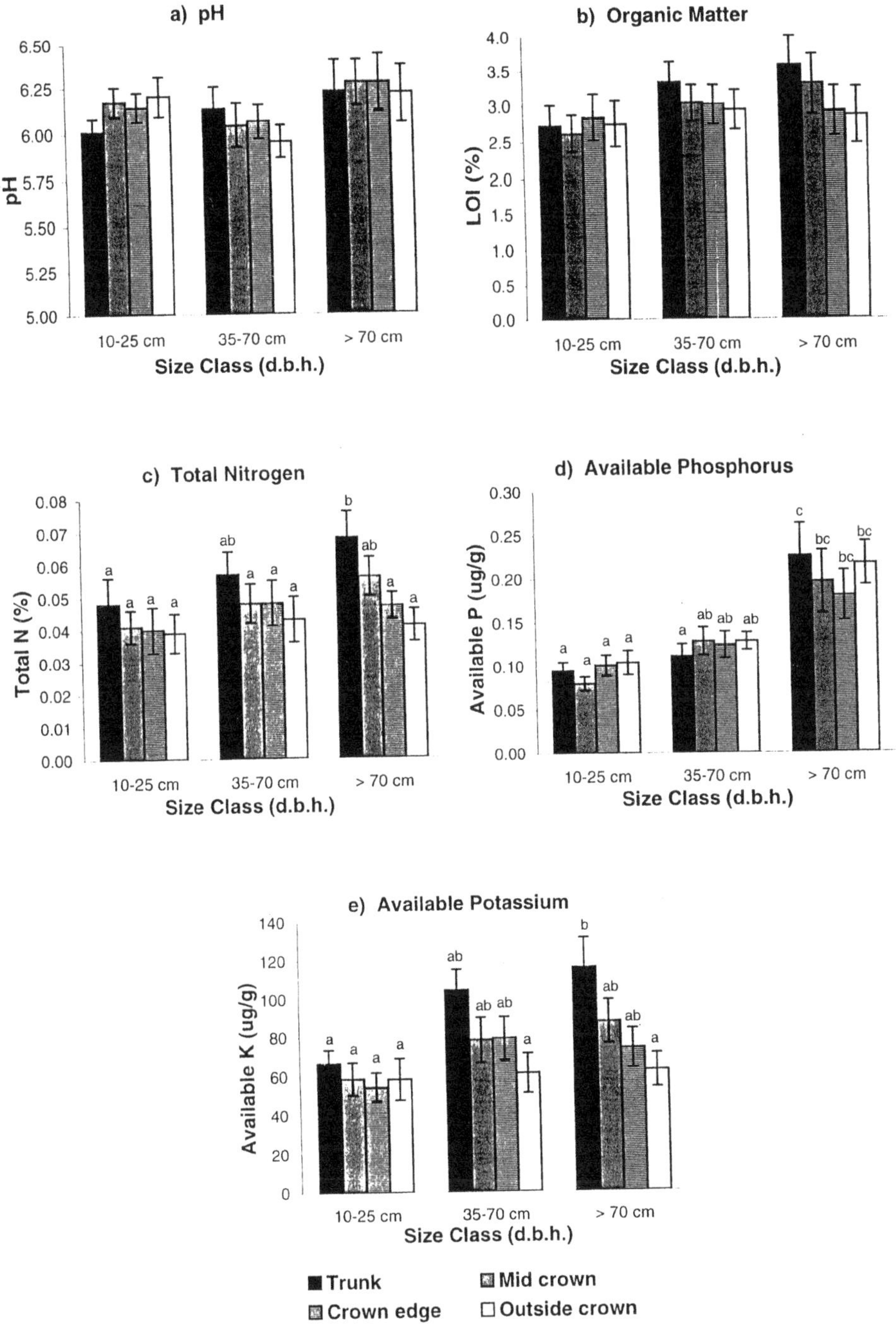

Fig. 2. Soil properties according to tree size class and distance from *Parkia biglobosa* trees (n = 18); values followed by different letters are significantly different at $p = 0.01$.

at each location and detailed information on the management and history at the different sites is lacking. Therefore, it is only possible to make general assumptions regarding the potential for soil amelioration under *P. biglobosa* over time. In order to make conclusive statements, it would be better to monitor changes in soil properties with time on the same site [Sanchez, 1987].

Rhizobial inoculation

There were no nodules present on the roots of any of the *P. biglobosa* trees examined in the field, although other plants in the vicinity e.g., *Erythrina senegalensis* and groundnut (*Arachis hypogaea*) had active nodules. In the greenhouse experiment, none of the seedlings grown in soil collected from beneath mature *P. biglobosa* trees, or the seedlings inoculated with the cowpea miscellany rhizobial culture, showed any indication of nodulation.

These results would indicate that either there is no suitable inoculum available in the soils studied, or that *P. biglobosa* is not capable of nodulation. *Parkia* species in Asia are known to nodulate but no nodules have been found on South American species in Brazil [de Faria et al., 1984]. Recent attempts to find nodules on *Parkia biglobosa* in West Africa have also been unsuccessful [Workman, 1986] (J. Sprent, pers. comm.). The trees were growing in soils with a very low phosphorus content, which may have been insufficient for nodulation [Högberg, 1986]. In the greenhouse experiment, phosphorus levels in the soils were augmented with nutrient solutions but still nodulation did not occur. The initial low phosphorus concentrations may have been a critical factor in the survival of rhizobia in the soil [Virginia, 1986], as suitable inoculum is unlikely to persist where conditions are unfavourable. The inoculation with cowpea miscellany was also unsuccessful, but this may have been related to host-strain specificity.

It is possible that the nodules may be ephemeral and only persist for a few weeks before senescing. Nodules on *Acacia* species grown in Sudanese soil were in an advanced state of decay nine weeks after inoculation [Deans et al., 1993]. To determine if this were the case with *Parkia biglobosa* it would be necessary to conduct destructive harvests over a very short time course. In another study, nodules were found on seedlings of *Faidherbia (Acacia) albida* but not mature trees. It has been proposed that amelioration of nitrogen under mature trees is likely to be a result of some mechanism other than N-fixation, possibly through redistribution by the root system [Dunham, 1992]. Nodules were also reported to be absent from the roots of mature *Acacia senegal* trees [Bernhardt-Reversat, 1982].

The indication that *P. biglobosa* may not nodulate, and therefore may not fix atmospheric nitrogen, does not imply that the tree is of little of value in agroforestry systems. It may be more worthwhile for a farmer to plant a tree with multipurpose uses rather than one whose sole benefits are N-fixation, because of the economic benefits generated. On many sites, investment in an extensive root system may increase tree growth and survival more efficiently than investment in nodules which can only alleviate nitrogen deficiency [Sprent, 1985].

Mycorrhizal infection

There was no evidence of ectomycorrhizal infection in the roots of *Parkia biglobosa*. Approximately 85% of the roots were infected with endomycorrhizae, but there were no significant variations with location and tree size class. These results are in keeping with reports from other Mimosoideae [Högberg, 1986; Högberg, 1992]. Non-nodulated endomycorrhizal species are highly competitive where the availability of phosphorus is low, which at the same time is a severe restriction for N-fixing species [Högberg, 1986]. Mycorrhizae increase the tree's absorptive capacity by enlarging the exchange surface available for uptake of mineral elements such as phosphorus and potassium [Berthelin et al., 1991], thereby increasing the efficiency of the root system. They promote tolerance to drought and salinity and bind soil particles together, thus forming and maintaining soil structure [Young, 1986]. They also play an important role in the conversion of nutrients within the nutrient cycle through their ability to take up nitrogen and phosphorus directly from litter. It is possible that in the case of *P. biglobosa*, endomycorrhizae enhance nitrogen uptake to a sufficient level to compensate for the lack of nodulation.

Root distribution

It has been suggested that the main mechanism by which legumes alter the fertility of the superficial soil layers is through root activity [Virginia, 1986]. Deep rooting is one possible mechanism by which this is achieved. However, rooting patterns in arid lands are likely to be predominantly shallow because of limited soil depth and soil wetting. Most data on West African root systems indicate that the roots are concentrated in the upper 20–50 cm of soil [Kessler and Breman, 1991]. Thus, redistribution of soil nutrients as a consequence of the lateral spread of roots is likely to be more important in concentrating nutrients under the tree crown compared with uptake from deep soil layers. Results from a study of root distribution of *Parkia biglobosa* in Burkina Faso (H. Tomlinson, unpub.), where roots were excavated to a depth of 50 cm using a logarithmic spiral trench [Huguet, 1973], showed that the majority of roots (approximately 60%) were concentrated in the upper 20 cm of soil but no attempt was made to assess the depth of the tap root. The study also suggested that the roots of *P. biglobosa* trees with an average crown radius of 7 m could extend up to 20 m from the trunk. Thus, the potential area for resource capture by the tree's roots is eight times that of the crown area. This is similar to data from other researchers. The root systems of *Faidherbia (Acacia) albida* in Zambezi riverine woodlands extended to two and a half times the crown radius, with a potential area for root exploitation of six times the crown area

[Dunham, 1992]. *Burkea africana* roots exceeded crown spread by seven times and consequently an accumulation of nutrients in the sub-crown habitat was predicted [Rutherford, 1982]. An extensive root system such as that of *P. biglobosa*, aided by mycorrhizae, could effectively mine nutrients over a large area which would then accumulate beneath the tree crown.

Other mechanisms of amelioration

Other external inputs to the soil beneath the crown could be derived from atmospheric deposition or animals. Dust deposition in Kano has been reported at a rate of 1.2 kg ha^{-1} phosphorus annually from the Sahara, 1000–1200 km away [McTamish and Walker, 1982]. Nutrients in aerially transported dust could reach the area below the crown via throughfall and stemflow. Nitrogen input in rain was recorded as 0.04 g m^{-2}, compared with 0.57 g m^{-2} in throughfall beneath *Acacia senegal* [Bernhardt-Reversat, 1982]. However, it was not stated whether this was as a result of leaching (recycling) or atmospheric deposition (genuine input). It is possible that the large umbrella-shaped crown of *Parkia biglobosa* would be efficient in trapping dust particles, but no studies have been conducted to evaluate this. The quantity of gross rainfall accounted for by stemflow may also vary considerably according to tree architecture. Stemflow in *Acacia aneura* accounts for 40% of gross rainfall compared with 5–10% for most tree species [Leyton, 1983]. It is likely that stemflow will be important for *P. biglobosa* as a result of the spreading nature of the tree's crown, and this may account for the increase in potassium close to the trunk.

Inputs from animals may be derived from defecation by cattle resting in shade beneath the trees or from birds roosting in the crown [Belsky et al., 1989]. There is some evidence of this at Pama, where increased phosphorus concentrations are likely to be a consequence of animals sheltering by the trees. However this would need to be explored further, and the role of nitrogen and potassium present in the faeces and urine determined more clearly.

Other possible mechanisms for soil amelioration include input of leaf litter and modification of the microclimate by the tree. Trees provide a range of carbon substances whose nutrients are released over different periods of time [Young, 1986]. Accumulation of carbon beneath the tree canopy may enhance the availability of soil nutrients as suitable carbon substrates can provide energy for soil micro-organisms to carry out mineralisation. The tree canopy also provides shade and moderates surface temperatures thus making conditions more favourable for microbial activity.

Implications for intercropping

Traditional practices (G. Ky, pers. comm.) and information from other studies [Kater et al., 1992; Kessler, 1992] suggest that planting crops beneath the crown of *P. biglobosa* trees is not a viable farming application. On the other hand, under conditions of marginal soil fertility it may be important to maximise the potential of these 'islands' of enhanced nitrogen and potassium. Advantage could be taken of these by growing secondary crops, e.g., vegetables and cowpea for fodder, or dwarf sorghum, as has been suggested by a survey of farming practices in Nigeria (H. Tomlinson, unpub.). Information would then be required on the root distribution of the tree and crops in order to obtain maximum benefits from such intercropping systems. Methods of pruning could also be established to reduce the negative effects of shading. This could mean a compromise between arable production and *P. biglobosa* seed production, although pruning may stimulate flower production. Similar conclusions to these were drawn from a study of *P. biglobosa* in South Mali [Kater et al., 1992].

Conclusions

The results from the tree-open land soil transects indicate that soil nitrogen and potassium increase with proximity to *Parkia biglobosa* trees. However, it is likely that the majority of this amelioration takes place as a result of the redistribution of nutrients by an efficient laterally spreading root system. Therefore, the increased nutrients are present under the tree crown at the expense of treeless areas [Kessler and Breman, 1991]. N-fixation through rhizobial symbiosis would be a genuine contribution to soil enrichment, but there is no evidence of this. The presence of endomycorrhizal symbioses would, however, enable the root system to exploit the existing soil nutrients more effectively. Aerial and animal inputs are also likely to provide a contribution to soil fertility beneath the crown but it would require further experimentation to confirm this.

In areas of marginal soil fertility it may be worth exploiting the area beneath the crowns of *P. biglobosa* trees, but appropriate farming practices would need to be adopted.

Acknowledgements

We wish to acknowledge staff at the Centre National de Semences Forestières (CNSF), Burkina Faso, and at the Forest Research Institute of Nigeria (FRIN), Nigeria, for assistance in collecting the soil and root samples. The technical assistance of Andy Owen is gratefully acknowledged for all chemical analyses. This research was financed by the European Commission Directorate General XII, under the 'Life Sciences and Technologies for Developing Countries' programme.

References

Anon (1990) Étude pédologique de la forêt classée à Gonse (parcelles CNSF). Burkina Faso, Ministère de l'Agriculture et de l'Élevage, Secrétariat Général, Bureau National des Sols, Ouagadougou, Upper Volta

Ahn PM (1970) West African Agriculture. Vol 1. West African Soils. 3rd Edition. Oxford University Press, London, UK

Allen ON and Allen EK (1936) Root nodule bacteria of some tropical leguminous plants: 1. Cross-inoculation studies with *Vigna sinensis* L. Soil Sci 42: 61–77

Belsky AJ, Amundson RG, Duxbury JM, Riha SJ, Ali AR and Mwonga SM (1989) The effects of trees on their physical, chemical and biological environments in a semi-arid savanna in Kenya. J Appl Ecol 26: 1005–1024

Bernhardt-Reversat F (1982) Biogeochemical cycle of nitrogen in a semi-arid savanna. Oikos 38: 321–332

Berthelin J, Laheurte F and De Guidici P (1991) Involvement of roots and rhizosphere microflora in the chemical weathering of soil minerals. In: Atkinson DJ (ed) Plant Root Growth: An Ecological Perspective, pp 187–200. BES Special Publication no. 10, Blackwell Scientific Publications, Oxford, UK

Booth FEM and Wickens GE (1988) Non-Timber Uses of Selected Arid-Zone Trees and Shrubs in Africa. FAO Conservation Guide 19, Rome, Italy, 176 pp

Brewbaker JL, Haliday J and Lyman J (1984) Economically important nitrogen fixing tree species. N Fix Tree Res Rep 11: 38–45

Campbell-Platt G (1980) African locust bean (*Parkia* species) and its West African fermented food product: dawadawa. Ecol Food Nutr 9: 123–132

Dancette C and Poulain JF (1969) Influence of *Acacia albida* on pedoclimatic factors and crop yields. African Soils 14: 143–184

Deans JD, Ali OM, Lindley DK and Nour HOA (1993) Rhizobial nodulation of *Acacia* tree species in Sudan: soil inoculum potential and effects of peat. J Trop For Sci 6(1): 56–64

De Faria SM, Franco AA, De Jesus RM, De Menandro MS, Baitello JB, Mucci SF, Dobreiner J and Sprent JI (1984) New nodulating trees from south-east Brazil. New Phytol 98: 317–328

Dunham KM (1992) Comparative effects of *Acacia albida* and *Kigelia africana* trees on soil characteristics in Zambezi riverine woodlands. J Trop Ecol 7: 215–220

Gijsbers HJM, Kessler JJ and Knevel MK (1994) Dynamics and natural regeneration of woody species in farmed parklands in the Sahel region (Province of Passoré, Burkina Faso). For Ecol and Manage 64: 1–12

Högberg P (1986) Soil nutrient availablility, root symbioses and tree species composition in tropical Africa: a review. J Trop Ecol 2: 359–372

Högberg P (1992) Root symbioses of trees in African dry tropical forests. J Veg Sci 3: 393–400

Hopkins HC (1983) The taxonomy, reproductive biology and economic potential of *Parkia* (Leguminosae:Mimosoideae) in Africa and Madagascar. Bot J Linn Soc 87: 135–167

Hopkins HC and White F (1984) The ecology and chorology of *Parkia* in Africa. Bull Jard Bot Nat Belg 54: 235–266

Huguet JG (1973) Nouvelle méthode d'étude de l'enracinement des végétaux pérennes à partir d'une tranchée spirale. Ann Agron 24: 707–731

Isichei AO and Muoghalu JI (1992) The effects of tree canopy cover on soil fertility in a Nigerian savanna. J Trop Ecol 8: 329–338

Kater LJM, Kante S and Budelman A (1992) Karité (*Vitellaria paradoxa*) and néré (*Parkia biglobosa*) associated with crops in South Mali. Agroforestry Systems 18: 89–105

Keay RWJ (1959) Vegetation Map of Africa South of the Tropic of Cancer. Oxford University Press, Oxford, UK, 24 pp

Kellman M (1979) Soil enrichment by neotropical savanna trees. J Ecol 67: 565–577

Kessler JJ (1992) The influence of karité (*Vitellaria paradoxa*) and néré (*Parkia biglobosa*) trees on sorghum production in Burkina Faso. Agroforestry Systems 17: 97–118

Kessler JJ (1994) Agroforestry in Burkina Faso: a careful balance. Ag-siev 6(3): 4–5

Kessler JJ and Breman H (1991) The potential for agroforestry to increase primary production in the Sahelian and Sudanian zones of West Africa. Agroforestry Systems 13: 41–62

Koske RE and Gemma JN (1989) A modified procedure for staining roots to detect VA mycorrhizas. Mycol Res 92: 486–505

Le Houérou HN (1980) Chemical composition and nutritive value of browse in tropical West Africa. In: Le Houérou HN (ed) Browse in Africa, the Current State of Knowledge, pp 261–289. ILCA, Addis Ababa, Ethiopia

Leyton L (1983) Crop water use: principles and some considerations for agroforestry. In: Huxley PA (ed) Plant Research and Agroforestry, pp 379–500. ICRAF, Nairobi, Kenya

McTamish GH and Walker PH (1982) Nature and distribution of Harmattan dust. Z Geomorph 26(4): 417–435

Mazzarino MJ, Oliva L, Nunez J, Nunez G and Buffa E (1991) Nitrogen mineralisation and soil fertility in the dry Chaco System. Soil Sci Soc Am J 55: 515–522

Okigbo BN (1983) Plants and Agroforestry in Land Use systems of West Africa In: Huxley PA (ed) Plant Research and Agroforestry, pp 25–41. ICRAF, Nairobi, Kenya

Phillips JM and Hayman DS (1970) Improved procedures for clearing roots and staining parasitic and versicular abuscular mycorrhizal fungi for rapid assessment of infection. Trans Brit Mycol Soc 55: 158–160

Rutherford MC (1982) Woody plant biomass distribution in *Burkea africana* savanna. In: Huntley BJ and Walker BH (eds) Ecology of Tropical Savannas, pp 120–142. Ecological Studies, Springer Verlag, Berlin, Germany

Sabiitti EN and Cobbina J (1992a) Initial agronomic evaluation *of Parkia biglobosa* in the humid zone of Nigeria. Agroforestry Systems 17: 271–279

Sabiitti EN and Cobbina J (1992b) *Parkia biglobosa*: a potential multipurpose fodder tree legume in West Africa. Int Tree Crops J 7: 113–139

Sanchez PA (1987) Soil productivity and sustainability in agroforestry systems. In: Steppler HA and Nair PKR (eds) Agroforestry: A Decade of Development, pp 205–223. ICRAF, Nairobi, Kenya

Sprent JL (1985) Nitrogen fixation in arid environments. In: Wickens GE, Goodin GR and Field DV (eds) Plants for Arid Lands, pp 215–229. Allen and Unwin, London, UK

Virginia RA (1986) Soil development under legume canopies. For Ecol Manage 16: 69–79

Von Maydell HJ (1987) Agroforestry in the dry zones of Africa. In: Steppler HA and Nair PKR (eds) Agroforestry: A Decade of Development, pp. 89–116. ICRAF, Nairobi, Kenya

Workman SW (1986) Nodulation trials with selected arboreal legumes of Burkina Faso, West Africa. N Fix Tree Res Rep 4: 7–8

Young A (1986) Effects of trees on soils. In: Prinsley RT and Swift MJ (eds) Amelioration of Soil by Trees, pp 10–19. Commonwealth Science Council, London, UK

Young A (1989) Agroforestry for Soil Conservation. CAB International, ICRAF, Wallingford, UK, 276 pp

Agroforestry Systems **30**: 161–173, 1995.

Root architecture in relation to tree-soil-crop interactions and shoot pruning in agroforestry

M. VAN NOORDWIJK and P. PURNOMOSIDHI
ICRAF-S.E. Asia, P.O. Box 161, Bogor 16001, Indonesia

Key words: competition, fractal, multipurpose trees, root methods, root pattern

Abstract. Desirable root architecture for trees differs between sequential and simultaneous agroforestry systems. In sequential systems extensive tree root development may enhance nutrient capture and transfer to subsequent crops via organic pools. In simultaneous systems tree root development in the crop root zone leads to competition for resources.

Fractal branching models provide relationships between proximal root diameter, close to the tree stem, and total root length or surface area. The main assumption is that a root branching proportionality factor is independent of root diameter. This was tested in a survey of 18 multipurpose trees growing on an acid soil in Lampung (Indonesia). The assumption appeared valid for all trees tested, for stems as well as roots. The proportionality factor showed a larger variability in roots than in stems and the effects of this variability should be further investigated. A simple index of tree root shallowness is proposed as indicator of tree root competitiveness, based on superficial roots and stem diameter.

Pruning trees is a major way to benefit from tree products and at the same time reduce above-ground competition between trees and crops. It may have negative effects, however, on root distribution and enhance below-ground competition. In an experiment with five tree species, a lower height of stem pruning led to a larger number of superficial roots of smaller diameter, but had no effect on shoot:root ratios or the relative importance of the tap root.

Introduction

Competition for below-ground resources between trees and food crops may mask or surpass many of the advantages trees may provide for the long term sustainability of agroforestry systems. Competition depends on above-ground demand for water and nutrients, as well as possibilities for uptake. Here initial tests of a new method to characterize the competitive strength of trees, as far as it is based on root distribution, are described.

Agroforestry systems can be biologically more productive than either pure crop or pure tree systems provided that trees and crops are at least partly complementary in the use of resources (above- and/or below-ground). If the tree and crop component represent approximately equal direct value to the farmer relative to the area occupied, a weak complementarity is sufficient to justify agroforestry; if trees have less or no direct value, the complementarity has to be pronounced [Van Noordwijk, in press].

Tree-(food) crop interactions in agroforestry can be spatial, temporal or both (Fig. 1). The lower left corner of Fig. 1 does not classify as agroforestry as the interactions are too weak, the upper right corner is only viable for shade-

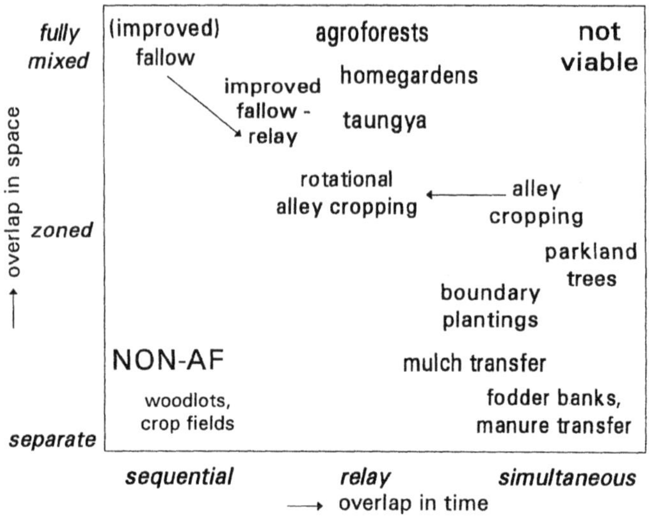

Fig. 1. Classification of agroforestry systems based on trees and food crops, by the degree of spatial and temporal overlap of trees and crops [Van Noordwijk, in press].

tolerant crops, as the interactions are strong [Van Noordwijk, in press]. Five main types of agroforestry systems based on trees and crops can be distinguished in this classification:

1. Upper left: *improved fallow* systems, where the trees and crops are completely separated in time, but spatially fully mixed;
2. Mid upper: *upper canopy tree* based systems, which allow for crop growth only in the early successional stages (taungya, homegarden, agroforests [sensu De Foresta and Michon, 1994];
3. Mid right: *hedgerow intercropping* systems or widely spaced upper canopy trees, which are spatially zoned, but where the tree and crop phase are fully integrated (e.g., coffee gardens with 'shade' trees);
4. Lower right: *mulch transfer* systems, where trees and crops are spatially separated;
5. The centre of the graph can be approached from upper left by improved fallow systems with relay cropping, or from mid right by rotational forms of hedgerow intercropping.

For complementarity of below-ground resource use, the major distinction is between simultaneous and sequential systems [Van Noordwijk et al., in press]. In sequential systems all biophysical interactions occur via soil conditions (i.e., its physical, chemical and biological fertility properties) at the end of the tree phase. On soils with sufficient nutrient reserves in the subsoil, deep rooted trees can act as 'nutrient pumps', bringing nutrients to the surface via litterfall. In simultaneous systems, competition for water and/or nutrients is of major importance, but if the tree roots occupy the zone just underneath the crop roots, they can capture *current* losses from the crop root zone and can therefore act as 'safety-nets'. The safety net function is especially important for nutrients which, depending on rainfall surplus and apparent nutrient adsorption constant, leach so fast that they cannot be reached by a subsequent deep rooted (fallow) vegetation. Under high rainfall conditions the safety net function may yield at least as much N to companion crops as N fixation [Van Noordwijk et al., 1991a].

Rooting patterns that maximize root length density appear attractive for sequential (or spatially fully separated) agroforestry systems, as they probably increase nutrient capture by the tree and nutrient mobilization for the crop through organic residues. A dense tree root system probably also enhances biological formation of soil structure [Kooistra and Van Noordwijk, 1995]. Deep rooted trees in such systems may help to create the 'old tree root channels' which can help subsequent crops [Van Noordwijk et al., 1991a].

For simultaneous systems, which normally are spatially zoned, complementarity in root distribution between tree and crop is desirable to reduce competition. How effective the spatial zonation is in reducing below-ground interactions depends on the lateral spread of the (tree) root system. Trees which rely mainly on superficial root systems are probably highly competitive with shallow rooted crops, unless their above-ground phenology leads to reduced demand for water and nutrients in the period most critical for the crop.

The available evidence in the literature suggests that trees with a well-developed tap root but few laterals extending in the top soil will generally have slow initial growth [Coster, 1932; Van Noordwijk et al., in press]. The notion of 'fast-growing, non-competitive trees' may be a contradiction in terms, at least where the initial establishment phase is concerned. A further obstacle for managing simultaneous systems may be that above-ground pruning, to harvest tree biomass and/or to reduce shading, will have below-ground consequences. It probably enhances root turnover, which may be a positive effect, but may also induce death of root apices and formation of new adventitious roots which tend to grow in the upper soil layers. The latter effect was first described for *Peltophorum dasyrachis* [van Noordwijk et al., 1991b; erroneously identified as *P. pterocarpum* in the original publication). As stem height left at pruning decreased, the number of proximal roots increased while their average diameter decreased. An experiment to test this effect for a range of tree species is described here.

Current root research methods on trees are laborious and cannot be directly related to farmers' criteria for selecting and judging the performance of trees. If root research stays in the domain of 'experts', it will not contribute to the development of agroforestry systems in the real world. The validity of generalizations about deep- or shallow-rooted trees, competitive and beneficial ones, is likely to be vastly overestimated, unless simple, non-destructive observation tools are developed that can be used to check root development for potentially any tree. As a first step in that direction, relations can be sought between the root diameter at the stem base (proximal roots) and functionally important root parameters [Van Noordwijk and Brouwer, 1995] such as the total length or surface area of the fine roots. Soumaré et al. [1994] found in a study in Mali that for *Acacia seyal* and *Sclerocaryea birrea* the root diameter at 1 m from the stem is a reasonable predictor of horizontal root spread.

In a 'fractal' branching pattern the same rules govern branching at each subsequent level [Fitter and Stickland, 1992; Van Noordwijk et al., 1994; Spek and Van Noordwijk, 1994]. The initial size (diameter) and the essential branching rules thus contain the information required to construct the whole pattern [Prusinkiewicz and Lindenmayer, 1990]. If root branching patterns have fractal characteristics, measurement of the proximal root diameter at the stem base and the branching rules as observed anywhere in the root system, would be enough to predict total root length, root diameter distribution and root length per unit dry weight (specific root length). Leonardo da Vinci claimed that the cross sectional area of a tree's main stem is equal to the sum of the cross sectional areas of its branches [Mandelbrot, 1983]. The same rule might apply to rivers, and may be based on the approximately constant volume of water passing through the river system from the sum of all sources to the final sink. A constant sum of squared diameters in trees might indicate a constant resistance to longitudinal water flow, if individual xylem cells have a constant diameter (the maximum of which is determined by the risk of forming air bubbles in large cells). For tree stems, stability and strength requirements [Coutts, 1983] may be as relevant as water transport capacities in determining stem diameters. The 'constant cross sectional area' or 'pipe-stem model' [Shinozaki et al., 1964] at least forms a valuable point of reference in studying the above-ground architecture of actual trees. A similar rule might apply to roots, but this assumption requires testing. For realistic representations of above-ground tree architecture many parameters are needed which go beyond the pipestem model [De Reffye et al., this volume]. Empirical relationships between stem diameter and root biomass [Kuiper et al., 1990] may be re-interpreted in the light of the pipe-stem model.

An algebraic model has been developed [Van Noordwijk et al., 1994] to derive the relation between total root system size and proximal root diameter for two extreme branching patterns: a dichotomous and a herringbone pattern. These patterns represent the wider classes of proportionate and determinate branching, respectively. The equations show that to predict total root length from proximal root diameter, information is needed on at least the minimum root diameter, the link ('internode') lengths as a function of root diameter and the proportionality factor α between total cross sectional areas before and after branching (Fig. 2). A further parameter, q, is defined as the ratio of the largest root diameter to the sum of all branch root diameters; in dichotomous systems q is 0.5, in herringbone patterns it gradually decreases from nearly 1 to 0.5. The parameter q is important for the length of the longest root and for the specific root length. Spek and Van Noordwijk [1994] presented an algorithm which can be used to explore these relationships for branching patterns intermediate between dichotomous and herringbone patterns. Here

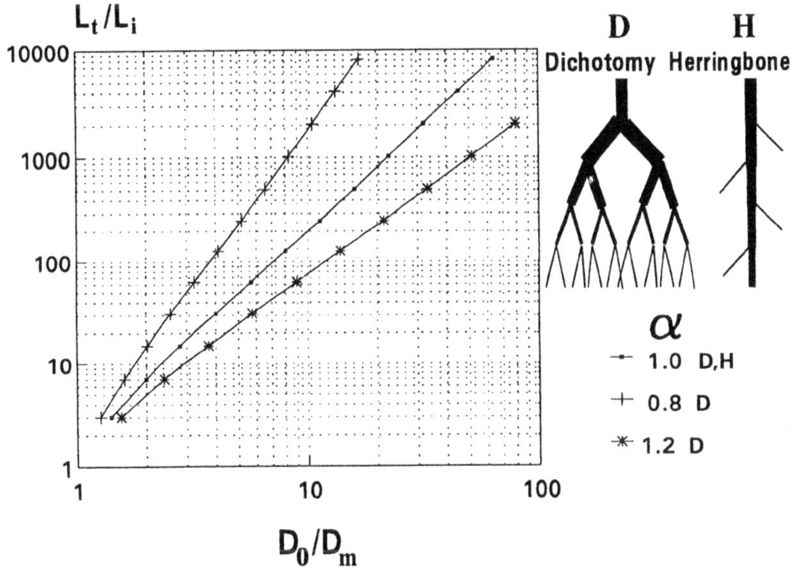

Fig. 2. Relation between the ratio of proximal (D_o) and minimum root (D_{min}) diameter and the number of links in a branched system (L_t/L_i, or the ratio of total and average link length) for a dichotomous (D) and herringbone (H) pattern, for various values of the proportionality factor α [Van Noordwijk et al., 1994].

initial tests of the validity of the main assumptions underlying these methods are described.

Material and methods

Site

Observations were made in North Lampung (4° 31′ S, 104° 55′ E) on a grossarenic kandiudult; soils are well drained, (very) low in CEC (4–10 cM_c kg^{-1}) and with a $pH(H_2O)$ of 4.5–5.0 and a pH(1 M KCl) of 3.8–4.2. Al toxicity is common, especially below a depth of 15 cm. Typical soil organic matter contents in the topsoil are 1.7–1.9%, root development is possible down to a plinthic layer at 1–1.5 m depth. Annual rainfall is about 2200 mm on average, with 1–4 months having less than 50 mm of rainfall. Further site description is given by Van Noordwijk et al. [1992].

Root survey

A number of multipurpose/fruit trees (see list in Table 1, one or two trees per species) were selected for a root survey from a homegarden close to the site of a long term soil fertility experiment, and the neighbouring village. Most trees were about six years old. The basal part of the roots at the stem base

Table 1. Root patterns of multipurpose trees on acid soil, North Lampung (Indonesia); preliminary survey of five to seven-year-old trees (mostly in homegardens).

Species	$Root_{vert}$:ΣRoot basal area ratio, A	Stem:ΣRoot basal area ratio, B	Index of shallow-rootedness (1-A)/B
Leucaena leucocephala	0.17	0.39	2.13
Parkia speciosa	0.23	0.46	1.67
Annona muricata	0.28	0.44	1.64
Perunema canescens	0.04	0.78	1.23
Psidium guajava	0.11	0.74	1.20
Nephelium lappaceum	0.15	0.77	1.10
Pithecolobium jiringa	0.10	1.00	0.90
Gliricidia sepium	0.18	0.94	0.87
Paraserianthes falcataria	0.17	0.86	0.97
Eugenia aquea	0.17	1.43	0.58
Ceiba pentandra	0.45	0.95	0.58
Gnetum gnemon	0.15	1.53	0.56
Anacardium occidentale	0.45	1.20	0.46
Calliandra calothyrsus	0.75	0.64	0.39
Peltophorum dasyrachis	0.74	0.69	0.38
Artocarpus heterophyllus	0.69	0.89	0.35
Durio zibethinus	0.75	1.14	0.22
Artocarpus integer	0.84	0.86	0.19
Mangifera indica	0.85	1.04	0.14

was carefully excavated, in a half sphere of 0.5–1 m radius. Root diameter and angle from a horizontal plane were measured for all proximal roots, i.e., roots originating from the stem base or as laterals from the top part of the tap root. Stem diameter D_{stem} was measured at breast height, or as 'root collar' diameter for pruned trees; stem basal area was calculated as πD^2_{stem}. The basal area of roots with an angle from the horizontal plane of less than 45° was calculated as $\pi \Sigma D^2_{hor}$, and for those with vertical orientation as $\pi \Sigma D^2_{vert}$. An *index of root shallowness* was calculated as $\Sigma D^2_{hor}/D^2_{stem}$, or as ratio of basal area of all horizontally oriented roots and the stem.

Test of fractal branching

To test the independence of branching rules from root diameter (a key assumption for the fractal model), parts of the root system were exposed, by tracing roots from the stem base. For each branching point where both the previous and subsequent 'links' had been exposed, root diameter before and after the branching point was recorded and the α parameter calculated as $D^2_{before}/\Sigma D^2_{after}$. The allocation parameter q was calculated as $\max(D^2_{after})/\Sigma D^2_{after}$. Regression of α, q and link length on root diameter was analyzed per tree with linear and non-linear models in Genstat-V. The null-hypothesis of 'no relation' was tested with a 5% probability level. Similar measurements were made on above-ground branching patterns of the same tree.

Effects of tree pruning height on root distribution

For five tree species (*Calliandra calothyrsus*, *Senna (Cassia) siamea*, *Gliricidia sepium*, *Paraserianthes (Albizia) falcataria* and *Peltophorum dasyrachis*) the effects of repeated tree pruning at 0.5, 0.75 or 1 m height on proximal root diameter and orientation were studied. Tree species were used as main plots, pruning height as a split-plot factor. Trees were planted at a distance of 2 × 0.5 m and were not intercropped. Pruning consisted of removing all above-ground biomass except for the stem, to the height indicated, at three-month intervals for two years. Observations were made on six trees for each pruning height. Further details of the experiment are given by Hairiah et al. [1992]; proximal root data were re-analyzed for this publication.

Results

Proportionality factors α *and q in relation to diameter*

In none of the 29 trees tested was a significant regression of α or q on D found, for stems or roots. When data for all tree roots and stems were combined (n = 4607 and 5112 branching points, respectively), again no significant regression was found.

Figure 3 shows that the α values for shoots where mostly in the 0.85–1.05

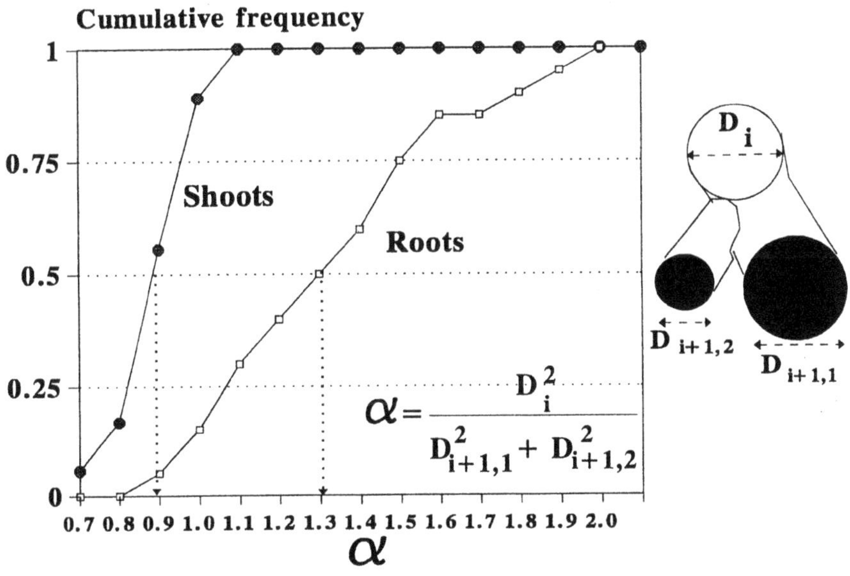

Fig. 3. Cumulative frequency of the α parameter of shoots and roots (averaged per tree) for multipurpose trees in Lampung.

range, while for roots the mean α per tree ranged from 0.9–2.0. The overall mean value for α for roots in the combined data set was 1.328 with a standard deviation of 0.5377. The α of shoots cannot be used to predict the α of roots (Fig. 4).

The allocation parameter q (by definition restricted to the range 0.5–1) had a mean of 0.765 for the stems of all trees combined (5112 branching points), with a standard deviation of 0.144, with an even spread over the 0.5–1 interval (Fig. 5). For roots the mean value was 0.8552, with a standard deviation of 0.1475, and a clear over-abundance of values in the range 0.9–1 (Fig. 5).

Link length was not correlated with diameter for the roots; above ground a weak relationship was found, mostly based on the unbranched portion of the main stem.

Stem basal area at breast height was 40–150% of the total root basal area (Fig. 6). The contribution of vertically oriented roots to the total root basal area ranged from 0.11–0.85. The values obtained for the 'index of tree root shallowness' ranged from 0.14–2.13 (Table 1), with the highest value for the shallow rooted *Leucaena leucocephala* and the lowest value for the relatively deep rooted *Mangifera indica.*

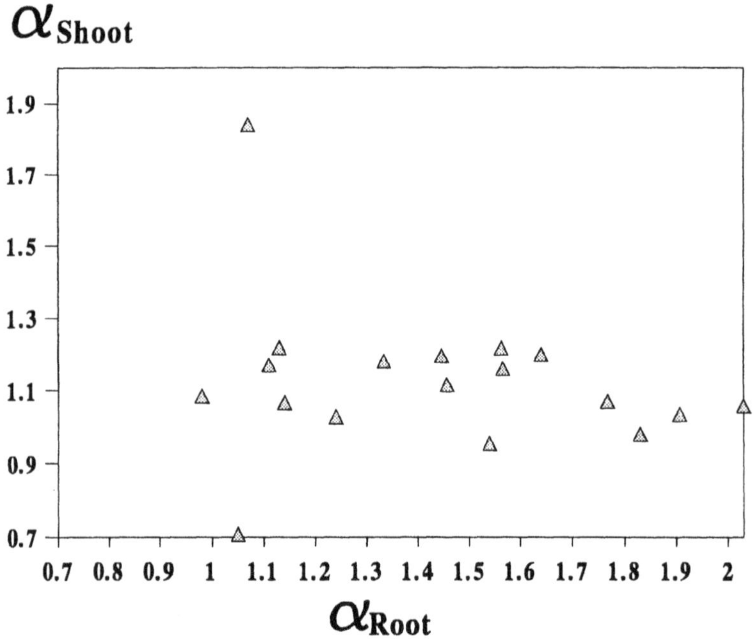

Fig. 4. Relation between average α of shoots and roots per tree; no statistically significant regression was obtained.

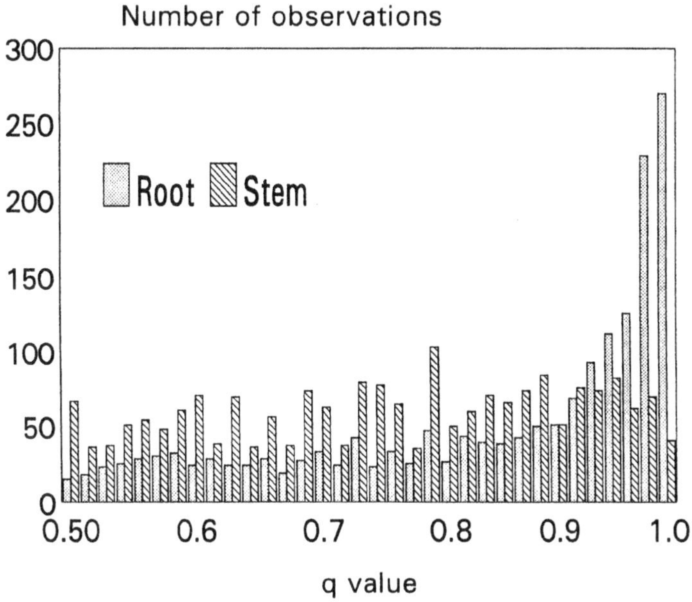

Fig. 5. Frequency distribution of q for roots and stem (based on 4607 and 5112 branching points, respectively, for all trees combined).

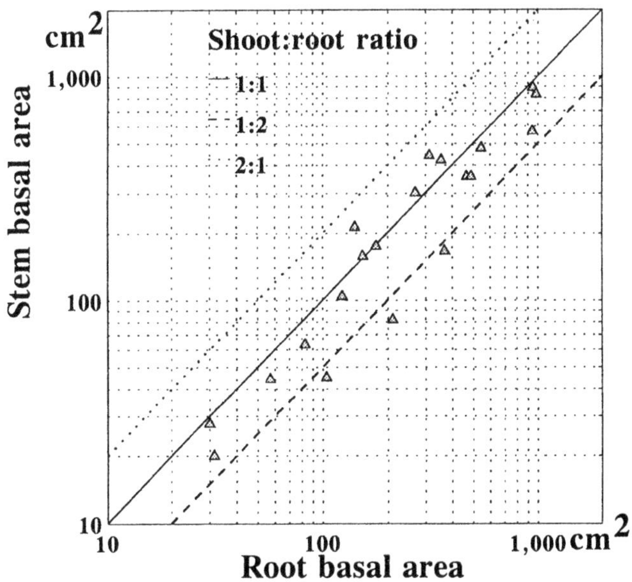

Fig. 6. Relation between the basal area of all proximal roots and stem basal area for 18 trees.

Pruning height experiment

A low pruning height did indeed lead to more but thinner adventitious roots for four of the five species tested. *G. sepium* was an exception as it made thick fleshy storage roots at low stem pruning height. Table 2 gives the main effects of tree species and pruning height. Pruning height had no effect on the ratio between cross-sectional area of the tap-root and all adventitious roots combined (A). No consistent effect was found on the index of root shallowness (last column). The average value of this index for these pruned trees was higher than that reported in Table 1 for *C. calothyrsus*, similar for *P. dasyrachis* and lower for *G. sepium.*

Discussion

Our preliminary results show that the assumptions of fractal branching models hold in that α and q are independent of root diameter. Thus, fractal models can be applied, as least as an approximation. Caution is needed, however, as the considerable variability in α and q within each root system may affect the relationships based on an average value, as the algebraic relations are non-linear.

Le Roux [1994] described root development of rubber trees and applied the SARAH model [Pagès and Aries, 1988]. The model follows root development from the primary axis and adventitious roots into secondary, tertiary and quaternary branch roots. For each class of root a different length of bare root tip and density of next order branching was found, suggesting that a fractal (scale-independent) description can only give a first approximation [Pagès et al., 1995]. Secondary thickening of roots is described as a function of time in the model of Le Roux [1994], and is not explicitly linked to branch root development. Both branch root development and diameter increase, however, are exponential functions of time and thus the model is not in contradiction with the fractal model description tested here.

Total basal area of stems and roots is clearly related across the tree species tested here, with a narrow range of shoot:root ratios. No architectural relationship was found, however, between shoot and root systems of the tree species in the sample, and the α and q parameters for roots cannot be predicted from the corresponding parameters for shoots, which are easier to measure.

The height of stem pruning had no consistent effect on the index of root shallowness, as the relative importance of the tap root remained constant. Possible differences in orientation of finer and thicker adventitious roots were not quantified. The hypothesis that a lower above-ground tree pruning height induces a shallower root system could not be confirmed by the present data. A lower pruning height does have an effect, however, on the number and diameter of proximal roots. The physiological basis for this effect may be

Table 2. Effect of tree pruning height on number and proximal diameter of adventitious roots; the last three columns are based on cross sectional area (diameter^2) ratios; only main effects of pruning height and tree species are given, although a species × treatment interaction was indicated, with *G. sepium* roots responding less than those of other trees [data re-analyzed from Hairiah et al., 1992].

	Avg. diam. of advent. roots, mm	Number of adventitious roots per diam. class (mm)			$Root_{vert}$: $\Sigma Root_{hor}$ basal area ratio, A	ΣRoot:Stem basal area ratio, B	$\Sigma Root_{adv}$:Stem basal area ratio, (1-A)/B
		< 5	5–20	20–40			
Grand mean	6.1	12.2	5.8	0.8	0.58	0.72	0.58
Pruning height (cm)							
50	5.2	15	6	0	0.59	0.86	0.48
75	6.2	11	6	1	0.55	0.63	0.71
100	6.8	10	6	1	0.60	0.67	0.60
Tree species							
Paraserianthes falcataria	5.9	11	6	1	0.66	1.01	0.34
Gliricidia sepium	8.5	7	5	1	0.72	0.60	0.47
Peltophorum dasyrachis	5.9	10	6	0	0.61	0.82	0.48
Senna siamea	6.7	7	4	1	0.72	0.60	0.70
Calliandra calothyrsus	5.1	20	8	1	0.37	0.66	0.96

found in the loss of apical control by the meristems of the main roots, as the recovery from pruning takes longer at reduced pruning height. Whether this is caused by reduced carbohydrate storage in the shorter stems or by a smaller number of buds resprouting, supplying fresh carbohydrates as well as hormonal signals, should be investigated further.

The results of this study indicate that the proximal root method and the 'index of tree root shallowness' as an indicator of competitiveness do require further testing. They may, however, provide a bridge between farmer-level observations ('trees with visible surface roots are competitive'), through the allometrics of root branching patterns to the functioning of roots in uptake of nutrients and water and thus to one of the major components of tree-soil-crop interactions. Combination of the fractal geometry and direct measurements of water flow through individual roots [Ong and Khan, 1993] may help to make this link.

Acknowledgements

Thanks are due to Mang Uyen for allowing us to dig around his homegarden trees, to Mrs Subekti Rahaya and Betha Lusiana with help in analyzing the data and to Dr Chin Ong, Dr Roger Leakey, Mr Richard Coe and two anonymous reviewers for comments on the manuscript.

References

Coster Ch (1932) Wortelstudiën in de tropen (Root studies in the tropics). I. De jeugdontwikkeling van het wortelstelsel van een zeventigtal boomen en groenbemesters (Early development of the root system of seventy trees and green manure species). Korte Meded vh Boschb Proefst no. 29

Coutts MP (1983) Root architecture and tree stability. Plant and Soil 71: 171–188

De Foresta H and Michon G (1994) Agroforests in Sumatra, where ecology meets economy. Agroforestry Today 6(4): 12–13

De Reffye P, Houllier F, Blaise F, Barthelemy D, Dauzat J. and Auclair D (1995) A model simulating above- and below-ground tree architecture with agroforestry applications. Agroforestry Systems (this issue)

Fitter AH and Stickland TR (1992) Fractal characterization of root architecture. Functional Ecology 6: 632–635

Hairiah K, Van Noordwijk M, Santoso B and Syekhfani MS (1992) Biomass production and root distribution of eight trees and their potential for hedgerow intercropping on an ultisol in Lampung. AGRIVITA 15: 54–68

Kooistra MJ and Van Noordwijk M (1995) Soil architecture and distribution of organic carbon. In: Carter MR (ed) Structure and Organic Carbon Storage in Agricultural Soils. Advances in Soil Science (in press)

Kuiper LC, Bakker AJJ and Van Dijk GJE (1990) Stem and crown parameters related to structural root systems of Douglas fir. Wageningen Agric Univ Papers 90-6: 57–67

Le Roux Y (1994) Mise en place de l'architecture racinaire d'*Hevea brasiliensis*, étude comparée du semis et de la microbouture. Thesis doctorat, Université d'Aix Marseille III, Aix Marseille, France

Mandelbrot BB (1983) The Fractal Geometry of Nature. Freeman, New York, NY, USA

Ong CK and Khan AAH (1993) The direct measurement of water uptake by individual tree roots. Agroforestry Today 5(4): 2–4

Pagès, L, Le Roux Y and Thaler P (1995) Modélisation de l'architecture racinaire. Plantations, Recherche, Développement 1995(1): 19–34

Pagès L. and Aries F (1988) SARAH: Modèle de simulation de la croissance, du développement et de l'architecture des systèmes racinaires. Agronomie 8: 889–896

Prusinkiewicz P and Lindenmayer A (1990) The Algorithmic Beauty of Plants. Springer-Verlag, Berlin, Germany, 302 pp

Shinozaki K, Yoda K, Hozumi K and Kira T (1964) A quantitative analysis of plant form – the pipe model theory. Japanese Journal of Ecology 14: 97–105

Soumaré A, Groot JJR, Koné D and Radersma S (1994) Structure spatiale du système racinaire de deux arbres du Sahel: *Acacia seyal* and *Sclerocaryea birrea*. Rapport PSS no. 5, Wageningen, The Netherlands, 45 pp

Spek LY and Van Noordwijk M (1994) Proximal root diameters as predictors of total root system size for fractal branching models. II. Numerical model. Plant and Soil 164: 119–128

Van Noordwijk M, Widianto, Heinen M and Hairiah K (1991a) Old tree root channels in acid soils in the humid tropics: important for crop root penetration, water infiltration and nitrogen management. Plant and Soil 134: 37–44

Van Noordwijk M, Hairiah K, Syekhfani M and Flach EN (1991b) *Peltophorum pterocarpa*: a tree with a root distribution suitable for alley cropping. In: Persson H and McMichael BL (eds) Plant Roots and Their Environment, pp 526–532. Elsevier, Amsterdam, The Netherlands

Van Noordwijk M, Widianto, Sitompul SM, Hairiah K and Guritno B (1992) Nitrogen management under high rainfall conditions for shallow rooted crops: principles and hypotheses. AGRIVITA 15: 10–18

Van Noordwijk M, Spek LY and De Willigen P (1994) Proximal root diameters as predictors of total root system size for fractal branching models. I. Theory. Plant and Soil 164: 107–118

Van Noordwijk M (in press) A simple model to quantify mulch and shade effects. In: Huxley PA and Ong CK (eds) Tree-Crop Interactions – a Physiological Approach, Chapter 3. CAB International, Wallingford, UK

Van Noordwijk M and Brouwer G (1995) Roots as sinks and sources of carbon and nutrients in agricultural systems. Advances in Agroecology (in press)

Van Noordwijk M, Lawson G, Groot JJR and Hairiah K (in press) Root distribution in relation to nutrients and competition. In: Huxley PA and Ong CK (eds) Tree-Crop Interactions – a Physiological Approach, Chapter 10. CAB International, Wallingford, UK

Agroforestry Systems **30**: 175–197, 1995.

A model simulating above- and below-ground tree architecture with agroforestry applications

P. DE REFFYE[1], F. HOULLIER[2], F. BLAISE[1], D. BARTHELEMY[1], J. DAUZAT[1] and D. AUCLAIR[1]
[1] *Unité de Modélisation des Plantes, CIRAD/GERDAT (Centre de Coopération Internationale en Recherche Agronomique pour le Développement), B.P. 5035, 34032 Montpellier cedex 1, France;* [2] *ENGREF, 14 rue Girardet, 54042 Nancy, France*

Key words: plant architecture, plant growth, plant modelling, stochastic processes

Abstract. Modelling plant growth and architecture requires two consecutive and complementary approaches. The first is a qualitative botanical analysis, in which the development sequence of a tree is studied by the identification of various levels of organisation and of homogeneous subunits. All of these – architectural unit, axis, growth unit – follow particular growth processes which can be described by using the second approach, the quantitative analysis. Modelling of the functioning of meristems based upon stochastic processes has been carried out since 1980, in combination with a large amount of experimental work on temperate and tropical plants. Calculations involved in tree simulations from field data are based upon the probabilistic Monte Carlo method for the topological part and on analytical geometry for the morphological part. Computer graphics methods are then used to visualise the computed plant. Several sectors in agroforestry are concerned with application of such plant architecture modelling: tree growth and yield, radiative transfers, timber quality and mechanics, simulation of competition, interaction between plant morphology and physiology.

Introduction

Agroforestry involves several components within an agricultural system, of which trees are one main element. These components interact with one another: trees influence growth (quantity and quality) of the crop or fodder species, they influence the microclimate and thus the energy balance and the production of animals [Sibbald et al., 1994], and conversely herbaceous plants and animals influence the growth of trees [Sibbald and Sinclair, 1990]. All the interactions are related to the spatial layout of the system.

In designing agroforestry combinations, knowledge of the spatial development of the various components is extremely important in order to optimise the crop-tree and/or the tree-tree interactions. The occupation of space by a tree and its evolution with time, as a result of its architecture [Hallé et al., 1978], influence physiological functions and interactions, such as light interception and transmission, water and nutrient uptake, and competition. Interactions between components are the main features of interest in agroforestry research, and particularly in agroforestry modelling. The knowledge and understanding of tree architectural development is therefore of utmost importance.

Although in many agroforestry systems tree development is often considered as approaching 'free growth', or at least with no strong above-ground interactions between trees, this is far from being the general situation and trees do very often interact, not only with herbaceous components but also with other trees (i.e., in alley cropping, in all rotational systems, and at the end of the rotation of the tree components) [Nair, 1991]. The architectural development of trees strongly influences the possibility of growing crops in association during the whole length of the rotation.

It is therefore most important to have a detailed description of tree architectural development, both qualitatively, with a botanical-type approach, and quantitatively for use in computer simulations.

The tree is a living organism, with an architecture being constantly transformed during growth [Hallé et al., 1978; Edelin, 1984; Barthélémy et al., 1991]. However, it is unusual to observe a plant structure from a dynamic point of view. Development takes place too slowly to be noticed and the eye only sees the global form reached at the time of observation. The description of this form is usually made with very general qualitative notions, such as 'columnar' or 'weeping'.

Only a trained eye can evaluate the complexity of the architecture of a tree and recognise in winter without difficulty a walnut or an ash, following characteristic branching patterns. The architect Le Corbusier said: 'The tree is a pure mathematical function', but he had no means of proving it.

The connection between tree architecture and growth was the object of very few studies until relatively recently, when a more rational study of plant structures started under the impulse of the botanists Hallé and Oldeman [1970]. This first qualitative study led to the concept of 'architectural model'.

The numeric approach, in its modelling and simulation aspects, began with the arrival of computers with increasing capacity and speed. Using this new potential, computer scientists developed algorithms capable of generating complex branching patterns (fractals and binary trees) and Lindenmayer's school produced computer grammars (L-systems) specifically developed for plant simulation [Prusinkiewicz et al., 1988]. These studies showed how simple transformation rules can generate very complex and apparently realistic forms, especially for simple and small plants. Although the first applications remained mainly theoretical and were very little subject to experimental validations, new developments of the L-systems have been pursued on bigger plants requiring specific growth rules [Kurth, 1994].

The plant modelling unit of CIRAD (in short 'AMAP': Atelier de Modélisation de l'Architecture des Plantes) focuses on the development of an experimental approach to modelling and simulation of plants for botanical and agronomic applications [De Reffye et al., 1993]. Its objective is the definition of a rigorous and faithful model of plants and plant growth, based upon botanical knowledge, and validated through observation and measurements on real plants. Françon [1991] and Françon and Lienhardt [1994] define this approach as a 'validated theory'. It integrates previous results of botanists and computer scientists and combines them with the measurement of real plants, in order to deduce the parameters of their growth processes.

The method was initially developed by De Reffye [1979] with the definition of a mathematical and macroscopic model of above-ground plant growth, first applied to *Coffea robusta*. It was then extended to a wide variety of tropical plants, followed by temperate species, in particular environments. Further developments still in progress concern the incorporation of interactions between plants, of plant-environment interactions, of below-ground architecture, and of physiological processes within the plant, in addition to many agronomic applications. These have been made possible by the use of a specific protocol: measurement of the plant in the field; coding of the data; computation of the laws and parameters of the growth and branching processes; and final growth simulation on a graphic workstation in the form of a three-dimensional structure giving an image close to that of the original plant.

Botanical and mathematical aspects of plant architecture

The architectural model

Taking into account both the architecture and the dynamics of tree growth, Hallé and Oldeman [1970] identified several '*architectural models*', based upon general classification criteria. The '*leafy axis*' is the basic element of every plant architecture. It generally consists of successive shoots called *growth units* (GU). A GU is produced by a terminal meristem, and is usually composed of a series of internodes and nodes, which carry leaves (Fig. 1). The set composed by the internode and the node above, axillary bud and leaf is named a '*metamer*' [White, 1979].

A tree species is classified as belonging to one architectural model or another, on the basis of its growth and branching pattern, the flowering position, and the orthotropic or plagiotropic nature of its axes. The main interest of this qualitative description is to show that there are strong endogenous constraints in the development of a plant, independent of the environment: botanical rules are more important than disturbances due to the environment in the production of new axes, and thus in producing a given architectural model. Of course, the environment affects the plant's physiology, which in turn affects its growth, but it expresses itself within the limits of the architectural model. The physical environment only modulates the quantitative expression of the architectural sequence.

Architectural unit and physiological age

However, even though the schematic character of the classification of Hallé and Oldeman [1970] has a high theoretical interest, it remains insufficient

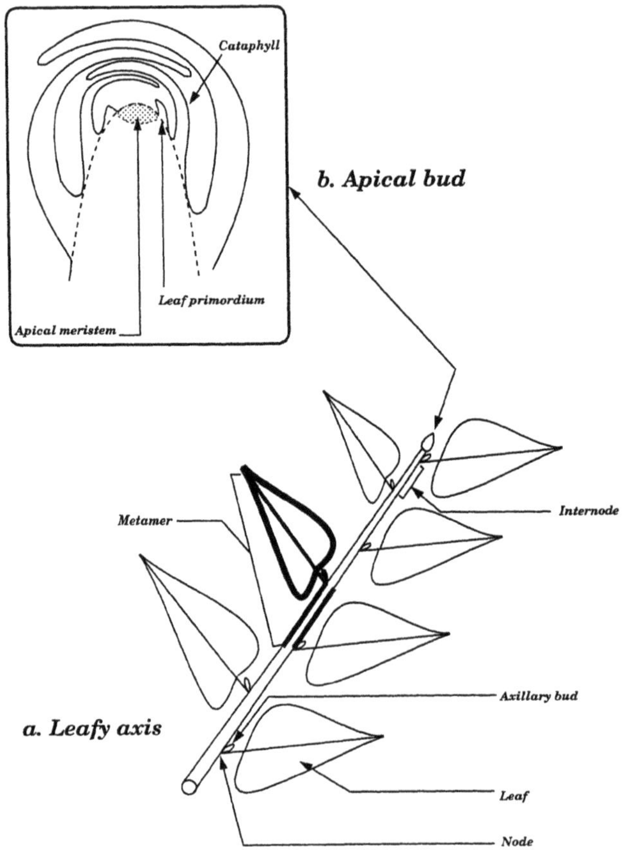

Fig. 1. A leafy axis (a) is composed of a succession of metamers (thick line), and is built up by the functioning of an apical meristem (b).

for detailed plant descriptions. That is why Edelin [1977] and Barthélémy et al. [1989] later defined the concept of '*architectural unit*', which characterises the fundamental elementary architecture of trees of a given species. The architecture of a plant can be seen as a hierarchical branched system in which the axes can be grouped into categories according to their morphological and functional characteristics (the stem is order 1 axis and bears order 2 axes, which in turn bear order 3 axes, and so on). For each species, the number of categories of axes is finite, and the identification of the architectural unit is achieved by the complete diagnosis of the functional and morphological features and the relations between all these categories of axes (Fig. 2).

Generally, the axes which constitute this architectural unit are not all homogeneous, but they are composed of a series of entities (annual shoots, growth units), with different morphological characteristics according to their location, expressing the '*physiological age*' of the meristems which produced them [De Reffye et al., 1991a; Barthélémy et al., 1995]. A meristem can be

AXIS 1 (Trunk)	AXIS 2	AXIS 3	AXIS 4
Rhythmic monopodial growth	Rhythmic monopodial growth	Rhythmic monopodial growth	Rhythmic monopodial growth
Indeterminate growth	Indeterminate growth	Short-term determinate growth	Short-term determinate growth
Orthotropic	Orthotropic	No precise growth direction	No precise growth direction
Spiral phyllotaxis: leaves with a radial arrangement	Spiral phyllotaxis: leaves with a radial arrangement	Spiral phyllotaxis: leaves with a more or less secondary bilateral arrangement	Spiral phyllotaxis: leaves with a more or less secondary bilateral arrangement
No sexual characters	No sexual characters	Female sexuality (terminal)	Male sexuality (lateral)

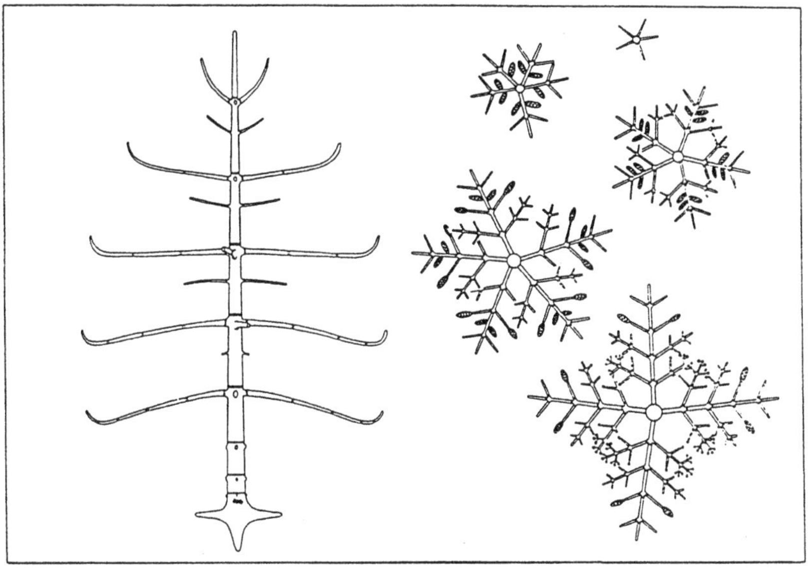

Fig. 2. Architectural unit of *Picea abies* [after Edelin, 1977]. Above: table describing the architectural unit; each category of axis displays its own series of characters. Axes 1 and 2 form the supporting structure. Axes 3 and 4 form the uptake and reproductive systems; they are short shoots with very short-term determinate growth. Below: architectural diagram. Left: side view. Right: view from above showing different tiers of branches. The leaves are not represented; female cones are indicated by oval hatched areas and male cones by large black dots. The transverse lines crossing axes indicate the resting period of an axis.

characterised by two ages: the chronological age, and the degree of differentiation of the meristem or physiological age. The latter is estimated by a series of qualitative and quantitative criteria. For example the short axis of fruit-trees is characteristic of a physiologically-aged element: growth units are short, bear flowers and have a short lifetime. These axes are born '*physiologically old*'! On the contrary, the forks (or '*reiterations*') of stems characterise vigorous '*physiologically young*' productions (Fig. 3).

Between these two extremes, there is a gradient of morphological expression ('*morphogenetic gradient*'). This can be observed in the progressive

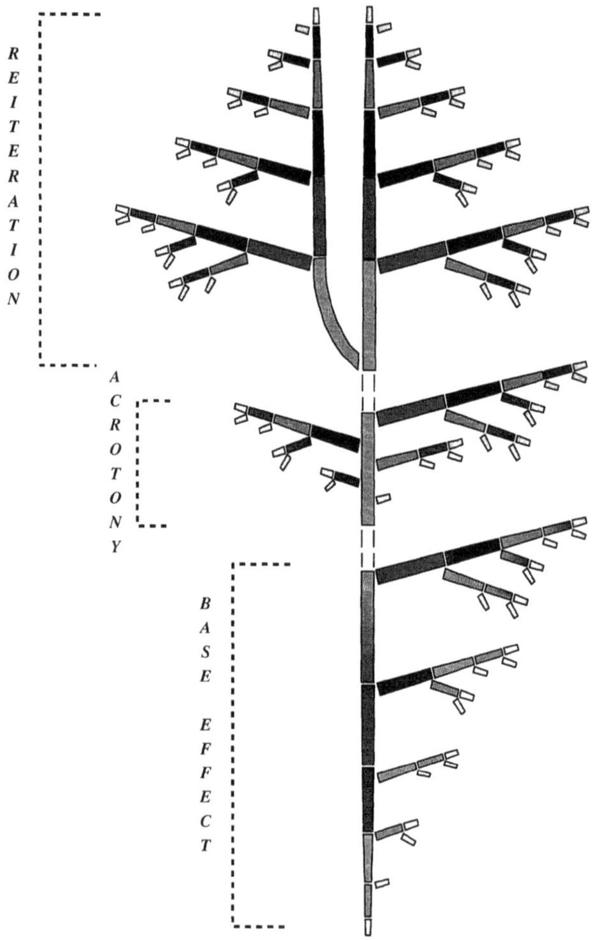

Fig. 3. Diagrammatic representation of the structure of subunits according to their position within the tree architecture [Barthélémy et al., 1995]. This may be seen as a hierarchical system composed of subunits (such as GUs) with different morphological characters, represented here by different sizes and shades. Within the architecture of the whole plant, subunits may be located in various positions according to morphogenetic gradients such as acrotony, succession of GUs along one axis, or events such as the 'base effect' characteristic of the first stages of development, or 'reiteration' which duplicates the architectural unit of a tree.

ageing of the apical meristem of an axis, in the *acrotony* within a growth unit (decrease in length and number of internodes of axillary shoots according to their position from top to base of a GU), and in the progressive increase of tree vigour during the juvenile phase. Plant development is thus characterised by a continuous change in the functioning of the meristems which causes a progressive change ('*metamorphosis*') in the architecture.

This qualitative architectural analysis of trees leads to a realistic representation of plant structure, and prepares the ground for a quantitative analysis.

The processes of growth and interruption

The meristem, which is located within the bud (Fig. 1), produces metamers at an embryonic stage during its active period: this is '*apical growth*' or *organogenesis*. A reserve of future metamers is thus constituted within the bud, invisible externally, with no elongation. Only the lengthening of the metamers which set up new leaves with their associated internode is visible: this is '*internodal growth*' or *elongation*. The metamers which develop in this way are replaced by new embryonic ones. A 'fifo' queue (first in first out) is then constituted.

In the case of temperate trees, the same apical bud can follow two different types of growth processes. In spring, a package of meristems coming from the reserve undergoes the lengthening process and constitutes the '*preformed part*' of the GU. Afterwards, a continuous growth process may be initiated and the metamers extend one by one immediately after their birth more or less regularly: they constitute the '*neoformed part*' of the GU [Caraglio and Barthélémy, 1995].

The formation of the GU is complete when the lengthening process stops for a long period of time. This *interruption* can take place almost simultaneously for all the buds of a tree (e.g., *Prunus avium*, *Fagus sylvatica*) or it can be delayed from one bud to an other (e.g. *Prunus armeniaca*, *Populus* spp.).

Stochastic interpretation of growth processes

It has been experimentally observed that the number of metamers produced by growth units of the same physiological age is variable, resulting from the compound processes of growth and interruption [De Reffye et al., 1991b]. If the internodal growth of a homogeneous population of leaf axes is observed during a given period, the 'number of internodes' follows a distribution which characterises the random growth of the shoots. The distributions obtained in this manner can be unimodal (*Fagus sylvatica*, *Hevea brasiliensis*) or bimodal (*Prunus avium*, *Populus nigra*), depending on whether or not the meristem produces a neoformed part during its growth (Fig. 4).

Within a tree, equivalent vegetative axes (i.e., those of the same branching order and of the same physiological age) follow similar distributions, which can be fitted statistically (Figs. 4 and 5). The processes of growth and inter-

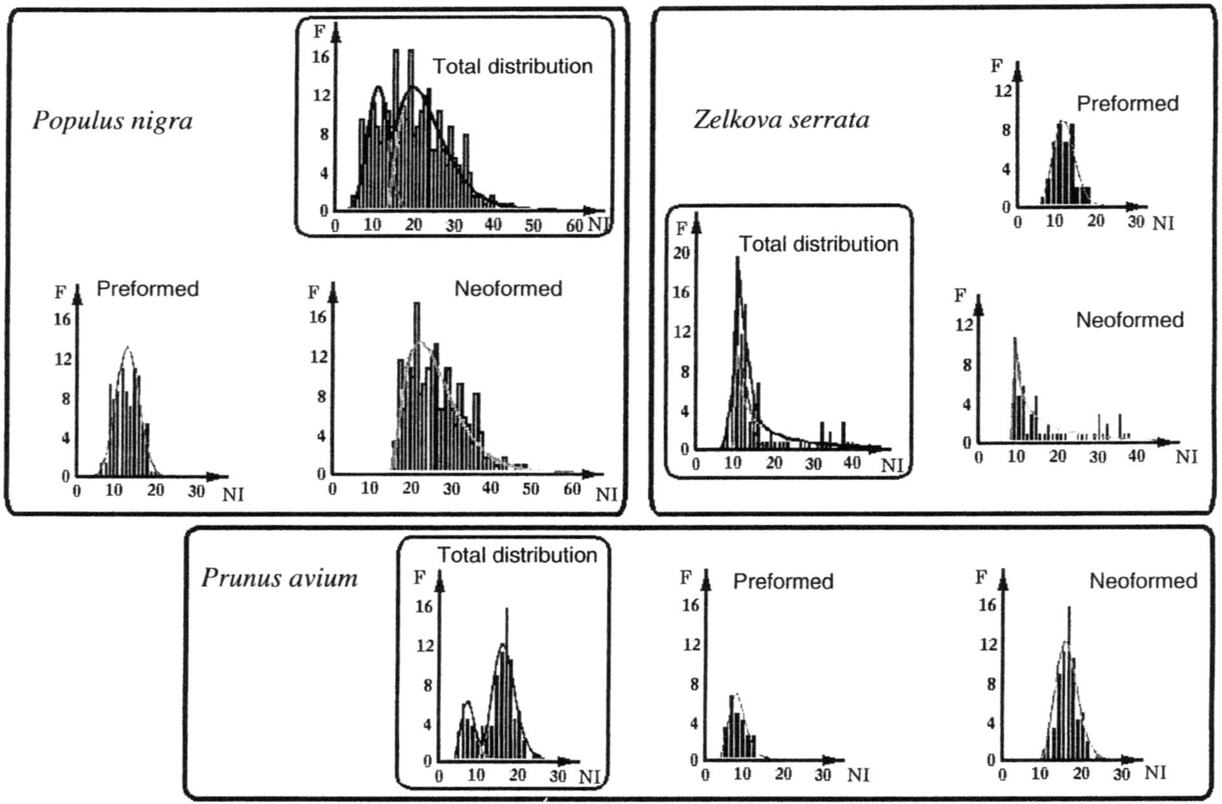

Fig. 4. Examples of diagrams showing the frequency of number of metamers per growth unit. The total frequency distributions can be decomposed into a first distribution corresponding to preformation and a second distribution corresponding to neoformation.

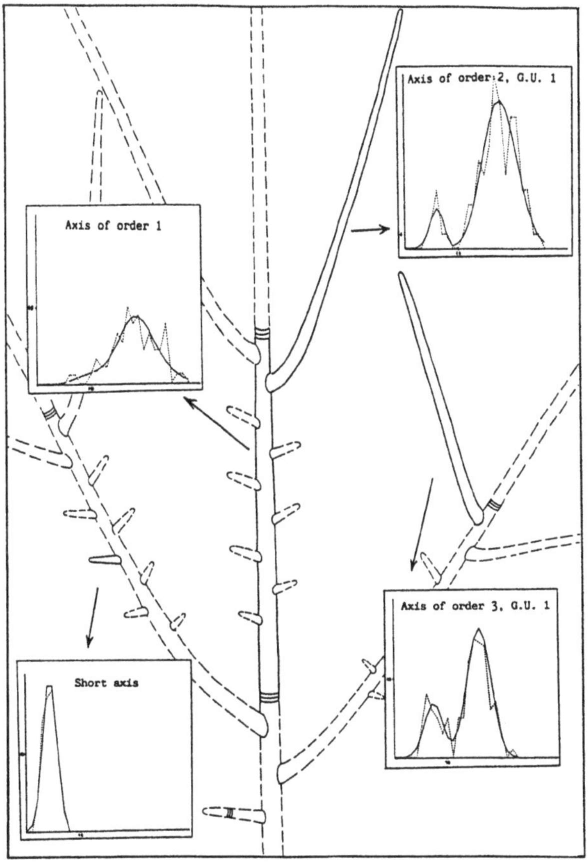

Fig. 5. Statistical evolution of number of internodes per GU according to tree architecture. Example of *Prunus avium* in southern France [De Reffye et al., 1991b]. Graph axes as in Fig. 4.

ruption which drive these distributions incorporate a number of factors which are up to now not fully predictable, and the outcome of these processes can therefore be given in terms of probabilities. It is thus considered here that growth follows stochastic processes, which can be very simply modelled by the probability theories of '*renewal*' and '*reliability*' [Cox, 1962; Feller, 1957]. The distributions observed at a macroscopic level are simple and do not necessarily depend on the complexity of the underlying phenomena.

Experimental observation has shown that the nature (or the form) of the observed distributions is independent of environmental conditions [De Reffye et al., 1991b]. The numeric parameters are modified when the same tree species grows in a different environment, but the simple underlying rules do not change and can be used to predict the form of distributions under new sets of circumstances. As we have already noted, the physiological age of shoots varies within the tree according to morphogenetic gradients (Fig. 3); the influence of these gradients also modifies numerically the parameters without changing the nature of the distributions.

A very high endogenous stability of the growth processes has been shown [De Reffye et al., 1991b]. The same distribution of metamers can be reached after a longer or shorter time of meristem activity, depending on environmental conditions. Using this approach, it is possible to characterise a physiological stage of architectural development, although the exact age of the tree remains unknown. An average architectural growth pattern can therefore be determined in a particular situation. Although exceptionally unfavourable years can disturb annual growth, trees usually receive enough cumulated energy and nutrients to develop according to their architectural model: architectural development occurs in a first stage and is little affected by limiting factors; it is followed by flowering and fructification processes which are much more dependent on the environment.

The branching process

The terminal meristem produces metamers with axillary buds at the leafy axes, which each correspond to at least one potential branch. However, branching can either occur immediately, or occur with a certain delay, or not occur at all. There is therefore a budding probability for dormant buds, which, although driven by (up to now) unpredictable external factors, can be described by simple probabilistic distributions.

Branches can be grouped in packets along an axis, resulting in an alternation between branched and unbranched areas. In an individual GU, branching can occur in a particular region: terminal region (*acrotony*), median region (*mesotony*), lower region (*basitony*). Therefore, the branching pattern cannot be simply described by only one parameter which would indicate the proportion of nodes carrying a branch, but it needs to be characterised by distributions of branched and unbranched zones: most of the time, the nature of these distributions is close to a geometric law and the branching pattern is well modelled by a Markov chain of order 1 with two states, the branched and the unbranched state (Fig. 6) [Guédon, 1995; De Reffye et al., 1993].

Field measurements

Modelling tree architecture requires both a qualitative and a quantitative assessment. The first step is the qualitative aspect, which consists of botanical observations, in order to provide an accurate description of the architecture and growth pattern of a tree. With this approach, homogeneous groups of GU and/or axes can be identified within the tree, according to morphogenetical gradients.

Quantitative measurements are then necessary for building the model. In

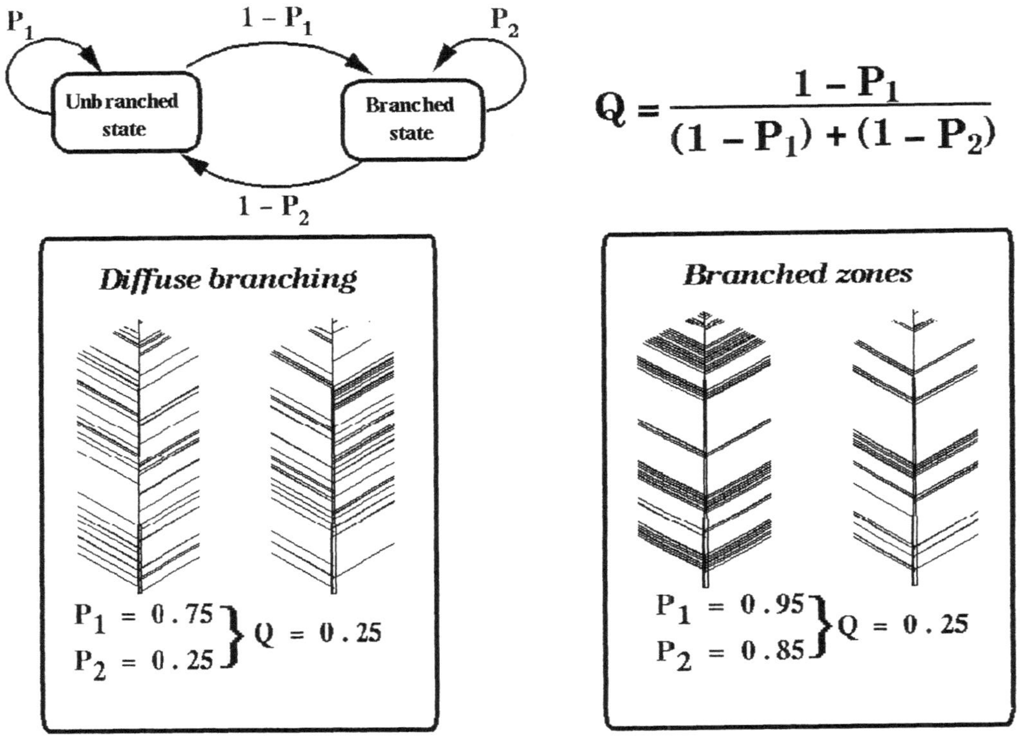

$$Q = \frac{1 - P_1}{(1 - P_1) + (1 - P_2)}$$

Fig. 6. Distribution of branched and unbranched zones within a growth unit. For a global branching probability Q, if one node is unbranched the next node has a probability P_1 of being unbranched, and if a node is branched the next node has a probability P_2 of branching. The homogeneous Markov process can simulate diffuse branching (left) or zones with more branches than others (right), according to the parameters of the distributions [De Reffye et al., 1993].

order to facilitate the subsequent coding, a strict order has to be respected. The following measurements are recorded:

i) for growth processes:
   - length of the growth units
   - number of internodes in each GU
   - polycyclism
   - state of the apex at each GU (living, dead, broken)
   - nature of the axis (dominant/dominated/forked)

ii) for branching processes:
   - nature and position of lateral buds, axes or floral buds

iii) for geometric aspects:
   - branch diameter at the base of each GU
   - angle of insertion

The number of samples to be measured depends on the objective of a given study, on the variability and the form of the distributions, on the degree of complexity of the architecture, as well as on the existing knowledge of the architectural model of a given tree species. It can generally vary from 20 to 100 growth units for homogeneous groups of GU or axes.

Computer simulation of tree architecture

Statistical analysis and fitting of the distributions

The software AMAPmod [Guédon, 1995; Guédon and Costes, 1995] is schematically described in Fig. 7. From the botanical observation of GUs and their classification into different classes (axis order, physiological age, long/short axis), the static frequency distributions of number of internodes per GU are computed. Theoretical distributions are then fitted as in Fig. 4, and their parameters calculated. According to the different types of GU observed, different distributions are fitted (Poisson, binomial, geometric) [Jaeger and De Reffye, 1992]. The dynamic growth and interruption processes which generate these distributions are estimated and validated with the help of mathematical routines. The parameters of branching processes (Markov chains) are then calculated to fit the distributions as in Fig. 6.

Simulation of primary growth

In order to build the architecture of a tree it is necessary to know how the successive GUs are stacked. As meristems age, the GUs which they produce can be characterised according to their physiological age. It is therefore possible to establish a classification of the GUs, by grouping them into classes representative of each particular type of functioning: in each class the GUs have the same physiological age and the same type of growth and branching

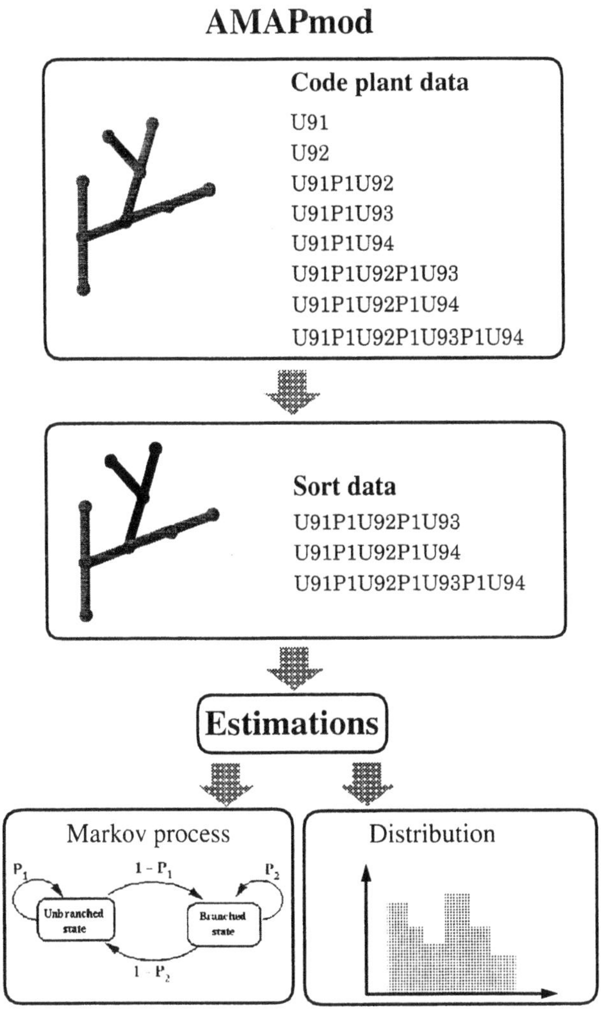

Fig. 7. Schematic description of the AMAPmod software.

process. A single theoretical graduated axis can thus be defined, which successively takes all the possible morphological differentiation stages of GUs according to their physiological age. This '*reference axis*' [De Reffye et al., 1991a] represents the changes in meristem functioning (Fig. 8).

The laws governing bud activity are expressed as progressions along the reference axis. These are generally step by step forward progressions, but they can take various forms according to the state of differentiation of a given axis. According to the morphogenetical gradients, the apical meristem of an axis can have different behaviours relative to the reference axis: a mature order 1 shoot may have the same behaviour as an order 3 or 4 shoot, or conversely an axis with order 3 or 4 behaviour can appear on an order 1 axis. Short shoots

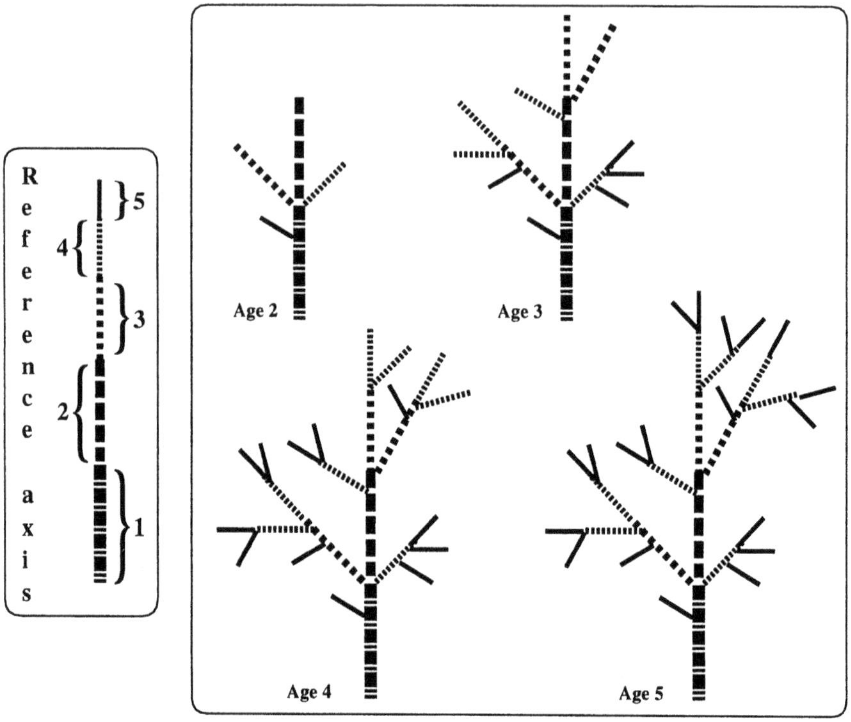

Fig. 8. The reference axis is a single theoretical graduated axis, which successively takes all the possible morphological differentiation stages of GUs according to their physiological age. During growth, the apical meristem of each axis can have different behaviours relative to the reference axis [Blaise et al., 1995].

and flowering are direct jumps up to the terminal stage. This concept of reference axis, fundamental for the description of bud activity, can also be extended to geometric properties of the elements resulting from growth (angle, phyllotaxy, length), and can be numerically deduced from architectural data taken on a given tree.

AMAPsim software [Barczi et al., 1995], schematically described in Fig. 9, manages simultaneously the two ages (temporal and physiological) during the simulation of tree development, according to the reference axis. It integrates i) the data of the growth process of the GU, where bud activity is simulated by the Monte Carlo process in which random numbers are combined with stochastic laws of mortality, interruption and branching, and ii) the measurable geometry (angles, length) in order to allow the creation of realistic three-dimensional models [Jaeger and De Reffye, 1992]. The reference axis is simulated by a finite '*left to right automaton*', where each state is characterised by three sets of parameters: topological parameters (growth, branching, and death laws), geometric parameters (length, angle), and parameters of the transition function from one state to another [Blaise et al., 1995].

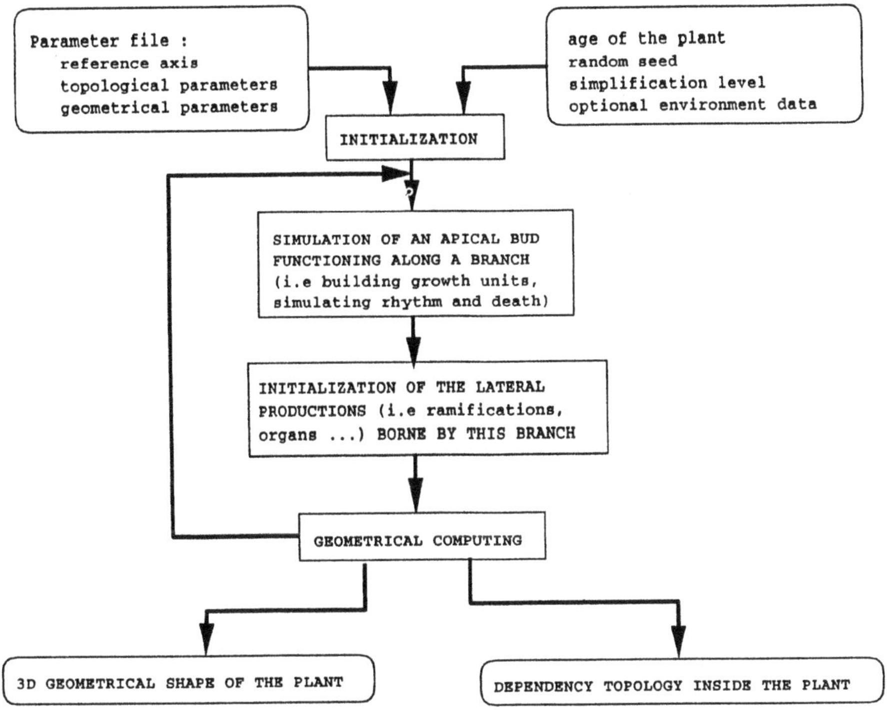

Fig. 9. Schematic description of the AMAPsim software.

This method has been successfully applied on tropical trees (*Coffea* sp., *Elaeis guineensis*, *Hevea brasiliensis*, *Litchi sinensis*) and on temperate trees (*Pinus pinaster* and *P. halepensis*, *Picea abies*, *Fagus sylvatica*, *Populus nigra*, *Zelkova serrata*). It is also suitable for annual plants (*Gossypium* spp., *Nicotiana* spp.). An example is shown in Fig. 10 for *Pinus halepensis*.

Simulation of radial growth

In most temperate woody plants, the apical meristems produce each year new leafy shoots and the diameter of vegetative axes is increased by one new growth ring. There will be a direct relationship between the radial structure of the growth rings making up an axis and the architecture of the tree, since the terminal leafy GU produces carbohydrates which will deposit along the pathway leading to roots according to an allocation law, thus building the outer growth ring.

The AMAPpara software [Blaise and De Reffye, 1994; De Reffye et al., 1995] simulates in a natural way the process of radial growth for any architectural model and any given allocation law (Fig. 11). Up to now, several theoretical carbohydrate allocation laws have been simulated, producing realistic results. In some simple cases the knowledge of the radial ring

Fig. 10. An example of simulation by AMAPsim of a 20 year-old *Pinus halepensis* tree in the French Mediterranean region (Y. Caraglio, unpub.).

sequence, measured on real trees, has led to very accurate representations (Fig. 12). The extension to more complex cases is at present under study.

Simulation of plant/plant and plant/environment interactions

The AMAPpara software manages the simultaneous growth of branches within a crown and of trees within a stand. With the use of a scheduler it is possible to simulate growth of buds which are subject to global physical constraints, even if they belong to different trees. The volume in which the forest stand is developed is discretised into elementary cubes ('*voxels*'), in which interactions can be modelled, leading to the simulation of competition for space between buds which grow within the same cube, whether they belong to the same tree or to neighbouring trees (Fig. 12) [Blaise, 1991; De Reffye et al., 1995].

It is therefore possible to simulate the result of silvicultural practices, because ring thickness depends on the leafy functional area of the crown,

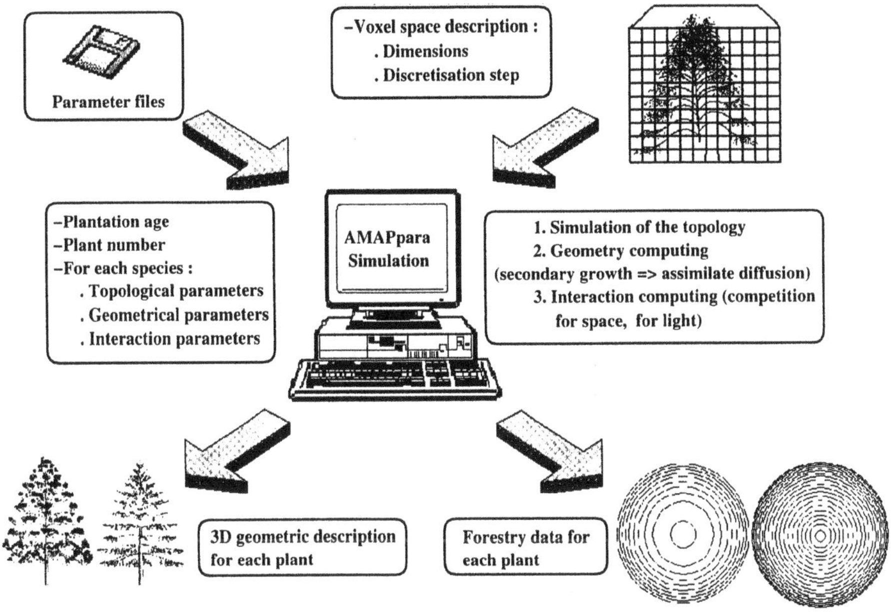

Fig. 11. Schematic description of the AMAPpara software.

and consequently on the self-pruning of branches caused by the contact between trees. This has been solved in some simple cases, but will still take some time to be applied to the whole stand level, principally on account of the limitations of computer power leading to prohibitive calculation times.

Simulation of below-ground architecture

A similar qualitative and quantitative approach can be applied to the below-ground system. Thus Atger [1992] and Atger and Edelin [1994] gave a qualitative description of root architecture, as Hallé and Oldeman [1970] with above-ground architecture. Despite some differences, close structural analogies with above-ground architecture can be found. Below-ground systems bear several categories of root axes, organised hierarchically with 'branching' orders, which form a functional and morphological unit variable from one species to another. As for above-ground systems, root reiteration can occur in different ways (sylleptic or proleptic, partial or total, etc.). However, although structures similar to architectural models can be clearly recognised, their expression is strongly dependent on the soil environment.

On the basis of these qualitative observations, the software developed for above-ground systems has been adapted to simulations of root systems, based on numeric data. Some such root simulations have been performed by Jourdan [1995], to describe *Elaeis guineensis* root axes (Fig. 13). Growth and mor-

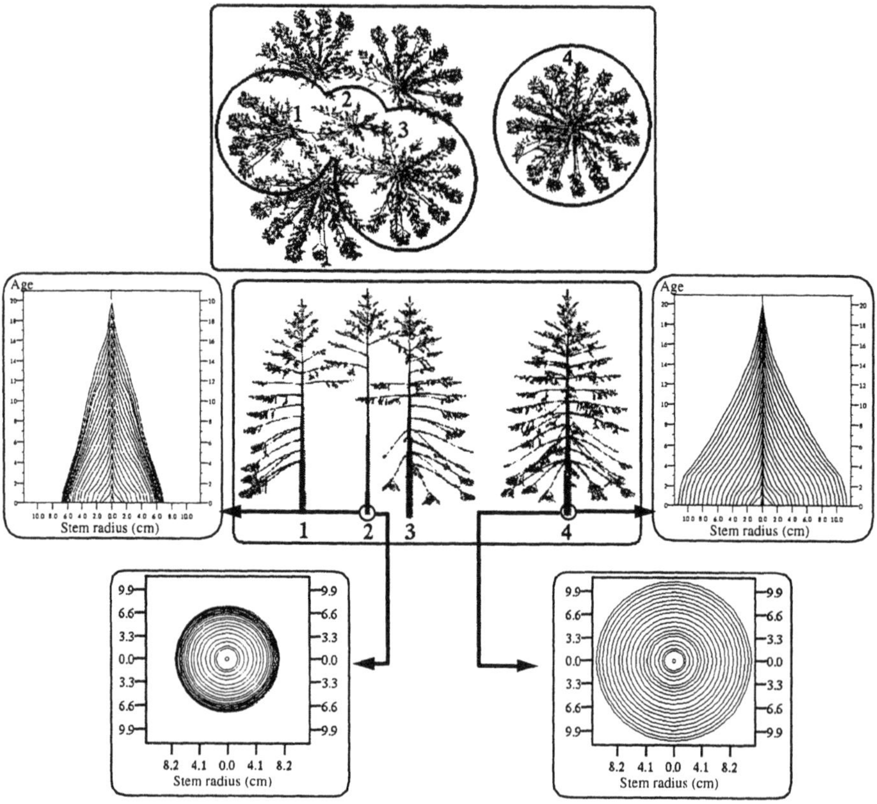

Fig. 12. An example of simulation from AMAPpara: simultaneous growth of several *Picea abies* trees within a stand in north-eastern France. Top: view from above; Centre: side view. Trees #1 and #3 are on the edge of the stand, and undergo self-pruning on the inside. Tree #2 is in the centre and suffers competition on all sides. Tree #4 is open grown. The effect of competition on radial growth appears at the bottom (basal disk), as well as along the stem profile (on either side of the figure) [De Reffye et al., 1995].

tality processes have been fitted in a similar manner to those of above-ground systems, although some differences have been observed. As roots are not built according to the same pattern as shoots (metamers with nodes and axillary buds), the concept of 'internode' is not practicable, and has been replaced by that of '*elementary length unit*': this is defined as the smallest observed length between two successive lateral roots. In this way the branching process of *Elaeis guineensis* roots has been modelled by a Markov chain with branched and unbranched states for each elementary length unit, similar to the branching of above-ground growth units [Jourdan et al., 1995].

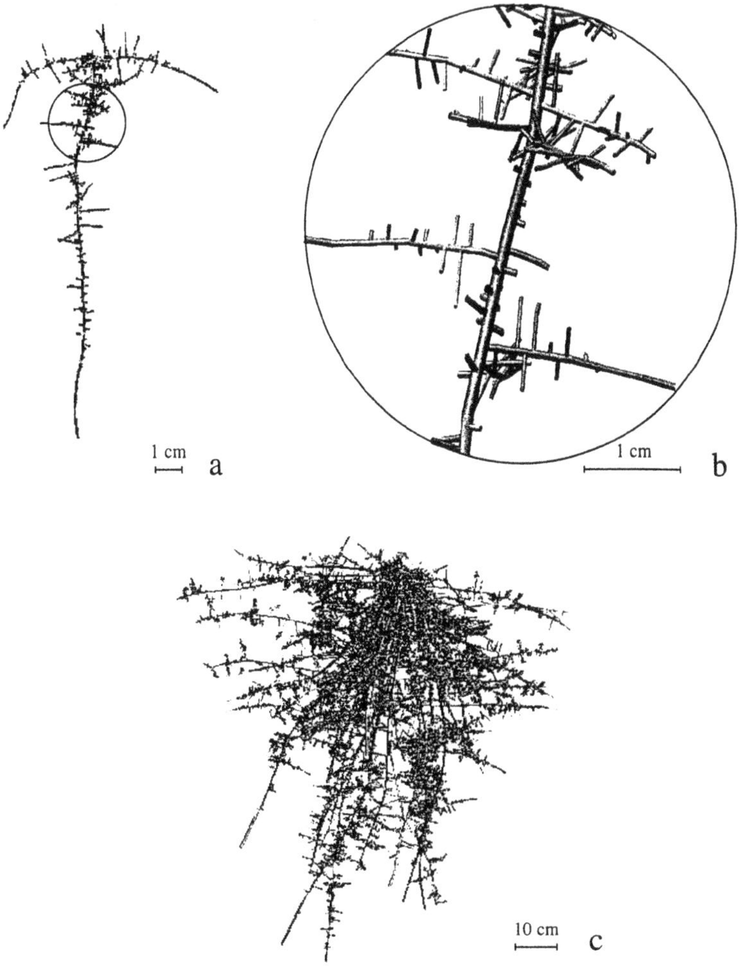

Fig. 13. An example of simulation of *Elaeis guineensis* root growth in Ivory Coast [Jourdan, 1995]. (a) 50-day-old root system; (b) detail; (c) the same root system after one year's growth.

Discussion

The modelling of plant architecture described here has many potential applications in agroforestry, as well as in traditional silviculture and agronomy.

For example, specific programmes have been developed for numerical simulations of radiative transfer. A ray-tracing programme [Dauzat and Hautecoeur, 1991] is used for the simulation of the directional reflectance of canopies. The MIR programme [Dauzat, 1994] simulates intercepted radiation (especially PAR for photosynthesis calculation) and maps the radiation transmitted through a canopy (Fig. 14). In addition, the biometric parameters of a canopy (LAI, LAD, etc.) can be extracted and used as an input for other

Fig. 14. Simulated map of transmitted PAR under a stand of *Elaeis guineensis* in Ivory Coast [Dauzat, 1994].

models. Problems concerning water transfer are also currently under investigation.

This has many obvious applications for agroforestry. The link with synthetic agroforestry models is currently under consideration [Auclair, 1995]. However a balance has to be found between relatively time-consuming detailed process-based models and operational economic models for use in decision-making. Simplified architectural models can be used as modules to produce input for more global models.

The mechanical properties of a standing tree have also been rigorously approached thanks to simulations involving both longitudinal and radial growth [Fourcaud and Lac, 1993; Fourcaud, 1995].

Simulation of root systems has a number of potential applications which should help to reduce time- and labour-consuming below-ground investigations: estimation of root biomass production and distribution, location of the absorbing parts of the root system (leading to the estimation of the useful volume of soil for optimising fertilizer application), studies of root growth in disturbed or compacted soil or under water stress, root competition between plants of the same or of different species, and so on.

More generally, the modelling of tree architecture can provide the basis for numerical investigations which were up to now impossible. Such applications have concerned the development of software concerned with fire propagation in Mediterranean regions, taking into account the structure of the plant cover (Y. Caraglio, unpub.), or the fitting of radar pictures with the simulation of radiation transfer in a tree canopy (J. Dauzat, unpub.).

Finally, linking the methods of architectural modelling with physiological modelling is today one of the most interesting objectives, because the coupling of these two approaches should give quantitative and qualitative prediction models of crops. Work in this direction is currently under way on annual plants. Further developments are at present being considered, in particular by adapting the pipe-model theory [Shinozaki et al., 1964] to the architectural models described here: Rennolls [1994] discussed the limits of the pipe-model theory and proposed improvements which seem very promising in the present context.

Acknowledgements

The authors are greatly indebted to the two anonymous reviewers for their very fruitful comments and suggestions on the first version of the manuscript.

References

Atger C (1992) Essai sur l'architecture racinaire des arbres. Thesis, Université de Montpellier, France

Atger C and Edelin C (1994) Premières données sur l'architecture comparée des systèmes racinaires et caulinaires. Can J Bot 72: 963–975

Auclair D (1995) Scientific and technical basis for sylvopastoral systems in Europe. In: CIHEAM (ed) Sylvopastoral Systems – Environmental, Agricultural and Economic Sustainability. Proceedings of the 8th Meeting of the FAO Working Group on Mediterranean Pastures and Fodder Crops, 29 May–2 June 1995, Avignon, France

Barczi JF, De Reffye P and Caraglio Y (1995) Essai sur l'identification et la mise en oeuvre des paramètres nécessaires à la simulation d'une architecture végétale. Science Update. INRA, Versailles, France (in press)

Barthélémy D, Edelin C and Hallé F (1989) Architectural concepts for tropical trees. In: Holm-Nielsen LB, Nielsen I and Balsler H (eds) Tropical Forests: Botanical Dynamics, Speciation and Diversity, pp 98–100. Academic Press, London, UK

Barthélémy D, Edelin C and Hallé F (1991) Canopy architecture. In: Raghavendra AS (ed) Physiology of Trees, pp 1–20. John Wiley & Sons, Chichester, UK

Barthélémy D, Caraglio Y and Costes E (1995) Architecture, gradients morphologiques et âge physiologique. Science Update. INRA, Versailles, France (in press)

Blaise F (1991) Simulation du parallélisme dans la croissance des plantes et applications. Thesis, Université Louis Pasteur, Strasbourg, France

Blaise F and De Reffye P (1994) Simulation de la croissance des arbres et influence du milieu: le logiciel AMAPpara. In: Tankoano J (ed) Proceedings of the 2nd African Conference on Research in Computer Science (CARI'94), 12–18 October 1994, pp 61–75. Burkina Faso, INRIA, ORSTOM, Ouagadougou, Upper Volta

Blaise F, Barczi JF, Jaeger M, Dinouard P and De Reffye P (1995) Simulation of the growth of plants – modelling of metamorphosis and spatial interactions in the architecture and development of plants. In: Kunii TL and Luciani L (eds) Synthetic Worlds. John Wiley & Sons, Chichester, UK (in press)

Caraglio Y and Barthélémy D (1995) Caractéristiques morphologiques de la croissance et de la ramification des végétaux. Définition, description, écueils terminologiques. Science Update. INRA, Versailles, France (in press)

Cox DR (1962) Renewal Theory. Chapman and Hall, London, UK

Dauzat J (1994) Radiative transfer simulation on computer models of *Elaeis guineensis*. Oléagineux 49(3): 8–90

Dauzat D and Hautecoeur O (1991) Simulation des transferts radiatifs sur maquettes informatiques de couverts végétaux. In: Hunt JJ (ed) Physical Measurements and Signatures in Remote Sensing, 14–18 January 1991, pp 415–418. European Space Agency (ESA), Courchevel, France

De Reffye P (1979) Modélisation de l'Architecture des Arbres par des Processus Stochastiques. Simulation Spatiale des Modèles Tropicaux sous l'Effet de la Pesanteur. Application au *Coffea robusta*. Thesis, Université de Paris-Sud Orsay, Paris, France

De Reffye P, Blaise F and Guédon Y (1993) Modélisation et simulation de l'architecture et de la croissance des plantes. Rev Palais de la Découverte 209: 23–48

De Reffye P, Dinouard P and Barthélémy D (1991a) Modélisation et simulation de l'architecture de l'orme du Japon *Zelkova serrata* (Thunb.) Makino (*Ulmaceae*): la notion d'axe de référence. In: Edelin C (ed) 2ème Colloque International sur l'Arbre, 10–15 September 1990, pp 251–266. Naturalia Monspeliensia, N° hors série A7, Montpellier, France

De Reffye P, Elguero E and Costes E (1991b) Growth units construction in trees: a stochastic approach. Acta Biotheor 39: 325–342

De Reffye P, Houllier F, Blaise F and Fourcaud T (1995) Essai sur les relations entre architecture d'un arbre et la grosseur de ses axes végétatifs. Science Update. INRA, Versailles, France (in press)

Edelin C (1977) Images de l'Architecture des Conifères. Thesis, Université de Montpellier, Montpellier, France

Edelin C (1984) L'Architecture Monopodiale: l'Exemple de Quelques Arbres d'Asie Tropicale. Thesis, Université de Montpellier, Montpellier, France

Feller W (1957) An Introduction to Probability Theory and its Applications. John Wiley & Sons, New York, NY, USA

Fourcaud T (1995) Analyse du comportement mécanique d'une plante en croissance par la méthode des éléments finis. Thesis, Université de Bordeaux, Bordeaux, France

Fourcaud T and Lac P (1993) Modélisation mécanique de la croissance des végétaux. In: Colloque National en Calcul des Structures, 11–14 May 1993, pp 131–148. INRIA, Giens, France

Françon J (1991) Sur la modélisation informatique de l'architecture et du développement des végétaux. In: Edelin C (ed) 2ème Colloque International sur l'Arbre, 10–15 September 1990, pp 231–247. Naturalia Monspeliensia, N° hors série A7, Montpellier, France

Françon J and Lienhardt P (1994) Basic principles of topology-based methods for simulating metamorphoses of natural objects. In: Magnenat Thalmann N and Thalmann D (eds) Artificial Life and Virtual Reality, pp 23–44. John Wiley and Sons, Chichester, UK

Guédon Y (1995) Modélisation de séquences d'événements décrivant la mise en place d'éléments botaniques. Science Update. INRA, Versailles, France (in press)

Guédon Y and Costes E (1995) Modélisation de la croissance d'un axe végétatif. Science Update. INRA, Versailles, France (in press)

Hallé F and Oldeman RAA (1970) Essai sur l'architecture et la dynamique de croissance des arbres tropicaux. Masson & Cie, Paris, France

Hallé F, Oldeman RAA and Tomlinson PB (1978) Tropical Trees and Forests – An Architectural Analysis. Springer Verlag, Berlin, Germany

Jaeger M and De Reffye P (1992) Basic concepts of computer simulation of plant growth. J Biosci 17: 275–291

Jourdan C (1995) Modélisation de l'architecture du système racinaire du palmier à huile. Thesis, Université de Montpellier, Montpellier, France

Jourdan C, Rey H and Guédon Y (1995) Architectural analysis and modelling of the branching process of the young oil palm root system. Plant and Soil (submitted)

Kurth W (1994) Morphological models of plant growth: Possibilities and ecological relevance. Ecological Modelling 75/76: 299–308

Nair PKR (1991) State of the art of agroforestry systems. In: Jarvis PG (ed) Agroforestry: Principles and Practice, pp 5–29. Elsevier, Amsterdam, The Netherlands

Prusinkiewicz P, Lindenmayer A and Hanan J (1988) Developmental models of herbaceous plants for computer imagery purposes. Computer Graphics 22: 141–150

Rennolls K (1994) Pipe-model theory of stem profile development. For Ecol Manage 69: 41–55

Shinozaki K, Yoda K, Hozumi K and Kira T (1964) A quantitative analysis of plant form – the pipe model theory. I. Basic analysis. Jpn J Ecol 14: 97–105

Sibbald AR, Griffiths JH and Elston DA (1994) Herbage yield in agroforestry systems as a function of easily measured attributes of the tree canopy. For Ecol Manage 65: 195–200

Sibbald AR and Sinclair FL (1990) A review of agroforestry research in progress in the UK. Agroforestry Abstracts 3: 149–164

White J (1979) The plant as a metapopulation. Ann Rev Ecol Syst 10: 109–145

Agroforestry Systems **30**: 199–221, 1995.

The tree-crop interface: representation by coupling of forest and crop process-models

G. J. LAWSON[1], N. M. J. CROUT[2], P. E. LEVY[3], D. C. MOBBS[1], J. S. WALLACE[4], M. G. R. CANNELL[1] and R. G. BRADLEY[2]
[1] *Institute of Terrestrial Ecology, Edinburgh, EH26 0QB, UK;* [2] *Tropical Crops Research Unit, University of Nottingham, LE12 5RD, UK;* [3] *Institute of Ecology and Resource Management, University of Edinburgh, EH9 3JU, UK;* [4] *Institute of Hydrology, Wallingford, OX10 8BB, UK*

Key words: agroforestry models, crop models, forest models, root growth

Abstract. Three process-based approaches to agroforestry modelling are described. These are (a) coupling a continuous-canopy forest model (Hybrid) and tropical crop model (PARCH); (b) coupling an individual-tree model (MAESTRO) with a crop model (PARCH); and (c) incorporating a combined model of evaporation and radiation interception by neighbouring species (ERIN).

The coupled Hybrid/PARCH was parameterised for maize and eucalyptus, and run in five contrasting weather-types. As expected, shade is the most important factor limiting yield in wet sites; water in dry sites. Year-to-year variability in crop yield is increased by light and water competition. MAESTRO/PARCH was run with similar assumptions, and gave comparable yield predictions, except at the driest site where it allows small areas distant from the tree sufficient water to produce a modest yield. Hybrid/PARCH predicted total crop failure in the same climate. Yields on drier sites were higher in the shade, but water competition was severe close to the tree.

ERIN is simpler than the above models, but is unique in including the transfer of heat and water vapour between the two canopies. Transpiration from a moist understorey can humidify air in the overstorey, and reduce its transpiration; whilst a dry understorey will give off sensible heat, which increases the vapour pressure deficit in the overstorey and causes its transpiration to increase. Changes in overstorey transpiration due to fluxes from the understorey may approach 15–20%.

Introduction

Agroforestry models can fulfil several roles:

- to conceptualise existing knowledge on the mechanisms of tree-crop interactions;
- to identify key areas of bio-physical or socio-economic uncertainty;
- to focus and co-ordinate programmes of experimentation and dissemination; and
- to assist managers and extension workers in decision-making.

The importance of models in agroforestry has often been stressed, and is comprehensively reviewed in a study commissioned by the UK Overseas Development Administration's 'Forestry Research Programme' (FRP) [Muetzelfeldt and Sinclair, 1993]. The present paper arises from another FRP commissioned project [The 'Agroforestry Modelling Project' (AMP) – Cannell et al., 1994], which has two objectives:

- to evaluate and utilise existing models to understand the mechanisms involved in tree-crop interactions, and to identify those situations where agroforestry may offer benefits;
- to evaluate existing FRP-sponsored field experiments with respect to tree-crop interactions, their conceptual frameworks, measurement protocols and ranges known for different variables.

The AMP has developed a simple model of rotational fallows [Robertson, 1995; Mobbs and Cannell, 1995], and is currently integrating the Edinburgh Forest model [Thornley and Cannell, 1992] with the Hurley Pasture model [Thornley and Verberne, 1989]. This paper deals with three inter-related approaches to agroforestry modelling, which are intended to meet the following objectives:

- linking existing forest (Hybrid) and crop (PARCH) models on a broad scale, to enable predictions to be made over long-periods and wide-areas, and;
- linking a detailed tree photosynthesis and light-interception model (MAESTRO) with a crop growth model (PARCH), to enable short-term spatial interactions between trees and crops to be examined in detail around individual trees;
- using light interception and energy combination theory to describe the partitioning of energy between sensible heat and evaporation in plant mixtures (ERIN).

Whilst the individual tree and crop models described in this paper have been validated against real tree and crop yields, there has been no attempt, as yet, to validate the coupled models.

Model components

Hybrid

The model Hybrid is derived from a forest gap model, an ecosystem process model and a photosynthesis model [Friend et al., 1993]. Individual trees fix and respire carbon and lose water on a daily basis. Carbon partitioning and canopy growth currently occur annually, but it is possible to increment these on a smaller timestep.

For each forest plot, the number of trees of each species and the height and diameter at breast height for each tree is provided as input. Each individual is assigned a set of fundamental physiological parameters depending on the species, and these are used at a daily or annual timestep to calculate water and carbon dynamics. The canopy is divided into layers depending on number, size and species of trees in the plot, but it is currently assumed to be horizontally uniform across the plot. Each tree is treated separately with respect to daily transpiration, carbon fixation and respiration. Daily stomatal conductance of individual trees is calculated and used to predict fluxes of carbon dioxide and water between the foliage and the atmosphere. The flux of water is summed across individuals in each plot for each day then subtracted from the soil water. The flux of carbon dioxide is summed across days, to give annual productivity.

Four input files are required: a daily climate driver, a tree species library and two files containing details on the individual site. The meteorological data file contains daily values for precipitation, maximum air temperature, minimum air temperature and solar radiation. For this study, the daily meteorological data were predicted from monthly recorded data using a weather generator [Richardson and Wright, 1984]. Any number of different species can be grown in the plot, provided that physiological parameters for each species (31 parameters such as light extinction coefficient and foliage turnover rate), are listed in the tree library file. The remaining two input files contain 35 parameters describing site and soil conditions, including specific run details (number of years, size of plot, etc.) and assumptions regarding decompostion and nutrient availability. Full details are given elsewhere [Cannell et al., 1994]. For this study, 22 *Eucalyptus globulus* trees are grown in a single plot of 200 m^2. Each tree has a stem diameter at breast height of 4.72 cm, and the height to the base of the crown is 1 m. These values result in trees of 5 m height and a leaf area index (LAI) approximately equal to one. Soil potential water storage values for each site are taken from a global ecosystems database [Webb et al., 1992].

MAESTRO

The MAESTRO model is based on a model of radiative transfer in plant canopies, originating in the USA and New Zealand [Norman and Welles, 1983; Grace et al., 1987], and further developed in Edinburgh [Wang and Jarvis, 1990]. The model simulates the absorption of radiation, transpiration and photosynthesis of a specified stand of plants using meteorological data as an input. The co-ordinates, dimensions and leaf area of all plants in the stand are specified in an input file. At an hourly time step, the radiation regime within the crown of a single 'target tree' is calculated in detail, at up to 120 spatial points and in three wavebands (photosynthetically active radiation (PAR), near infrared, and thermal). At each of these points (representing a subvolume of the crown), stomatal conductance is calculated from the specified response to temperature, vapour pressure defecit (VPD), soil moisture and PAR. The conductance model is based on Jarvis [1976]. Transpiration and photosynthesis are then calculated for each subvolume. Values are summed to give outputs in moles per tree per day. Radiation is also calculated at points on the ground beneath the canopy.

Precipitation is partitioned between throughfall and intercepted components. Interception loss is calculated using the Rutter et al. model [1971], the remainder reaching the soil as stemflow and throughfall. The soil is represented as a number of layers, each with a given storage capacity. When this capacity is reached, water infiltrates to the layer below. When transpiration occurs, water is extracted first from the topmost layer, then from deeper layers as each layer dries to permanent wilting point.

The model differs from most others in that it represents canopy structure fully in three dimensions and can therefore simulate structurally complex, heterogeneous plant canopies. It is thus suitable for simulating agroforestry systems, which do not conform to simple, one dimensional canopy representations.

Input files

MAESTRO uses four input files:

- A Site File – with various parameters including site latitude and longitude, slope, leaf area and angle distribution functions, leaf optical properties, canopy and soil hydrological parameters.
- A Tree File – with co-ordinates and dimensions of all trees in the plot.
- A Physiology File – with parameters for stomatal conductance, leaf photosynthesis and phenology routines.
- A Meteorology File – with hourly values of windspeed, soil temperature, air temperature, relative humidity (RH), solar radiation, PAR and precipitation.

For the purposes of this study the model was configured to simulate a single tree of *Eucalyptus globulus* Labill. (Tasmanian blue gum) [Jarvis et al., 1989, treatment (F)]. It has the following dimensions: height = 5 m; height to base of crown = 1 m; crown radius = 1.25 m; leaf area = 25 m^2. This leaf area remains constant during the year. Crown shape is assumed to be conical. These are generally the same tree characteristices as used in the Hybrid simulation.

Meteorological inputs

Daily inputs required are solar radiation, maximum and minimum temperature, windspeed, humidity and precipitation. For this study, solar radiation, maximum and minimum temperature and precipitation were obtained using the Hybrid weather generator. A simple program was written to produce the necessary hourly input data for MAESTRO from these daily variables. Over the course of a day, air temperature is assumed to follow a sine wave with a maximum at 1400 h. Soil surface temperature and air temperature are assumed to be equivalent. Standard earth-sun geometry was used to calculate time of dawn and dusk. Solar radiation is assumed to follow a sine wave between dawn and dusk with the maximum at noon. A constant conversion factor of 2.1 is used to convert between solar radiation (W m^{-2}) and PAR (μmol m^{-2} s^{-1}). Above-canopy windspeed is assumed to be a constant 3 m s^{-1}.

Computation procedure

To calculate water use, the model is run using hourly meteorological data where the calculation points for below-canopy radiation are positioned on the ground around the tree. An hourly time step is relatively coarse resolution to capture the distribution of light below the tree, so the maximum possible spatial resolution of 120 points is used. The calculation of transmitance at these points is the most time consuming procedure in the model, but as transmittance changes only with canopy structure and solar position, it was considered unnecessary to recalculate for every hour of every day in each year. Instead, these are calculated only once for every daylight hour, for one day each month in a single year. By assuming that the diurnal solar track is constant within a month (and the same from year to year), the total transmitance computation time is reduced by a factor of around 30. The area around the tree is divided into 12 sectors based on azimuth angle and distance from the tree (Fig. 1), and transmitances are averaged within each sector. These

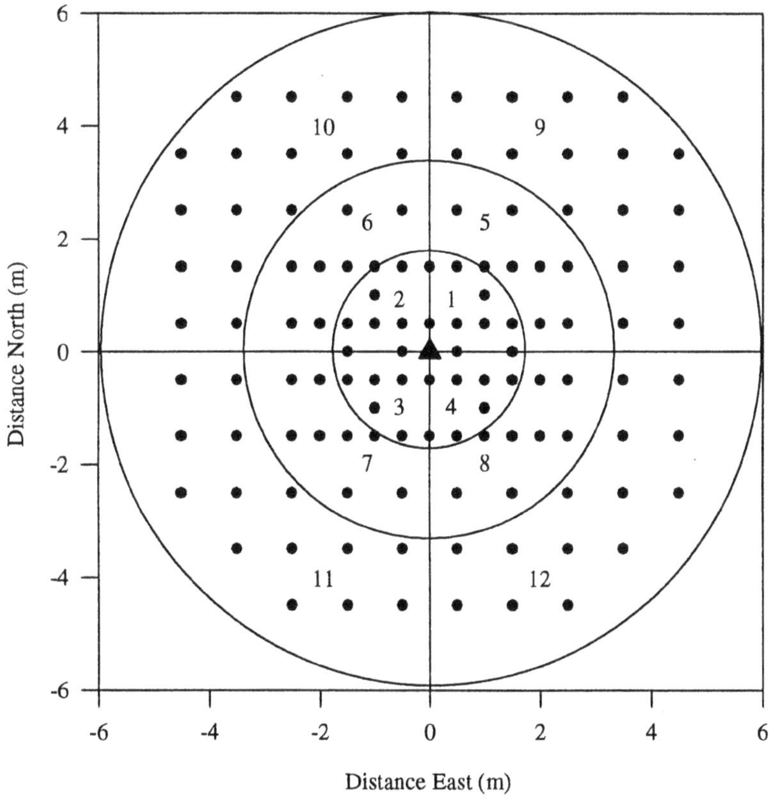

Fig. 1. Distribution of 120 points in a 10 × 10 m area on the soil surface under a *Eucalyptus globulus* tree where radiation receipt is predicted hourly for sample days each month. Values from these points are averaged into 12 zones and used to calculate the yield of an understorey crop of maize.

average transmitances can then be multiplied by incident radiation to give the radiation received in each sector every hour and summed to give daily values.

PARCH

PARCH is based upon a daily time step for interception of light and water resources, and assimilation of dry matter. Depending upon the availability of these resources, and the crop's ability to intercept them, growth is considered as either light or water 'limited'. An index of crop stress is calculated in terms of the ratio of light- to water-limited growth. This stress index is then used to control a number of physiological responses to stress, such as leaf rolling or increased partitioning to roots.

Partitioning of resources between plant organs is calculated by empirically derived fractions which are adjusted according to growth stage and level of stress. Resources partitioned to the leaf canopy and root system add to leaf area and root length, thereby feeding back into subsequent calculations for light interception and water uptake.

Although PARCH is a general model, it has been developed for applications in the dry topics, and considerable attention has been paid to the crop's water supply. A multi-layer water balance model is a key component of PARCH. This simulates vertical redistribution of soil water, infiltration, drainage, and soil evaporation. More information is provided in the user manual [Bradley and Crout, 1994].

Crop growth

Daily assimilation is calculated following PARCH's pre-cursor model RESCAP [Monteith et al., 1989], based upon conversion coefficients for plant light and water use.

Light limited growth, G_L (g m^{-2} d^{-1}) is given by:

$$G_L = f_i \varepsilon S F_{temp} \tag{1}$$

where

- f_i is the fractional light interception of the canopy calculated by a conventional Beer's law approach, but allowing for the effects of sparse canopies,
- ε is the light to dry matter conversion coefficient (g MJ^{-1}),
- S is daily solar radiation (MJ m^{-2} d^{-1}), and
- F_{temp} is a zero to unity factor which simulates the effect of temperature on photosynthesis.

Water limited growth, G_W (g m^{-2} d^{-1}), is given by:

$$G_W = X_{max} \frac{\Omega}{D} \tag{2}$$

where

- X_{max} is the maximum amount of water the root system can supply (mm d^{-1}). The calculation of this is described below,
- Ω is the crop's dry matter to transpiration ratio adjusted for atmospheric saturation deficit (g mm^{-1} kPa), and
- D is the daily average saturation vapour pressure deficit (kPa).

Daily crop growth, G (g m^{-2} d^{-1}), is then taken as:

$$G = \min(G_L, G_W) \tag{3}$$

Daily transpiration Q (mm^{-1} d^{-1}) is related to daily growth G via the transpiration equivalent Ω by:

$$Q = G\frac{D}{\Omega} \tag{4}$$

This amount of water is debited daily from the multi-layer soil water sub-model.

PARCH considers an 'unstressed crop' to be one where growth proceeds at a rate limited only by the availability of light. If water supply or temperature are such that growth is reduced then the crop is considered 'stressed'. PARCH contains a number of relationships in which the crop's behaviour is modified according to the level of stress it experiences. These relationships are framed in terms of a 'stress index', which is calculated on a daily basis as a proportion of the maximum water stress, and temperature stress.

The value of the stress index is used to dynamically control the plant's response to its environment. This is a key feature of the model as the crops studied have a high potential to modify their behaviour to meet a wide range of conditions. This takes place through: (a) day to day changes in effective leaf area by leaf rolling; (b) permanent changes in leaf area by advanced leaf senescence; (c) changes in the rate of thermal development; (d) changes in root to shoot partitioning of assimilates; and (e) changes in the plant's specific leaf area for new leaf growth.

Potential water extraction

Total potential water extraction is calculated each day by summing the potential extraction for each layer of the water balance sub-model. The calculations are based on a maximum uptake rate of water per unit depth of soil, U_{max}. This approach is empirical, but has been successfully applied in a number of crop simulation models. Although U_{max} will be dependent upon soil texture a single value of 0.1 mm (water extracted) mm^{-1} (of soil depth) has been found to be effective [Jansen and Gosseye, 1986; Robertson et al., 1993b].

U_{max} is assumed to apply to water extraction from a layer if it contains free water (i.e., at or above field capacity) and is 'saturated' with roots such that the addition of more roots will not increase water extraction significantly. If the water content of a layer, or the root density are below these values then potential extraction is reduced. Extraction is assumed to be proportional to $(aw/aw_{fc})^2$ (where aw is the plant's available water in the layer, and aw_{fc} is the plant's available water at field capacity). The power of two is used to approximately simulate the effect of a soil moisture release characteristic. This gives a value of 1.0 at field capacity, reducing rapidly to zero as soil moisture falls. This will give a poor approximation at high soil moisture, when water extraction declines only slowly as the soil becomes drier. However in these situations the model's estimate of water extraction is based upon light limited growth, and not soil water availability.

Water extraction is assumed to follow a 'diminishing returns' relationship proportional to $\sqrt{\rho/\rho_{max}}$ where ρ is root length density (m root m^{-3} soil), and ρ_{max} is maximum effective root length density.

Combining these ideas, and multiplying by the maximum uptake rate gives the potential extraction for each layer (X_{pot} in mm water mm^{-1} soil profile):

$$X_{pot,\,i} = \frac{aw^2}{aw_{fc}^2}\sqrt{\frac{\rho_i}{\rho_{max}}}\,U_{max},\ \rho \le \rho_{max} \tag{5}$$

This is then multiplied by the thickness of the layer and summed to give the total potential extraction of water by the roots, X_{max} (kg m^{-2}):

$$X_{max} = \sum X_{pot,\,i}\,\Delta Z_i \tag{6}$$

where ΔZ_i is the thickness of layer i in mm.

This is the maximum *potential* extraction from the soil profile. The amount of water actually extracted is equal to the plant transpiration Q calculated from Eq. (4). If the crop's growth is water limited this will be equal to X_{max}.

Root growth

Root growth is simulated in two stages.

- Assimilate partitioned to the root system is distributed to roots in each layer of the soil profile within the rooting zone.
- The root front is advanced down the profile, extending the rooting zone, although roots in this extended zone are not considered to extract water until the following day.

The expansion of roots in each layer is modelled first, as the 'new root front' is deemed to be extending ready for subsequent lateral expansion. This agrees with the hypothesis of Robertson et al. [1993a], who describe a situation where the first roots to penetrate a layer extract little water prior to the proliferation of further roots in that layer. The approach is a refinement of the framework given by Monteith [1986], where the extraction front closely corresponds to the rooting front, and roots in that layer are assumed to be fully expanded immediately the root front arrives.

The rate of root growth and expansion is currently controlled by developmental stage, temperature, stress, soil physical conditions, and assimilation rate.

ERIN

The Evaporation and Radiation Interception by Neighbouring species model (ERIN) [Wallace, 1995] provides a physically robust, yet comparatively simple theoretical framework which accounts for the coupled light interception and water use of two species grown as a mixture. In the agroforestry context it quantifies the competition for light between the crop and tree canopies, and calculates the evaporation from each canopy, and the interactions between them. There are two sets of equations: one for radiation interception and the other for transpiration.

Radiation interception

The light interception model only requires the individual values of extinction coefficient (K) and leaf area index (L) for each species along with their relative heights (h_1/h_2). The fraction of light intercepted by each component, f_i is given by:

$$f_i = f_i^s + \frac{h_i^2}{h_1^2 + h_2^2} [f_i^d - f_i^s] \tag{7}$$

where f_i^s and f_i^d are the amounts of light species i would intercept if its foliage was either completely above (f_i^d) or below (f_i^s) the other canopy. f_i^d and f_i^s are therefore given by the equations:

$$f_i^d = 1 - e^{-K_i L_i} \tag{8}$$

and

$$f_i^s = (1 - e^{-K_i L_i}) \sum_{j \neq i}^{2} e^{K_j L_j} \tag{9}$$

Figure 2 shows an example of the output of Eqs. (7), (8) and (9) for a mixture of two species both with a leaf area index of 3 and an extinction coefficient of 0.5. When species 1 is completely subordinate ($h_1/h_2 \sim 0$), species 2 intercepts the incident light first and the remainder falls on species 1. As h_1/h_2 increases, species 1 becomes more light dominant until, ultimately, it intercepts the incident light first, the remainder falling on species 2. This simple model applies to horizontally uniform species mixtures and needs further modification for use in agroforestry systems where the over-storey tree

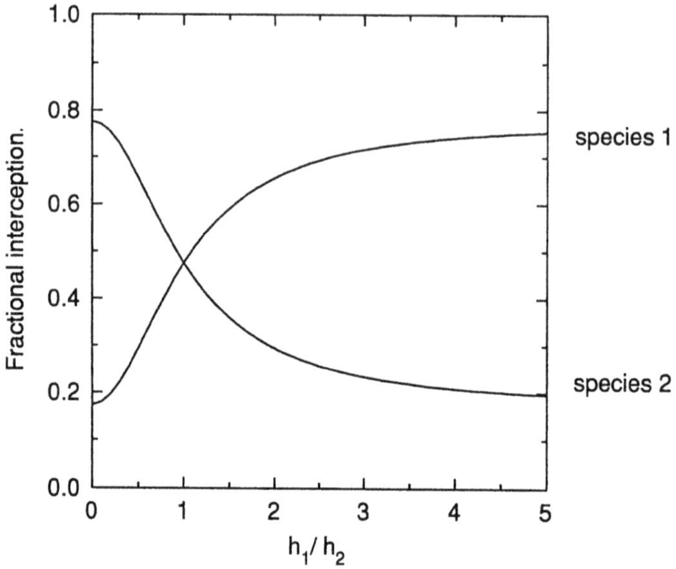

Fig. 2. The fractional light interception of two species in a mixture as a function of their relative heights estimated using the ERIN [Wallace, 1995] model. In this example each species has the same leaf area (3.0) and extinction coefficient (0.5).

canopy is spatially clumped. The *ERIN* model has a relatively simple way of dealing with canopy inhomogeneity by allowing the extinction coefficient of each species to vary with the degree of clumping in the canopy [Wallace, 1995].

Transpiration

The transpiration model in ERIN allows for interactions between the two canopies in an agroforestry system. It is based on the energy combination theory described by Shuttleworth and Wallace [1985] and is coupled to the radiation interception model described above. Shuttleworth and Wallace originally proposed a model to describe heat and water vapour transfer from a mono-cultural sparse crop. Wallace [1994] applied this model to a two component crop system, in which case the total evaporation, E can be written as:

$$\lambda E = C_1 PM_1 + C_2 PM_2 \tag{10}$$

where C_1 and C_2 are defined below, and PM_1 and PM_2 are terms similar to the Penman-Monteith formula which would apply to closed canopies of the constituent species. They have the form:

$$PM_i = \frac{\Delta A + [\rho c_\rho D - \Delta r_a^i(A - A_i)]/[r_a^a + r_a^i]}{\Delta + \gamma[1 + r_s^i/(r_a^a + r_a^i)]} \tag{11}$$

where A is the total energy available to the mixed canopy, A_i is the amount of energy available to the component species i, so that

$$A_i = f_i R_n \tag{12}$$

Since the values of f_i are generated by the radiation interception model, described in the previous section, the available energy and hence transpiration from each canopy are coupled to the above ground competition for light.

The various resistance terms in Eq. (11) are shown diagrammatically in Fig. 3. The bulk leaf boundary layer resistances of the two component species are represented by r_a^1 and r_a^2, and r_a^a is the aerodynamic resistance between the in-canopy mixing point (D_0) and the reference height (D). The coefficients C_1 and C_2 in Eq. (10) are given by Shuttleworth and Wallace [1985]. Equations (10) and (11) provide the means to calculate evaporation from a two species canopy when only the above canopy net radiation, temperature and vapour pressure deficit are known.

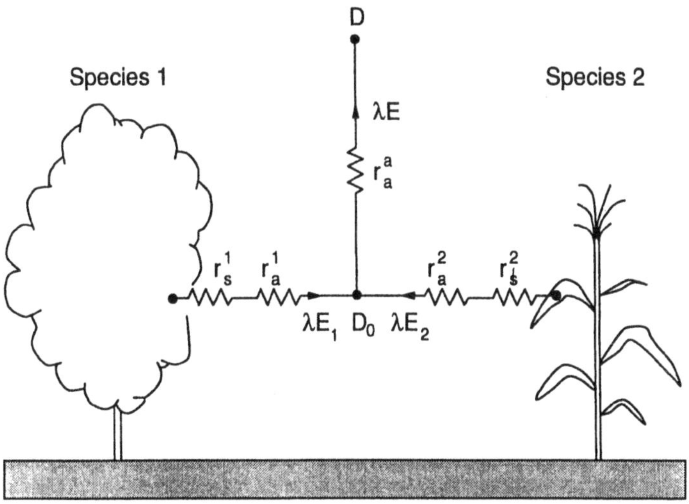

Fig. 3. A schematic representation of the transpiration from a two species mixture. Each species has its canopy (r_s^1 and r_s^2) and boundary layer resistance (r_a^1 and r_a^2). Component transpiration fluxes, E_l and E_2, converge within the mixed canopy where the vapour pressure defecit is D_0. The total tanspiration, E, mixes with the air at screen height via the aerodynamic resistance r_a^a.

Linking Hybrid and PARCH: long-term, regional comparisons

Linking the models

As an initial approach, the daily solar radiation and rainfall values used by PARCH were replaced by daily estimates from Hybrid of radiation and throughfall reaching the ground beneath a canopy of *Eucalpytus globulus*, with a leaf area index of 1.

Hybrid's representation of below-ground processes is rather simple, and soil water is modelled as a single compartment. The linked model assumes that the daily transpiration value calculated by Hybrid is abstracted from the different soil-layers of PARCH according to the available water in each layer and the distribution of tree roots. Thus the water extraction from each layer is calculated from Eq. (5) modified by a factor (F) expressing the distribution of tree roots as a declining exponential function of depth (z):

$$F(z) = ae^{-az} \tag{13}$$

where:

$$a = \frac{\ln(0.5)}{Z_{0.5}}$$

and $Z_{0.5}$ is the 'half depth' of the function (m); Z is the depth of the layer in the profile (m).

Simulation runs

Five sample sites were selected from a global database of mean monthly meteorological data [Mueller, 1982], and daily values were interpolated for a 30-year period using the weather generator in Hybrid. The sites span a wide-range of rainfall, with contrasting unimodal and bimodal rainfall patterns (Table 1). Two types of simulation were carried out.

Table 1. Rainfall and temperature characteristics of the 5 sites selected from the Mueller [1982] database, and used to test the combined models.

Site	Latitude (degrees)	Longitude (degrees)	Total annual rainfall (mm)	Rainfall pattern	Mean max. temp. (°C)	Mean min. temp. (°C)
Townsville, Australia	–19.2	146.8	1217	Continuous	28.0	20.5
Bujumbura, Burundi	–3.4	29.3	848	Bimodal	28.7	18.7
Bogota, Colombia	4.6	–74.1	939	Unimodal	18.9	9.9
Lilongwe, Malawi	–14.0	33.7	840	Unimodal	26.7	13.0
Mogadishu, Somalia	2.0	45.3	398	Unimodal	30.2	24.0

Multi-year simulations

The full meteorological data set was used for each of the five sites, and competition between trees and maize was examined in three scenarios:

1. maize growth with no trees present;
2. maize growth under trees, but with tree water uptake turned off (ie. trees compete for light only);
3. maize growth under shallow-rooted trees, with trees competing for both light and water (tree root depth parameter $Z_{0.5}$ = 1 m).

The tree-growth functions in Hybrid were disabled, and the canopy of *Eucalyptus globulus* was reset to a LAI of 1 at the end of each year of the 30 year. Maize was assumed to be planted at each site the day after plant available soil water rose to 40 mm. Naturally, maize may not be an appropriate crop for all these sites, and different cultivars of maize would be optimal for different locations. However our intention here is not to provide site specific yield predictions, but to make simplified comparisons between different environments. This is a necessary first step, but its limitations should not be neglected.

Australia, Colombia and Malawi showed the highest yields (Fig. 4a) in the crop only case (Scenario 1). Australia has a particularly reliable climate, and yield coefficient of variation (CV) is only 27%. Burundi has intermediate yields, and there is significant limitation due to water supply. Somalia has the lowest mean yield, as might be expected for the driest site. Of particular importance is its very high inter-year variability (with a CV of 112% – Fig. 4b).

In all cases, yields are reduced by the presence of the tree canopy. This can be separated into the effects of light and water competition by comparing Scenario 2 with Scenario 3. For Somalia, yield is only slightly reduced by light competition alone (relative mean yield 93.6%). For the other sites light competition has significant effects on crop yield, and relative mean yields range from 41 to 59%. For all sites the inclusion of water competition further reduces mean yields, although the effect for Australia is very small. Clearly Somalia is highly water limited, Australia is light limited, and the other sites are intermediate.

The effect of trees on crop yield variability is considerable. In Scenario 3 (water and light competition), trees clearly increase the variability of yield. Comparison with the case of Scenario 2 (light competition only) shows that this increased variability is due mainly to water competition. The drier the environment the greater the increase in variability.

Whilst the evaporative demand of trees reduces the soil water available for the crop, it was not possible to include a feedback in the reverse direction during the multi-year simulations. This feedback would have reduced the growth and water uptake of the trees. Thus, the simulations in this section may significantly over-estimate of the competitive effect of trees on crops in terms of water use, particularly on drier sites.

(a)

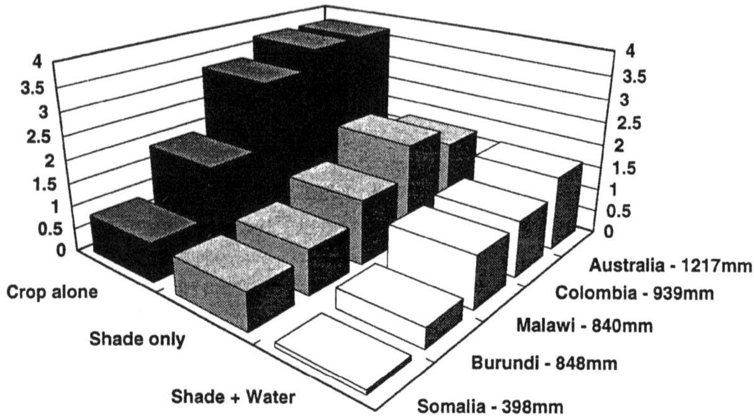

(b)

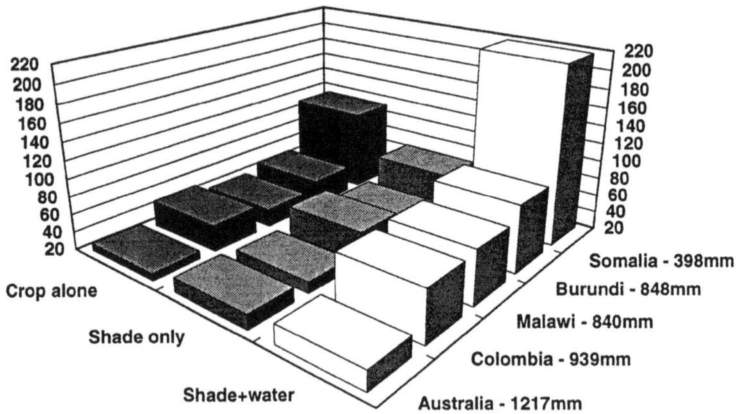

Fig. 4. (a) Predicted 30-year average maize yields in t ha^{-1} yr^{-1} assuming Scenario 1 (no trees), Scenario 2 (trees provide shade only), Scenario 3 (trees provide shade and compete with crops for water – assuming that the root volume declines exponentially from the surface with a $z_{0.5}$ value of 1 m. (b) Predicted coefficeints of variation for 30-year average predicted maize yields, with the same scenarios as in Fig. 4a.

Single-year simulations

A limited feedback of crop water use on tree growth was introduced into the Hybrid/PARCH simulations, by using daily outputs of crop evaporation from PARCH to reduce the soil water pool available to Hybrid. This is a laborious and temporary expedient, but has been used for sample years from the 30-year dataset to demonstrate: that competition for water does not significantly reduce daily net primary productivity for trees at the wettest site (Fig. 5d); and that the largest reductions in net primary productivity are observed late in the crop growing-season, due to depleted soil-water pools which are not replenished by rain (Fig. 5c).

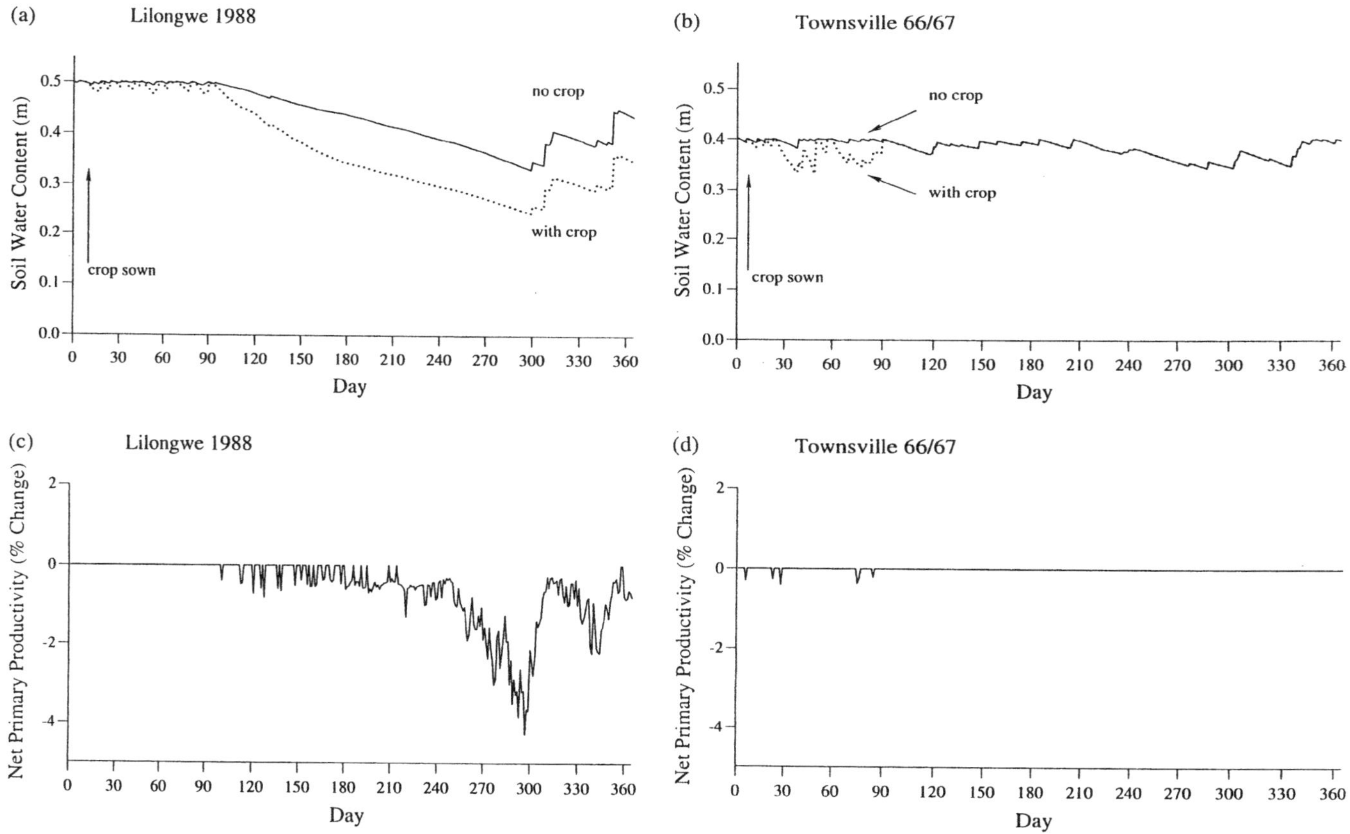

Fig. 5. The soil water content and percent change in tree net primary productivity (NPP) caused by the introduction of maize beneath LAI-1 *Eucalpytus globulus* growing at Lilongwe (1985–1986, 840 mm rainfall) and Townsville (1966–1967, 1217 mm rainfall).

Linking MAESTRO and PARCH: spatial variation around a single tree

Linking the models

MAESTRO gives daily light and throughfall values in 12 zones beneath the standard LAI-1 *Eucalyptus* tree. PARCH is run separately for each of these 12 zones, and it is currently assumed that there is no horizontal flow of water between zones. The water uptake by tree-roots is assumed to decline exponentially with radial distance from the trunk. The relative proportion of the tree's water that is extracted at radius R (m) is given by a factor F_{radial}.

$$F_{radial} = be^{-bR}R \qquad (14)$$

where:

$$b = \frac{\ln(0.5)}{R_{0.5}}$$

and $R_{0.5}$ is a 'half radius' parameter (m).

Water extraction is normalised with radius and depth to ensure that the transpiration calculated by MAESTRO is satisfied.

Simulation runs

To permit direct comparison with the Hybrid/PARCH single-year simulations, MAESTRO was run for the same years (30-year median rainfalls) and sites (Australia, Malawi and Somalia). Two scenarios are presented here:

1. Light competition only – with PARCH run for each of the 12 MAESTRO zones, and tree water uptake set to zero. This compares to Hybrid scenario 2.
2. Light and water competition included – with water uptake from each zone calculated assuming $Z_{0.5}$ = 1 m and $R_{0.5}$ = 2 m.

Predicted maize yields from each of the 12 zones were spatially averaged over the same 100 m^2 area as used by Hybrid. The half-radius setting allows the tree to receive virtually all its water from the modelled 10 × 10 m region, and therefore be as comparable as possible with Hybrid.

Preliminary results show reasonable agreement between the average crop yields predicted by PARCH under the spatially uniform canopy of Hybrid, or the 3-D tree canopy of MAESTRO (Table 2). In Somalia however Hybrid/PARCH predicts complete crop failure, whilst MAESTRO/PARCH calculates and average yield of 0.5 t ha^{-1} yr^{-1}, which is a respectable yield in many low input systems.

SHADE ONLY

TOWNSVILLE (1217 mm)

	West					East
	2.02					2.02
		1.91			1.89	
North			1.46	1.31		
South			1.05	1.15		
		1.64			1.57	
	1.96					1.91

LILONGWE (840mm)

	West					East
	0.84					0.85
		0.67			0.66	
North			0.30	0.27		
South			0.29	0.39		
		0.68			0.68	
	0.82					0.85

MOGADISHU (398mm)

	West					East
	1.57					1.59
		1.49			1.49	
North			1.07	0.97		
South			0.73	0.88		
		1.27			1.33	
	1.54					1.55

SHADE AND WATER COMPETITION

TOWNSVILLE (1217 mm)

	West					East
	1.79					1.79
		1.1			1.1	
North			0.29	0.27		
South			0.22	0.23		
		1.01			0.98	
	1.74					1.7

LILONGWE (840mm)

	West					East
	0.84					0.85
		0.66			0.65	
North			0.29	0.26		
South			0.28	0.37		
		0.66			0.67	
	0.81					0.84

MOGADISHU (398mm)

	West					East
	0.83					0.82
		0.0			0.0	
North			0.0	0.0		
South			0.0	0.0		
		0.0			0.0	
	0.82					0.82

Fig. 6. Predicted maize yields (t/ha) for 12 zones (Fig. 1) around a *MAESTRO Eucalyptus globulus* tree for three sites, with two scenarios for water uptake competition.

Further examination of the distribution of yield around the MAESTRO tree for each of the three sites (Fig. 6) shows:

- a general pattern of reduced crop yield close to the tree due to competition for light which is exacerbated when competition for water is included;
- higher crop yields on the shaded side of the tree (north for Somalia, south for Australia and Malawi), particularly when water competition is excluded.

The respectable yields in zones furthest from the MAESTRO tree in Somalia, explain the average yield of 0.5 t ha^{-1} yr^{-1} in Table 2, and provide an instance where the assumptions in Hybrid/PARCH that both the light-interception and tree-water extraction are uniform may be unacceptable.

Table 2. MAESTRO/PARCH predicted maize yields (t ha^{-1} yr^{-1}) for median rainfall years underneath a *Eucalyptus globulus* tree of LAI 1. Yields are spatially adjusted averages of 12 zones at different positions around the tree. Comparisons are made with appropriate HYBRID/PARCH predictions for tree water use.

Scenario		Predicted maize yield (t/ha) at each site		
		Townsville ('66/67)	Lilongwe ('85/86)	Mogadishu ('86)
No water uptake by the tree	MAESTRO	1.84	0.74	1.45
	Hybrid	2.21	0.70	1.24
Trees have priority for available water	MAESTRO – shallow roots ($R_{0.5}$ = 1 m)	1.09	0.69	0.51
	Hybrid – shallow roots	1.27	0.65	0.00

Linking light interception and evaporation (ERIN)

The effect of canopy transfer on heat and water vapour is illustrated in Fig. 7 for the case of a dominant but sparse main canopy (species 1) with a leaf area index of 1 and a surface resistance of 200 sm^{-1}. The diagram shows the ratio of species 1 transpiration calculated using the Penman-Monteith [Monteith, 1965] formula to that calculated using Eq. (10) as a function of the sub-canopy (species 2) surface resistance. The calculations are performed for average weather conditions and for three different values of dominant canopy leaf area index. Since Eq. (10) allows for flux interactions between the two canopies and the Penman-Monteith formula does not, their ratio indicates the degree to which this is important under the conditions illustrated. When the sub-canopy resistance is very low, as it would be for non-stressed vegetation with a high leaf area, or if the sub-canopy was wet just after rain, the Penman-Monteith Equation overestimates transpiration from the dominant canopy by as much as 15%. Conversely, when the sub-canopy resistance is high, as would be the case if it were water stressed or had a low leaf area,

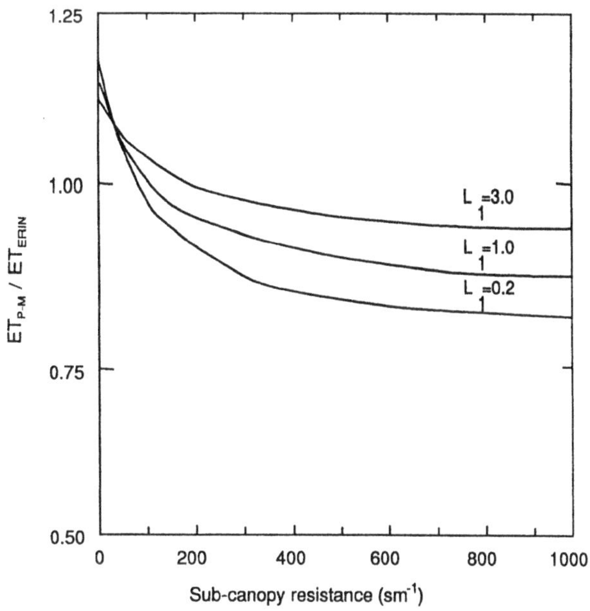

Fig. 7. Variation in the ratio of transpiration calculated using the Penman-Monteith formula and Eq. (10), ($ET_{P\text{-}M}/ET_{ERIN}$) with sub-canopy surface resistance. The three lines are for a dominant over-storey with leaf area indices of 0.2, 1.0 and 3.0 under typical mid-day conditions.

the Penman-Monteith Equation under-estimates the over-storey transpiration by nearly 20%. Clearly, the net effect of flux interactions from the two canopies over a season will depend on the constituent species surface resistances and how these are affected by rainfall and soil moisture stress.

The mechanism by which the over-storey canopy transpiration is altered is via the modification of the within-canopy vapour pressure deficit (D_0) because of fluxes of heat and water vapour from the sub-canopy. When the sub-canopy resistance is low, transpiration from the sub-canopy is large enough to humidify the air around the over-storey and hence reduces its transpiration. When the sub-canopy resistance is high, the sensible heat flux from the sub-canopy is sufficient to increase the vapour pressure deficit around the over-storey, thereby increasing its transpiration rate. The magnitude of this effect is greatest when the over-storey leaf area is low, because the amount of energy reaching the sub-canopy, and hence the degree of modification of D_0 is largest.

Discussion

Present problems

The results presented here are sensible, with competition for light and water represented at broad (Hybrid/PARCH) and detailed (MAESTRO/PARCH) levels. The effect of fluxes of heat and water on tree-canopy transpiration have also been explored (ERIN). However the four models described here are being continuously developed and integrated. This paper records an initial stage of integration where several deficiencies exist.

1. Hybrid and PARCH are linked by the exchange of input and output files, and are executed in different platforms and in different languages. Trees are assumed to take all available soil-water, rather than reducing their uptake as a result of competition for water from crop roots.
2. The Hybrid/PARCH link assumes horizontally-uniform light absorption and water uptake. This is not realistic for widely-spaced hedges or individual trees, and does not allow management strategies such as tree spacing or pruning to be investigated.
3. Hybrid requires 74 input parameters (31 tree, 23 soil and 20 site or run specific). There are few, if any, tropical agroforestry species for which sufficient physiological information is available. This is even more the case for MAESTRO, which requires 112 parameters (55 tree, 4 soil and 53 site or run specific).
4. Both Hybrid and PARCH contain rudimentary routines to allow soil and tissue nitrogen levels to control growth, but the routines are not compatible.
5. The MAESTRO/PARCH link represents tree-root distribution in three-dimensions, but tree-roots were assumed to be fixed through the growing season, and do not respond to areas of high water-content or permeability. The drying front around a tree will expand horizontally as well as vertically, and by ignoring dynamic root growth the competitive effect of trees is overestimated close to the stem, and underestimated further out.
6. MAESTRO is one of the most sophisticated three-dimensional models of a tree canopy. It has high spatial and temporal resolution, and provides a sophisticated approach to light interception and photosynthesis. However, it does not yet partition photosynthate or grow the tree. This is not a problem in current simulations, where leaf area has been held constant, but will be a limitation in the future when it is desired to test the effect of a developing tree canopy on crop yield.
7. PARCH has so far only been fully validated for sorghum [Bradley et al., 1994], although data exists for both millet and maize.
8. No effects of the fluxes of water vapour, heat or CO_2 from the crop on tree photosynthesis or transpiration (some of which are included in ERIN) are yet included in MAESTRO or Hybrid.

9. ERIN has yet to include the effects of spatially heterogeneous canopies and needs to be linked to a complete agroforestry water balance model.

Future development: HyPAR

The main aim is to develop the Hybrid/PARCH model to become a comprehensive physical agroforestry model, incorporating all the essential processes required to design optimum tree-crop systems which maximise crop and timber yields in particular locations. A new fully-coupled model is already in development, and will be called HyPAR.

Hybrid already couples the soil-plant-atmosphere system, and will be developed within HyPAR to:

(i) enable the canopy to be disaggregated to simulate individual trees and hedgerows (simplifying MAESTRO by using approaches suggested in Jackson and Palmer [1979] and Thornley and Johnson [1990]);
(ii) include more than one soil stratum, enabling different patterns of water and nitrogen abstraction to be examined in response to different vertical and horizontal rooting patterns, (using techniques in PARCH and FUSSIM2 [Heinen and De Willigen 1992]); and
(iii) allow essential micro-climatic feedbacks to occur between the crop and tree canopies (using techniques in ERIN).

PARCH has been translated into FORTRAN and partially integrated with Hybrid to form an initial version of HyPAR which allows feedbacks between tree water uptake, crop water uptake and soil water pools on a daily time-step. MAESTRO is being used in conjunction with PARCH to examine the effect of canopy pruning on tree photosynthesis, crop-growth and overall water consumption. Simple techniques to deal with canopy clumping are being developed within ERIN.

Once HyPAR is complete, its predictions will be compared with observations of biomass and crop yield, both regionally and at specific sites. Different tree-crop combinations will be parameterised, and management systems explored in conjunction with bio-economic models.

Acknowledgements

The ODA Forestry Research Programme (FRP) Grant R5651 has sponsored integration of Hybrid, MAESTRO, PARCH and ERIN. Development of PARCH has been funded by the Agronomy & Cropping Systems and the Resource Assessment Programmes of ODA. Work on ERIN has been supported by the ODA-FRP Project R4853.

References

Bradley RG and Crout NMJ (1994) PARCH – User Guide. Tropical Crops Research Unit, University of Nottingham, Nottingham, UK

Cannell MGR, Crout NMJ, Dewar RC, Lawson GJ, Levy PE, Mobbs DC and Robertson WH (1994) Agroforestry Modelling and Research Co-ordination. Annual Report June 1993–94. Institute of Terrestrial Ecology, Edinburgh, UK

Friend AD, Schugart HH and Running SW (1993). A physiology – based gap model of forest dynamics. Ecology 74: 792–797

Grace JC, Jarvis PG and Norman JM (1987) Description, sensitivity and field test of a radiative transfer model for widely-spaced stands. New Zealand Journal of Forest Science 17: 193–209

Heinen M and De Willigen P (1992) FUSSIM2: A Simulation Model for Two-Dimensional Flow of Water in Unsaturated Soil. DLO-Instituut voor Bodemvruchtbaarheid, Haren, The Netherlands

Jackson JE and Palmer JW (1979) A simple model of light transmission and interception by discontinuous canopies. Annals of Botany 44: 381–383

Jansen DM and Gosseye P (1986) Simulation of growth of millet (*Pennisetum americanum*) as influenced by water stress. Simulation reports CABO-TT no. 10. Centre for Agrobiological Research, Wageningen, The Netherlands

Jarvis PG (1976) The interpretation of the variations in leaf water portntial and stomatal conductance found in canopies in the field. Philosophical Transactions of the Royal Society of London B 773: 593–610

Jarvis PG, Wang YP, Borralho NMG and Pereira JS (1989) Simulation of the role of stress on radiation absorption assimilation transpiration and water use efficience of stands of Eucalpytus globus. In: Pereira JS and Landsberg JJ (eds) Biomass Production by Fast-Growing Trees, pp 169–179. Kluwer Academic Publishers, Dordrecht, The Netherlands

Mobbs DC and Cannell MGR (1995) Optimal tree fallow rotations: some principles revealed by modelling. Agroforestry Systems 29: 113–132

Monteith JL (1965) Evaporation and environment. In: Fogg GE (ed) State and Movement of Water in Living Organisms. 19th Symposium of the Society for Experimental Biology, pp 205–234. University Press, Cambridge, UK

Monteith JL (1986) How do crops manipulate water supply and demand. Philosophical Transactions of the Royal Society of London B 316: 245–259

Monteith JL, Huda AKS and Midya D (1989) RESCAP: a resource capture model for sorghum and pearl millet. In: Virimane SM, Tandon HLS and Alagarswamy G (eds) Modelling the Growth and Development of Sorghum and Perl Millet, pp 30–34. Research Bulletin 12. ICRISAT, Patancheru, India

Mueller MJ (1982) Selected Climatic Data for a Global Set of Standard Stations for Vegetation Science. Junk Publishers, The Hague, The Netherlands

Muetzelfeldt RI and Sinclair FL (1993) Ecological modelling of agroforestry systems. Agroforestry Abstracts 6: 207–247

Norman JM and Welles JM (1983) Radiative transfer in an array of canopies. Agronomy Journal 75: 481–488

Richardson CW and Wright DA (1984) WGEN: A Model for Generating Daily Weather Variables. ARS-8 ed. US Department of Agriculture, Washington, DC, USA

Robertson MJ, Fukai S, Ludlow MM and Hammer GL (1993a) Modelling root growth of grain sorghum using the CERES approach. Field Crops Research 33: 113–130

Robertson MJ, Fukai S, Ludlow MM and Hammer GL (1993b) Water extraction by grain sorghum in a sub-humid environment. II. Extraction in relation to root growth. Field Crops Research 33: 113–130

Robertson WH (1995) Modelling soil nitrogen levels under acacia/sorghum rotations. Agroforestry Systems 27: 283–292

Rutter AJ, Kershaw KA, Robins PC and Morton AJ (1971) A predictive model of rainfall interception in forests, 1. Derivation of the model from observations in a plantation of Corsican pine. Agricicultural Meteorology 9: 367–384

Shuttleworth WJ and Wallace JS (1985) Evaporation from sparse crops – an energy combination theory. Quarterly Journal of the Royal Meteorological Society 111: 839–855

Thornley JHM and Verberne ELJ (1989) A model of nitrogen flows in grassland. Plant, Cell and Environment 12: 863–886

Thornley JM and Johnson IR (1990) Plant and Crop Modelling. Clarendon Press, Oxford, UK

Thornley JHM and Cannell MGR (1992) Nitrogen relations in a forest plantation – soil organic matter ecosystem model. Annals of Botany 70: 137–151

Wallace JS (1994) Towards a coupled light partitioning and transpiration model for use in intercrops and agroforestry. In: Sinoquet H and Cruz P (eds) Ecophysiology of Tropical Intercropping, pp 153–162. INRA Editions, Versailles, France

Wallace JS (1995) ERIN: a coupled light partitioning and transpiration model for use in mixed plant communities. In: Agroforestry Modelling and Research Co-ordination, Second annual report to the UK Overseas Development Administration. Institute of Terrestrial Ecology, Edinburgh, UK

Wang YP and Jarvis PG (1990) Description and validation of an array model – MAESTRO. Agricultural and Forest Meteorology 51: 257–280

Webb RS, Rozenweig CE and Levine ER (1992) Global soil particle size properties In: Kineman JJ and Ohrenschall MA (eds) Global Ecosystems Database. US Department of Commerce, Colorado Springs, CO, USA

Agroforestry Systems **30**: 223–234, 1995.

A framework for a modular modelling approach for agroforestry

R. I. MUETZELFELDT
The University of Edinburgh, Institute of Ecology and Resource Management, Darwin Building, King's Buildings, Mayfield Road, Edinburgh EH9 3JU, Scotland, UK

Key words: modelling environment, simulation

Abstract. There are a number of problems with conventional methods for implementing models of agroforestry systems, including the difficulties of building, modifying and understanding them. One way of addressing these problems is to adopt a modular modelling approach, in which a model is constructed by linking a number of submodels, which in turn may consist of smaller computational units.

This paper presents a basic framework for the design of a modular modelling approach suitable for agroforestry modelling. This approach is based on a close correspondence between the notion of 'module' in modelling and of 'subroutine' in a programming implementation of a model, so that selecting and linking modules is analogous to choosing and issuing calls to subroutines. This approach has the potential for enabling the complex modelling structures that are required for agroforestry modelling, such as spatial aspects and the representation of individual trees, to be handled in a consistent and uniform manner.

The approach requires that subroutines should be held as symbolically-represented structures, rather than just source code in a programming language, so that they can be referenced as discrete units, and internally modified by symbol-processing operations. It is shown that the logic-programming language Prolog is well-suited to the task of representing subroutines in template form, ready to be fleshed out and combined into a single program in response to the selections made by a user. Prolog's suitability for the development of a user-friendly model design interface is also discussed, enabling models to be constructed simply by selecting options from a menu of design choices.

Introduction

The value of computer simulation modelling in agroforestry research and management is well recognised [Muetzelfeldt and Sinclair, 1993]. Models make it possible to test one's understanding of how agroforestry systems function; to predict the outcome of particular agroforestry systems; and to explore alternative management regimes.

Such models are usually implemented in a conventional programming language (such as Fortran), and their design is communicated to the rest of the research community in the 'Model Description' part of a report or journal paper. The use of a programming language for implementation brings with it a number of problems, including the length of time it takes to implement and debug the program, the difficulty of exploring alternative but equally plausible model formulations, and the difficulty of elucidating the logic of model design from the program listing. The use of English text for communicating model design is also unsatisfactory: such descriptions are frequently incomplete or ambiguous, and there are no agreed standards for describing model structure.

It would seem that there is a need for a computer-based environment to support the modelling process. Such environments are common-place in other domains, such as engineering and architecture, where they provide the modeller with a toolkit of model-building elements that correspond to the sorts of real-world elements appropriate to the domain, such as pumps and valves in chemical engineering. In contrast, the only alternatives to programming available to the agroforestry modeller are the use of simulation languages or diagram-based modelling software. None of these enables the modeller to design a model using agroforestry concepts, nor do they allow for all the model-design features required for agroforestry modelling.

The aim of this paper is to present a framework for the development of a modular modelling system for agroforestry modelling. The ideas in this framework underpin an ongoing project to implement such a system. This project uses the Prolog programming language both for representation of modules and models, and also for implementation of the various procedures involved in constructing and working with models. For those readers unfamiliar with Prolog, a number of introductory textbooks are available, for example [Malpas, 1987; Sterling and Shapiro, 1985], while an introduction to Prolog in the context of ecological modelling is given in [Muetzelfeldt et al., 1988].

Modelling in an agroforestry context

For present purposes, it is assumed that agroforestry is concerned with growing two types of plant on a given area of land. The larger and more permanent plant type is referred to as 'tree', recognising that in some contexts this plant might be more appropriately labelled as 'shrub'; and the smaller and less permanent plant type is labelled 'crop', recognising that in some contexts this might be (for example) grass which is grazed by sheep or cattle. Tree and crop might be separated temporally, in a rotational system, or might occupy the same area at the same time in a number of ways, such as in blocks, in rows, or with the crop growing amongst dispersed trees.

The management objective is, in general, to optimise (in some sense) the return from such a system, where the return is characterised in terms of the yield and service roles from the crop and the tree components, set against inputs of labour and money. Management choices include the choice of tree and crop species and varieties, temporal and spatial planting patterns, and the type, timing, location and intensity of management actions such as weeding, pruning and mulching.

It is implicit in the term 'agroforestry' that there will be interactions between the tree and crop components of the system. Individual interactions may be negative, such as shading in a light-limited environment, or positive, such as nitrogen fixation by the tree benefiting the crop. The adoption of an agroforestry system is based on the assumption that the positive effects will outweigh the negative ones, such that the net return to the farmer will be greater when the two components are grown together and these interactions occur.

The aim of modelling agroforestry systems is generally to explore alternative management options through simulation, in order to permit a reasoned assessment of which options are likely to be preferable. It follows that the model must produce, as output, estimates of those factors that are important to the farmer (such as yield); that it must permit the setting of management inputs corresponding to those that the farmer would be able to vary; and that the outputs of the model for given management inputs should faithfully mimic those that would be obtained for the corresponding real-life system managed in the same way. In order to achieve these objectives, agroforestry modelling must be capable of taking into account the key features of agroforestry systems identified above: that is, the spatial and temporal interactions between different types of component.

Reasons for adopting a modular modelling approach

There are a number of reasons for preferring a modular modelling approach over a conventional programming approach for agroforestry modelling. These are outlined below in terms of what can be achieved through the adoption of such an approach.

- Simplification of the task of designing and implementing models – Model construction becomes a design activity, achieved by selecting components and linking them together, rather than a programming activity.
- Simplification of the task of modifying model structure – The structure (design decisions) in models implemented in a programming language readily become ossified in the program used to implement the model. The need for sensitivity analysis on model parameters is widely recognised. A modular approach makes it much easier to undertake structural sensitivity analysis and to explore alternative but equally plausible model designs.
- Increased co-operation between modellers and researchers – An improved modelling environment will encourage modellers and researchers to interact to a greater extent than at present in the design of appropriate models.
- Increased reusability – Internationally, a huge amount of effort goes into determining modelling relationships, and in constructing models and submodels. The adoption of a common modular modelling approach would greatly facilitate the re-use of these modelling components by others, leading to a great reduction in overlap of effort. This is particularly pertinent in an interdisciplinary subject area such as agroforestry.
- Support for the modelling process – A modular approach, combined with standard ways for documenting modules, opens up the way for computer-

based guidance in the model-construction process. For example, the user could specify context (country, species) and required outputs, and the computer could suggest one or more suitable modules.

- Increased clarity of communication of model structure – A modular approach would make it much easier for others to understand the structure of a given model, since it lends itself to diagrammatic presentation, provides a disciplined and hierarchical format, and guarantees a direct correspondence between model description and implementation. This in turn is necessary if a model is to be considered as part of the scientific process, and again aids the appreciation of models amongst scientists across disciplines.
- The ability to investigate model components separately – It is difficult, using conventional methods of model implementation, to explore the behaviour of individual relationships or submodels; an activity which is crucial if one is to have confidence in the overall model behaviour. A modular approach means that all the tools provided for exploring model behaviour (run control, sensitivity or uncertainty analysis) can as readily be applied to individual components of the model.

Conceptual basis of a modular modelling approach

There are a number of real-world systems that provide useful analogies for thinking about modularity. For example, there are the familiar examples of modular electronic systems: a radio consists of a number of circuit boards, each of which performs some designated task and which contains lower-level electronic units, such as transistors and integrated-circuit chips. At any level, modules are connected though wires or copper track.

Computer-aided design (CAD) software has been developed in a number of subject areas, supporting a modular approach. For example, there is software that supports the process of designing electronic circuits, in which components are selected from 'bins', 'plugged into' a circuit board, and 'wired up', operations directly paralleling the real-world operations. A similar modular approach is used in CAD software in architecture. In engineering, a number of modelling systems based on block diagramming have been developed for systems analysis, in which the basic modules correspond to mathematical operations, such as multiplication and integration. These modules can in turn be combined together to give higher-level modules. These systems clearly demonstrate the effectiveness of graphical, mouse-based user interfaces for supporting the design process.

A generic modular modelling environment, SIMPLEX, has been developed [Eschenbacher and Schmidt, 1987], and applied in an environmental modelling context [Wittman, 1994]. This software represents an explicit commitment to a hierarchical modelling approach.

For people with programming experience, the concept of a 'subroutine' is a useful one for thinking about modularity: a program consists of some calls to subroutines, each of which performs some designated task and which may itself call lower-level subroutines. At any level, subroutines communicate through the passing of data structures, such as variables or arrays. A diagram showing the subroutine calling sequence of a model is frequently used to illustrate the submodel structure of the model: for example, PARCH [Bradley and Crout, 1992] and MAESTRO [Wang and Jarvis, 1990].

Indeed, modellers frequently respond to discussions on modular modelling by asserting that they work in this manner already, by the extensive use of subroutines in implementing a model. Why not, they argue, simply encourage good structured programming practices, and the development of a common subroutine library, so that models could be constructed simply by issuing the appropriate subroutine calls? However, with such an approach, model construction would still be a programming activity, and the programmer would have responsibility for ensuring correct matching of subroutine arguments. Moreover, modifying the functions called within library routines would be difficult. In addition, it would be hard to provide support for the process of selecting appropriate modules, and there would be no guarantee that model documentation would correspond to the design of the implemented model. These are, however, problems with current methods for working with subroutines, rather than with the subroutine concept itself.

The modular modelling approach presented here is based on the idea of developing a user-friendly interface for manipulating subroutine-like modules. Manipulation refers to the processes of selecting subroutines from a subroutine library, constructing subroutines in terms of calls to other subroutines, linking the input and output arguments of subroutines, and modifying the internal structure of previously-programmed subroutines. The development of an interface for supporting these tasks requires us to view a subroutine as a symbolically-represented structure, which can be referenced as a single unit, internally modified by symbol-processing operations, and reasoned with using some form of symbolic logic.

Conventional implementation of an agroforestry model

In order to provide a basis for discussion, it is useful to consider a model which is simple enough for purposes of exposition, but yet which captures a number of the significant features of agroforestry systems. This will be presented as an outline program, which will serve as a reference example for considering various ways in which concepts of modularity can be supported by an appropriate modelling environment.

The following pseudo-program represents a simple model of a tree-crop system, in which:

- the tree component is represented by the height of a set of individual trees;

- the crop component is represented as crop biomass on each of a number of grid squares;
- the interaction between the two is only in terms of tree shading reducing crop growth rate.

A row of three dots (. . .) represents parts of the programming (in this case, argument lists for subroutines) that are omitted for the sake of clarity.

1. call init (. . .)
2. for year = 1 to 20
3. call shading (. . .)
4. call tree_growth (. . .)
5. call crop_growth (. . .)
6. call tree_update (. . .)
7. call crop_update (. . .)
8. call display (. . .)
9. next year

There are several points to note about the programming features shown here. First, the program consists of calls to subroutines, plus some control structures (in this case, a loop over years). The extensive use of subroutines reflects a certain conventional view of 'modularity', and may be taken as representing a good structured-programming solution. It provides the basis for the approach to modular modelling explored in this paper.

Second, the tree component of the system is handled by not just one but two subroutines, one for calculating the growth (height increment) of each of the trees, the other for updating height with the new growth. The same is true for the crop. This reflects the requirement that, in general, all the rates in a system should be calculated before updating the state variables [Supit et al., 1994]. It is not uncommon to find published model flow-charts using a single subroutine call for each component with the main time loop, with both rate calculations and updating of state variables occurring during a single call, but this can lead to significant errors in modelling processes that operate in parallel.

Third, some of the sequencing of the subroutine calls is important, while some is not. 'shading' needs to be called before 'crop_growth', and the 'growth' routines need to be called before the 'update' routines; but the ordering of the two 'growth' routines and the two 'update' routines is irrelevant and has no effect on model behaviour.

Fourth, this apparently simple program can encompass a model of considerable complexity. For example, the subroutine 'shading', which takes as input data structures representing the size and location information for all the trees and returns a shading value for each grid square, could incorporate a sophisticated light interception model. Similarly, the tree growth routine could be based on a detailed process-based representation of the mechanisms of photosynthesis and partitioning of assimilates. Thus, even a simple main program such as this is capable of modelling complex spatial, organisational and geometric aspects of agroforestry systems.

Clearly, this example illustrates a certain notion of 'modularity': the one task of initialising and performing the calculations for a complex tree-crop model has been broken down into a number of subtasks. However, it might also be expected that other aspects of modularity might be indicated: for example, the notion of re-useability, with perhaps the 'shading' routine being copied from another program; or the notion of choice, with perhaps the program writer selecting one of several available tree growth routines.

Conventionally, working with these notions of modularity is a programming task: the programmer must edit a program to replace a call to one routine with a call to another; calls to additional routines must be edited into the program; and changes to the argument list in subroutine calls must be made as required to agree with the specified argument lists of the routines included. Moreover, this requirement applies equally to the subroutines called by the main program, to the subroutines that they call, and those called in turn at lower levels. These programming tasks require considerable skill, are error-prone and time-consuming activities, and are a long way removed from any notion of model construction through interactive module selection and linkage.

In order to move towards a subroutine-based modular modelling environment, it is necessary to view a subroutine as a structured item of information, not as text entered by a programmer using a text editor. Viewed in this way, it is then possible to develop software tools that can locate, select, insert and delete subroutines and – equally importantly – that can modify the internal structure of a subroutine to ensure consistency with the rest of the program. Conventional programming languages are ill-suited for this task. However, languages developed within Artificial Intelligence for reasoning with qualitative, symbolically-represented information are ideally suited. The logic-programming language Prolog is particularly appropriate, since it can be used for representing subroutines themselves as structured items of information, for guiding selection of appropriate subroutines given the goals of the model and the choices made so far, and for undertaking any re-writing of the subroutines to ensure program consistency.

Interactive model design

There are two ways in which the model design process can be supported within a subroutine-based modular modelling environment. The first is that the user can be given choices corresponding exactly to the modules (subroutines) themselves. The second is that the user can be asked to specify features of the system that are to be captured, and these can then be used behind-the-scenes to select appropriate modules.

Using the first approach, it is merely necessary to ensure that the user is presented at each stage with a choice of modules that are consistent with those already selected. There is an obvious hierarchy in the calling structure of subroutines, so it is natural to exploit this in interacting with the user.

Prolog lends itself very naturally to representing a set of choices that are contingent on those already made. For example, the first two of the following four Prolog 'clauses' (statements) assert that there are two subroutines ('growth1' and 'growth2') that are compatible with the 'tree_growth' slot in the program called 'main1' (which can be taken as being the program given above). The next two clauses assert that there are two subroutines ('photosyn5' and 'photosyn7') that are compatible with the photosynthesis call in the 'growth2' subroutine (a process-based submodel of plant growth).

```
possible_sub(main1,tree_growth,growth1).
possible_sub(main1,tree_growth,growth2).
possible_sub(growth2,photosyn,photosyn5).
possible_sub(growth2,photosyn,photosyn7).
```

Given such a representation, it is straightforward to write a Prolog program that can present the user with a menu of all the outstanding choices. With modern Prologs, this can be done in a GUI (graphical user interface) environment, enabling the user to build a model simply by selecting options from scrolling menus using the mouse. The user can be supported in making this choice by the provision of extra information on each module: again, it is a straightforward task in Prolog to associate some descriptive text with the symbolic name of each module.

The second approach involves the computer identifying appropriate modules in response to the user specifying desired features of the model. This task is directly analogous to that of identifying a plant from its attributes, or a disease from its symptoms, and can be handled by a variety of techniques developed for such purposes, such as look-up tables, decision trees and expert systems. The relational and rule-based features of Prolog make it well-suited for such tasks, and it has been used to develop a large number of expert-system-type programs.

Prolog representation of subroutines as modules

The argument for basing a modular modelling approach on the handling of subroutines as modules depends critically on the ability of software to manipulate the source code of the main program and of subroutine blocks. The aim of this section is to demonstrate the feasibility of such manipulation.

In order to provide a sufficiently concise example for presentation, it has been restricted to a model of just the tree component of an agroforestry system. The main program therefore does not include the shading and crop aspects of the previous example. Tree growth is modelled by a simple logistic growth term. Also, some small liberties have been taken with the programming aspects (for example, the exclusion of dimensioning information for arrays). These restrictions have been done only for purposes of clarity, and in no way detract from the central message of this section, that it is indeed possible to manipulate subroutine source code symbolically in ways that result in a consistent and coherent simulation program.

In this approach, each subroutine available for inclusion in a model is represented in template form in a Prolog clause headed by the predicate 'module'. In the Prolog program given below, there are two of these: a module called 'main1' (lines 2–15), and a module called 'grow1' (lines 16–24). Each 'module' clause is a rule, with the general form:

```
module(Name,Template) :– Conditions.
```

in which the symbol :– is read as the word 'if'. The Template is represented as a list (enclosed in [. . .]) of terms that are closely related to program statements. Thus, the Template for 'main1' is in lines 3–8, and that for 'grow1' is in lines 17–21. The task of the Conditions component is (despite its name) mainly to flesh out elements of the template with particular variable and array names, depending on the suite of modules selected by the user.

Line 1 is a clause which contains the list of modules selected by the user: this list is the result of interaction with the user interface, as described in the previous section. The 'compatible_with' predicate (line 25) indicates which subroutines can fit into a particular slot of a particular calling routine or main program, while the 'argument_list' predicate (line 26) indicates separately the state variable, parameter and rate variable arguments for each subroutine.

```
 1. user_selected([main1,grow1]).

 2. module(main1,
 3.   [(call init(Args1)),
 4.     (for year = 1 to 20),
 5.     (call Sub(Args2)),
 6.     (call update(Args3)),
 7.     (call display(State_vars)),
 8.     (next year)]):–
 9.       user_selected(L),
10.       member(Sub,L),
11.       compatible(main1,tree_growth,Sub),
12.       argument_list(Sub,State_vars,Params,Rate_vars),
13.       append(State_vars,Params,Args1),
14.       append(Args1,Rate_vars,Args2),
15.       append(State_vars,Rate_vars,Args3).

16. module(grow1,
17.   [(subroutine grow1(Args)),
18.     (for tree = 1 to ntree),
19.     (growth(tree) = r*height(tree)*(1-height(tree)/maxheight)),
20.     (next tree),
```

```
21.     (end sub)]):–
22.       argument_list(grow1,State_vars,Params,Rate_vars),
23.       append(State_vars,Params,Args1),
24.       append(Args1,Rate_vars,Args).

25. compatible(main1,tree_growth,grow1).

26. argument_list(grow1,[height(ntree)],[r,maxheight],[growth(ntree)]).
```

The following pseudo-code program for the selected modules is generated from the above set of templates with an extra 10 or so lines of Prolog, their main functions being to declare words such as 'subroutine' and 'call' as operators, and to step through the modules in the selected list. The correspondence between the program templates and the generated program can be seen by comparing the template for 'main1' in lines 3–8 above with the generated program in lines 1–6 below, and similarly for 'grow1' (lines 17–21 above with lines 7–11 below).

Generated program:

```
 1. call init([height(ntree),r,maxheight])
 2. for year = 1 to 20
 3. call grow1([height(ntree),r,maxheight,growth(ntree)])
 4. call update([height(ntree),growth(ntree)])
 5. call display([height(ntree)])
 6. next year

 7. subroutine grow1([height(ntree),r,maxheight,growth(ntree)])
 8. for tree = 1 to ntree
 9. growth(tree) = r*height(tree)*(1-height(tree)/maxheight)
10. next tree
11. end sub
```

As stated above, a number of simplifications have been made for presentation purposes. These include the convention that 'height(ntree)' is used to denote that 'height' is a one-dimensional array with 'ntree' elements; the absence of specification statements; and the inclusion of square brackets in argument lists. For the same reasons, the example does not demonstrate how the programming code for the subroutines 'init', 'update' and 'display' would be generated, though in fact that would be done using similar code-generation methods to those shown here. There are also numerous refinements to this system that can be envisaged, such as enabling the system to handle subroutines nested to any depth, and undertaking more rigorous type checking on the subroutine argument lists.

Discussion

In this paper, one possible approach to the problem of designing a modular modelling system for agroforestry has been presented. It has two key features. First, there is a close correspondence between the notion of 'module' in modelling and 'subroutine' in programming. In this approach, a library module corresponds to the subroutine itself, while an instance of a module in a model corresponds to a call to a subroutine, binding the subroutine arguments to particular variables or values. This correspondence enables the generation of a conventional program from the model specification, with a one-to-one relationship between a module and subroutine. This greatly simplifies the task of being able to generate a runable version of the model. It should also help in persuading practising modellers that the modelling environment works correctly, since they can check through the generated program.

Second, arbitrarily complex ways of representing the components in the system can be supported, such as grid squares and layers for soil water, or an individual-based tree component. This is achieved by allowing modules to contain arbitrary data structures (e.g., 2-D arrays), and arbitrary amounts of programming in a conventional programming language for performing calculations with these data structures. This is necessary in order to be able to make use of the large number of agroforestry-relevant submodels that already exist as program subroutines, and to capture important spatial interactions in agroforestry systems.

This approach results in a system that corresponds closely to everyday notions of modularity in the context of modelling, such as notions of 'submodel' and 'choice of light response function'. There is little doubt that the implementation and widespread adoption of such a modular modelling system would greatly enhance the effectiveness of agroforestry modelling: reusability of model components would be much simpler; alternative model formulations could be readily compared and evaluated; and it would be much easier to understand other people's models. A common objection to this approach – that it would constrain the modeller – has some theoretical merit, but in practice most modellers would probably have greater flexibility, rather than less, since they would have access to pre-programmed representations, and so would be less constrained by the limitations of their own programming ability or awareness of modelling approaches.

References

Bradley RG and Crout NMJ (1992) PARCH – User Guide. Tropical Crops Research Institute, University of Nottingham, Sutton Bonnington, Leicester, UK

Eschenbacher P and Schmidt B (1987) The model description language SIMPLEX-MDL. In: Adelsberger H and Broeckx F (eds) Discrete Event Simulation and Simulation and Operations Research. Proceedings of the European Simulation Multiconference, 7–10 July, 1987, Vienna, Austria, pp 44–50

Malpas J (1987) Prolog: A Relational Language and Its Applications. Prentice-Hall International, Inc, Englewood Cliffs, NJ, USA

Muetzelfeldt RI and Sinclair FL (1993) Ecological modelling of agroforestry systems. Agroforestry Abstracts 6: 207–247; Forestry Abstracts 54: 1103–1143

Muetzelfeldt RI, Robertson D, Bundy A and Uschold M (1988) The use of Prolog for improving the rigour and accessibility of ecological modelling. Ecol Modelling 46: 7–64

Sterling L and Shapiro E (1985) The Art of Prolog. The MIT Press, Cambridge, MA, USA

Supit I, Hooijer AA and Van Diepen CA (1994) System Description of the WOFOST 6.0 Crop Simulation Model Implemented in CGMS. Volume 1: Theory and Algorithms. Office for Official Publications of the European Communities, Luxembourg, Luxembourg

Wang YP and Jarvis PG (1990) Description and validation of an array model – MAESTRO. Agric and Forest Meteorol 51: 257–280

Wittman J (1994) The SIMPLEX II simulation system to solve problems in environmental protection. Ecol Modelling 75/76: 563–573

Agroforestry Systems **30**: 235–248, 1995.

Incorporation of indigenous knowledge and perspectives in agroforestry development

Part 1: Review of methods and their application

D. H. WALKER[1], F. L. SINCLAIR[2] and B. THAPA[3]

[1] *Division of Tropical Crops and Pastures, CSIRO, P.O. Aitkenvale, Queensland 4814, Australia;* [2] *School of Agricultural and Forest Sciences, University of Wales, Bangor, Gwynedd, LL57 2UW, UK;* [3] *Pakhribas Agricultural Centre, c/o BTCO, P.O. Box 106, Kathmandu, Nepal*

Key words: knowledge-based systems, participatory rural appraisal, rapid rural appraisal

Abstract. Calls for the effective integration of indigenous knowledge and perspective into agroforestry are increasingly familiar in agroforestry programmes. This is the result of a need to better target research, ethical concerns about participation and power and the recognition that indigenous knowledge is a potentially powerful source of understanding that may often be complementary to scientific knowledge.

Incorporating indigenous knowledge into development may be achieved through farmer participation in research planning and implementation, external survey of local needs as a basis for planning research or the active synthesis of indigenous and scientific knowledge (in addition to collaboration between the scientist and the farmer) in order to capitalise on their potential complementarity.

It is argued that active synthesis has received inadequate attention. In part this is because it demands rigorous analysis of indigenous and scientific knowledge. This is an exacting process and requires effective means of explicitly representing the knowledge concerned. An approach to this task is briefly introduced. A case study description and evaluation can be found in the accompanying paper [Thapa et al., 1995 (this issue)].

Introduction

Calls for more effective integration of indigenous knowledge and perspective into development are increasingly familiar to those working in agroforestry. Although some past research has placed considerable emphasis on local knowledge [see e.g., Richards, 1939], mainstream interest is relatively recent. Brokensha et al. [1980] provided a key impetus by bringing together a set of case studies and coherently arguing the need for enhanced understanding of local knowledge in planning and implementing development activities. Growth in the application of indigenous knowledge to rural research and development over the succeeding decade and a half is evident from the establishment of three international, two regional and nine national networked indigenous knowledge resource centres [CIRAN, 1993]. This ground-swell of activity can be attributed to:

- the demonstrable need to target research to client needs more effectively if the results of the research are to achieve high rates of adoption;

- a rise in the importance accorded to ethical issues relating to participation and power; and
- the recognition that indigenous knowledge, particularly in relation to agroforestry, may be a powerful resource in its own right and complementary to knowledge available from scientific sources.

Targeting research and development activities to client needs

It is widely acknowledged that many of the technological solutions that have been proposed to address problems in rural communities have been compromised by a failure to take into account local knowledge, perspective and practise [Chambers, 1983; Warren and Cashman, 1988; Chambers et al., 1989; Haverkort et al., 1991]. The role of farmers' perspectives in addressing this problem has been recognised by increasing numbers of researchers and development professionals, who point out that effective external interventions can best be achieved '*once we know what they already know, and what else might be most useful to add to their store of knowledge and tools*' [Rocheleau, 1987].

A variety of approaches to better targeting research to client needs have been espoused. The defining feature of all such approaches is some form of interaction with the target communities during project planning and implementation.

Ethical concerns

There is a growing awareness of the extent to which colonisers, administrators and social scientists in the nineteenth and early twentieth centuries established negative attitudes towards indigenous peoples and their knowledge [Slikkerveer, 1989; Warren, 1989]. Negative attitudes to indigenous knowledge persist and Hatch's assertion [1976] that '*even the supposedly enlightened development profession suffers from an entrenched superiority complex*' is still relevant. These negative values are often reinforced by the educated elite of developing countries who are strongly influenced by the current dominance of Western science [Compton, 1989] which is firmly linked to wealth, power and prestige [Chambers, 1983].

Development programmes are usually conceived within the framework of formal science and Western values. Until very recently the values of the rural agricultural practitioner were barely taken into account because societies that were perceived as primitive were assumed to have primitive modes of thought. The failure of many agricultural projects and technology packages imposed by the industrialised world appeared to reinforce these negative attitudes. The slow rate of assimilation of new technology and cropping schemes was attributed to local inertia rather than the failure to take account of the knowledge, experience and needs of local people upon whom they were imposed [Brokensha et al., 1980; Chambers, 1983; Richards, 1985; Rocheleau, 1987; Warren et al., 1989]. From this perspective, research into indigenous knowledge aims to validate the expertise of rural practitioners and to encourage and inform a problem-solving dialogue between them and the scientific community in order to establish a more constructive and equitable approach to development.

Indigenous knowledge as a resource for science

In agroforestry as a science, there remain many more questions than answers. Much scientific investigation of the ecology of agroforestry is intended to produce knowledge relevant to farmers in managing agroforestry practices [Gholz, 1987]. Indeed, recent reviews of research agendas in agroforestry indicate that future activities, particularly where process-based, will be increasingly productive of accurate, predictive understanding of interactions within agroforestry practices relevant to farmers [Anderson and Sinclair, 1993]. Nevertheless, research has yet to address many of the issues pertinent to the improvement of agroforestry practices [Nair, 1990].

Experimental research, although an important means of improving the information upon which farmers make decisions, cannot be relied upon to address all the pertinent issues over the next decade. With this in mind, it is worth considering whether relying on experimental scientific methodology alone is, in fact, the most expedient, effective and efficient way to fill gaps in current understanding about the functioning of agroforestry systems. In particular, it is worth seeking approaches that could complement experimental investigation, leading to a more rapid understanding of the functioning of agroforestry systems. One approach to filling the gap between what it is desirable to know about agroforestry practises in addressing development needs and what knowledge is currently held by research and extension staff, as a result of their professional training and from the scientific literature, is to mobilise and synthesise currently available knowledge from all sources more effectively. It is proposed that one of the richest but least utilised sources of knowledge available to the development professional is the indigenous knowledge of farmers who practice agroforestry.

For whatever combination of these different motivations, reference to indigenous knowledge and local participation is increasingly common in agroforestry-based project proposals. A considerable literature justifies this emphasis by arguing the case for making more effective use of indigenous knowledge in development [Brokensha et al., 1980; Chambers et al., 1989; Warren, 1991a; Warren et al., 1994]. This is supported by work that has characterised and explored indigenous knowledge in a range of contexts – including bibliographic reviews of indigenous knowledge in relation to private tree management [Mathias-Mundy et al., 1992], in sustainable agriculture [Warren, 1991b] and natural resources more generally [Warren, 1991a]. Much of this work remains anecdotal and descriptive but provides a persuasive basis for pursuing a more rigorous investigation of the nature and extent of indigenous knowledge.

Consideration of the literature demonstrates that indigenous knowledge and perspective can impact on agroforestry development at multiple levels and stages. Nevertheless, it is also apparent that making effective practical use of indigenous knowledge and perspective in designing and implementing development activities is not easily done.

In this paper, a range of approaches to the integration of indigenous knowledge and perspective into development activities are reviewed. Some strengths and weaknesses and appropriate contexts for application are indicated. It is argued that the active synthesis of indigenous ecological knowledge with scientific knowledge has a potentially important role to play but has received inadequate attention. An approach to this demanding process is briefly introduced. This approach is described and evaluated in more detail through case study evaluation in an accompanying paper [Thapa et al., 1995 (this issue)].

Approaches to incorporating indigenous knowledge and perspective

Three broad categories of approach to incorporating indigenous knowledge and perspective into development can be distinguished:

- approaches based on the active participation of local experts in development activities in which no understanding, per se, of indigenous knowledge is required to be sought by development professionals;
- interaction with local people to generate descriptions of actual practice and constraints on the system in order to improve development professionals' understanding of the target community's needs; and
- interaction with local people to investigate indigenous explanation of the ecological functioning of the system, in order to facilitate the integration of local and scientific knowledge and produce a more complete and useful resource base for use by professionals or local people in the incremental improvement of land-use systems.

Local participation in research

The integration of target communities into the research process is actively espoused and implemented by growing numbers of development professionals [Richards 1985; Chambers et al., 1989; Amanor 1989; Scoones and Thompson, 1994].

Rocheleau [1987] shows that local experts can often adapt new methods to local conditions better than outsiders, that local control of change often matters more than rapid transformation, and that new technologies are often most successful when built upon existing practice and knowledge. Furthermore, it has been argued that local participation can complement scientific manpower: local people, as competent observers, can extend the range of data channelled to the scientific decision maker [Howes and Chambers, 1980]. Indeed, many instances of active indigenous experimentation and consequent evolution of farming systems have been recorded [Geertz, 1963; Johnson, 1972; Richards, 1985; Bentley, 1994; Stolzenbach, 1994; Winarto, 1994]. With this in mind, effective integration of internal (local) and external ('scientific') efforts may help to avoid wasteful duplication of effort.

To a certain extent, farmers have long been participants in the externally driven research and development process, albeit at a late stage, in that they adapt extension recommendations according to detailed knowledge of interactions between crops, pests and soils under local climatic and edaphic conditions. Richards [1994] documents a long-established process of discovery amongst Mende-speaking communities in Sierra Leone involving screening of new rice germplasm, arising both from seed transport into the area and genetic variation. The farmers used an explicit methodology capable of yielding objective knowledge concerning crop-environment interactions both through intentional experimentation, such as observation plots of new rice varieties and comparative trials; and from drawing inferences from opportunistic observations. Richards [1994] asserts that the farmers are generating knowledge that is '*in conformity with general scientific principles, but which, because it embodies place-specific experience, allows better assessments of risk factors in production decisions*'. Recently, agricultural development initiatives have explicitly tried to incorporate farmer experimentation within the research and development cycle. Millar [1994] for example, describes how the results of farmer experimentation with cereals and tubers in northern Ghana (driven by curiosity, a problem-solving requirement, a need for adaptation and peer pressure) was incorporated into development in NGO projects through dialogue between extension staff and farmers. Bentley [1994] reports how scientists were able to stimulate farmers to experiment with biological pest control in Honduras by explaining ecological relationships, such as the existence and behaviour of parasitoids and entomopathogens, that were previously unknown to farmers because they are not easily observed. The involvement of farmer evaluation in on-farm research is now known to lead to the design of more appropriate technical interventions as, for example, in Zambia, where adaptive research has been institutionalised at a national level for over a decade [Drinkwater, 1994].

Effective participation evidently provides not only a means of integrating indigenous knowledge and perspective into the development process but also an environment that encourages better understanding through dialogue. However, without an independent record of the indigenous knowledge system, each new interaction with a local community effectively starts with a similar level of ignorance on the part of the professionals, and only those professionals who actively participate may learn what people know and be forced to re-evaluate their scientific knowledge accordingly. Much of the process of incorporating indigenous knowledge remains implicit, occurring because the balance of power shifts from imposing the perspective of the researcher to enabling the farmer to realise his or her objectives [Chambers, 1992]. Thus, while participation may empower farmers to pursue their own research, and make the activity of individual research and extension workers more relevant to farmer circumstances, it does not necessarily imply, nor result in, a sufficiently detailed and explicit evaluation of indigenous knowledge for this knowledge itself to be used as a basis for planning research and extension strategies.

Fuller understanding of target communities

Traditionally, communication in the development process has been one-way, from the researcher to the farmer. While this approach has had spectacular success [see e.g., Khush, 1993], it has frequently resulted in very limited adoption. Success may be more likely in circumstances where a relatively simple technology or new input can be fitted into relatively simple production systems in relatively homogeneous ecological zones to produce a high return. In the context of development programmes, these circumstances are the exception rather than the rule. Furthermore, it has been argued that while a one-way model of extension may be suitable for the environment in which it has evolved, the US and Europe, the one-way flow of information is less appropriate where there is a more serious element of cross-cultural translation between researcher and target community [Meehan, 1980]. As a result, traditional extension has been increasingly recognised as being of limited utility within development programmes. This has led to developments in approaches to farming systems research [Byerlee et al., 1982; Hildebrand, 1990; Biggs, 1985; Simmonds, 1985] and, separately [Ison, 1993], to the development of Rapid Rural Appraisal methodologies [Carruthers and Chambers, 1987] and specifically in agroforestry to the Diagnosis and Design methodology [Raintree, 1987].

Approaches to Rapid Rural Appraisal (RRA) are essentially qualitative survey methodologies and are primarily aimed at better understanding the needs of farming communities and the constraints under which they operate. As with approaches described in the previous section, these techniques are concerned with participation by the target communities in development activities. However, participation is essentially passive (i.e., responding to requests for information) rather than active. The emergence of Participatory Rural Appraisal (PRA) [Mascarenhas et al., 1991] has gone some way towards blurring this distinction.

The development and adoption of RRA and similar methodologies over the last fifteen years has had a profound and positive impact on the development process in many instances. However, resulting experience casts doubt on the prevailing view that the primary requirement in agroforestry is for adaptive research, aimed at the development of technology packages that are then extended to farmers [Beer et al., 1990; Van den Hoek and Bekkering, 1990]. The extension of prefabricated technology packages has certainly proved less effective in agroforestry development than incremental improvements that build on local practice [Bunderson et al., 1990]. The need to synthesise research outputs into technological packages as in the design phase in the Diagnosis and Design methodology [Raintree, 1990], has not been justified by field experience. In fact, there is evidence that farmers who do not adopt an integrated package may nevertheless be interested in elements of that package [Buck, 1990]. A need to support farmers in decision making about the incorporation of trees within their farming systems based on the results of more fundamental ecological research by the scientific community has been identified as a result [Anderson et al., 1993].

This recognition of the need to provide support to the farmer in the light of the farmer's role as the decision maker, demands means of assessing the ecological knowledge that farmers use in their decision-making processes, and the knowledge that farmers hold about how to make decisions on the basis of that ecological knowledge. Simply describing actual practice and the constraints contributing to actual practice is inadequate: an understanding of the knowledge used to generate that practice is also necessary.

Analysis of indigenous knowledge and synthesis with scientific knowledge

Given that farmers have much more intimate experience of their production practices than external professionals, it seems intuitive that they have developed an in-depth understanding about the ecology of those practices. There is clear evidence to support this suggestion [Conklin, 1957; Barker et al., 1977; Brokensha and Riley, 1980; Posey, 1985; Brosius et al., 1986; Chandler, 1991; Barrow, 1992].

In principle, the knowledge held by farmers can provide a resource for science irrespective of the participation of the farmers in the use of that knowledge. In other words, the ecological understanding developed by one farming community may be relevant to addressing the constraints encountered by other communities (with all the attendant issues of intellectual property rights). As a result, investigating local knowledge may be a powerful and efficient means of rapidly filling gaps in scientific understanding about agroforestry. So, amalgamating specific local knowledge and general scientific knowledge may be more powerful in designing appropriate land use practices than the use of either alone. Indeed, it has been claimed that agroforestry is in a unique position to learn from traditional knowledge and practice given the long history of traditional agroforestry practice as compared to a relatively recent scientific interest [Rocheleau, 1987]. The investigation of local knowledge is not reliant on that knowledge being 'accurate' [Mathias-Mundy et al., 1992]. Understanding local knowledge systems can expose inaccurate or technically outdated components and allow the identification of complementary external knowledge [Rocheleau, 1987], and incomplete areas of understanding [Bentley, 1994], allowing extension (where appropriate) to be targeted accordingly [Gladwin, 1980; Bentley, 1994].

It is argued that, by contrast with the two categories of approach already discussed, effective use of indigenous ecological knowledge demands rigorous analysis of indigenous explanation of system function. Evaluations of the current state of indigenous and scientific knowledge that rely on some implicit synthesis of the two knowledge systems, through extensive participant observation for example, may be effective in some instances but are limited in that the process and, therefore, the validity of analysis and synthesis cannot be evaluated. Explicit analysis provides an evaluable process. Furthermore, the process of explicit representation and analysis of knowledge helps to highlight assumptions and incongruities that are otherwise easily missed. Rigorous analysis essentially comprises an evaluation of what is currently known and its implications and an evaluation of gaps in current knowledge. Evaluation of what is known might focus on coherence and consistency. Gaps in indigenous knowledge might be defined in terms of information needs for 'optimised' management or by comparison with scientific knowledge. Rigorous analysis, in turn, depends fundamentally on means of getting farmers to articulate their knowledge and means of recording that knowledge in a form that is amenable to rigorous analysis and subsequent synthesis with scientific knowledge. In effect, a comprehensive and explicit representation or model of indigenous knowledge is required. Once this representation has been created, a comparative analysis in relation to a similar representation of scientific knowledge can be undertaken. This process leads to the possibility of a rigorous and effective synthesis of that scientific and indigenous knowledge.

In summary, the process of explicit representation, analysis and synthesis of indigenous ecological knowledge and scientific knowledge is methodologically demanding but can have a profound impact on development activities – as illustrated in the accompanying paper [Thapa et al., 1995 (this issue)]. It is argued that, in contrast to local participation in research and the use of survey methods to better understand target communities, these opportunities have not received adequate attention.

In the next section, one approach to the explicit representation and analysis of indigenous ecological knowledge and its synthesis with scientific knowledge is briefly introduced.

Formal representation and analysis of indigenous ecological knowledge

Previous work on indigenous knowledge systems has been undertaken largely by ethnographers. Although their studies are often rich in detail, the descriptive nature of this work renders it difficult to analyse systematically in terms of ecological processes. Simply recording indigenous knowledge collected in such research remains a challenging task [McKiernan, 1989].

One approach to addressing this problem is through explicit representation of knowledge as it is collected. This involves storing the knowledge in a form that permits repeated and variable analysis and interpretation of the knowledge for decision-making purposes. This can be achieved through the application of knowledge-based systems techniques for representing knowledge on computers [Walker et al., in press]. This allows the processing power of desk-top computers to be applied to the representation, analysis and use of large and complex sets of knowledge, much of which would be impractical without computing assistance to handle tedious and repetitive tasks.

The application of knowledge-based systems techniques also allows the application of automated reasoning enabled by the explicit and syntactically precise representation of knowledge and concomitant development of appropriate inferencing mechanisms. These may be used in: managing complex sets of knowledge; evaluating complex sets of knowledge; synthesising knowledge from different sources; producing clear justifications for research plans; and producing extension materials [Kendon et al., in press]. In practice, this type of approach builds on many of the features of participatory and RRA approaches but incorporates a more formal and rigorous procedure than previously possible, involving the analysis of what is and what is not known and the specification, in functional terms, of what knowledge new research aims to create.

The Agroforestry Knowledge Toolkit (AKT) methodology and software [Walker et al., 1994], is based on this knowledge-based systems approach to agroforestry research and extension. The methodology provides a means of iteratively creating knowledge bases about the ecology of agroforestry from multiple sources and of applying formal reasoning tools and less formal exploration of the content of the knowledge base (through the automated creation of, for example, systems diagrams) to the analysis of the content of the knowledge base. The knowledge base provides a central point of reference throughout the research. Knowledge may be formally represented on the computer from the first interaction with the community of interest. Formal and informal analysis of this prototype knowledge base then provides a basis for further knowledge elicitation. In this way a tight (typically daily) cycle of knowledge elicitation, knowledge representation and analysis helps target interaction with informants to create a 'complete', coherent and accurate representation of the knowledge of a target community. This process of knowledge base modification and further knowledge elicitation will continue throughout the useful life of the knowledge base.

Although the AKT toolkit and methodology have been designed primarily for the representation of knowledge elicited through semi-structured interview, a multi-method approach to knowledge elicitation can be used. Methods successfully applied include questionnaire surveys, slip sorting, interactive diagramming and sketch mapping. While the knowledge base will undergo continuous enhancement throughout its useful life, the process of knowledge base creation has a clear life cycle, different approaches to knowledge elicitation being appropriate at different stages. So, in the early stages of establishing the boundaries of a new knowledge base, extensive questionnaire survey across a community is appropriate. During the bulk of knowledge base creation, more intensive and interactive dialogue with targeted key informants is necessary. Prior to undertaking analysis of the content of the knowledge base in order to facilitate the planning of research or extension activities, questionnaire survey will be needed to establish how representative the knowledge obtained from key informants is of the knowledge in the broader community.

The creation of a knowledge base representing the indigenous ecological knowledge of a community of interest can have a significant utility in better understanding that community and thereby, making research and extension activities more relevant to that community. Moreover, comparable knowledge bases can be created representing scientific knowledge. The knowledge bases from different local and scientific communities can then be synthesised in an attempt to create a truly 'encyclopedic' knowledge base about focussed aspects of the ecology of particular agroforestry practices. While this process is exceedingly demanding, both practically and theoretically, it can be extremely revealing. In the accompanying paper [Thapa et al., 1995 (this issue)], an example of the impact of this process on research and extension activities is described.

While the AKT software and methodology has only begun to address issues relating to the analysis of indigenous knowledge and its synthesis with scientific knowledge, the experience such as that described in Thapa et al. [1995 (this issue)] provides a sound basis for arguing that this type of approach should receive more attention within agroforestry development.

Conclusions

In summary, it has been argued in this paper that:

- understanding indigenous knowledge is, in conjunction with understanding practice in the context of social, economic, cultural and environmental variables, an important means of understanding needs and constraints facing farmers;
- understanding indigenous knowledge is an important activity in improving the communication between target communities and development professionals in the development process;
- facilitating active partnerships between development professionals and local people in research can help to result in effective and well-targeted research;
- indigenous knowledge provides a useful source of understanding of the functioning of agroforestry practices that may be complementary to scientific knowledge.

In particular, the explicit representation of knowledge may have an important role to play in enabling:

- more thorough understanding of target farming communities through evaluation of the knowledge that they hold;
- more effective identification and prioritisation of research objectives, leading to more appropriate research;

- improved dissemination of knowledge to target farming communities; and
- improved monitoring and evaluation of research activities.

References

Amanor KS (1989) 230 Abstracts on farmer participatory research. Network paper, Agricultural Administration Research and Extension Network no. 5. Agricultural Administration Unit, Overseas Development Institute, London, UK

Anderson LS and Sinclair FL (1993) Ecological interactions in agroforestry systems. Agroforestry Abstracts 6(4): 207–247

Anderson LS, Muetzelfeldt RI and Sinclair FL (1993) An integrated research strategy for modelling and experimentation in agroforestry. Commonwealth Forestry Review 72(3): 166–174

Barker D, Oguntoyinbo JS and Richards P (1977) The utility of the Nigerian peasant farmer's knowledge in monitoring of agricultural resources. Chelsea College General Report Series 4. Monitoring and Assessment Centre, London

Barrow ECG (1992) Building on local knowledge: the challenge of agroforestry for pastoral areas. Agroforestry Today 3(4): 4–7

Beer J, Borel R and Bonnemann A (1990) On-farm agroforestry research planning in Costa Rica. In: Budd WW, Duchhard I, Hardesty LH and Steiner F (eds) Planning for Agroforestry. Developments in Landscape Management and Urban Planning, pp 58–78. Elsevier Science Publishing Co Inc, Amsterdam, The Netherlands

Bentley JW (1994) Stimulating farmer experiments in non-chemical pest control in Central America. In: Scoones I and Thompson J (eds) Beyond Farmer First, pp 147–150. Intermediate Technology Publications, London, UK

Biggs SD (1985) A farming systems approach: Some unanswered questions. Agricultural Administration 18: 1–12

Brokensha D and Riley BW (1980) Mbeere knowledge of their vegetation and its relevance for development: A case study from Kenya. In: Brokensha D, Warren DM and Werner O (eds) Indigenous Knowledge Systems and Development, pp 111–127. University Press of America, Lanham, MD, USA

Brokensha D, Warren DM and Werner O (eds) (1980) Indigenous Knowledge Systems and Development. University Press of America, MD, USA

Brosius JP, Lovelace GW and Marten GG (1986) Ethnoecology: An approach to understanding traditional agricultural knowledge. In: Marten GG (ed) Traditional Agriculture in South-East Asia: A Human Ecology Perspective, pp 187–197. Westview Press, Boulder, CO, USA

Buck L (1990) Planning agroforestry extension projects, the Care International approach in Kenya. In: Budd WW, Duchhard I, Hardesty LH and Steiner F (eds) Planning for Agroforestry. Developments in Landscape Management and Urban Planning, pp 101–131. Elsevier Science Publishing Co Inc, Amsterdam, The Netherlands

Bunderson TW, El Wakeel A, Saad Z and Hahim H (1990) Agroforestry practices and potentials in western Sudan. In: Budd WW, Duchhard I, Hardesty LH and Steiner F (eds) Planning for Agroforestry. Developments in Landscape Management and Urban Planning, pp 227–246. Elsevier Science Publishing Co Inc, Amsterdam, The Netherlands

Byerlee D, Harrington L and Winklemann DL (1982) Farming systems research: Issues in research strategy and technology design. American Journal of Agricultural Economics 62: 877–904

Carruthers I and Chambers R (1987) Rapid rural appraisal for development. Agricultural Administration 8: 433–450

Chambers R (1992) Rural appraisal: rapid, relaxed and participatory. IDS discussion paper 311. Institute of Development Studies, University of Sussex, Sussex, UK

Chambers R (1983) Rural Development: Putting the Last First. Longmans, London, UK

Chambers R, Pacey A and Thrupp LA (eds) (1989) Farmer First: Farmer Innovation and Agricultural Research. Intermediate Technology Publications, London, UK

Chandler PM (1991) The indigenous knowledge of ecological processes among peasants in the People's Republic of China. Agriculture and Human Values 8(1–2): 59–66

CIRAN (1993) Indigenous knowledge resource centres. Indigenous Knowledge and Development Monitor 1(2): 48

Compton JL (1989) Strategies and methods for the access, integration and utilisation of indigenous knowledge in agriculture and rural development. In: Warren DM, Slikkerveer LJ and Titilola SO (eds) Indigenous Knowledge Systems: Implications for Agriculture and International Development. Studies in Technology and Social Change, no. 11. Iowa State University, Iowa City, Iowa, USA

Conklin HC (1957) Hanunoo agriculture, a report on an integral system of shifting cultivation in the Philippines. Forestry development paper no. 12. FAO, Rome, Italy

Drinkwater M (1994) Knowledge, consciousness and prejudice: Adaptive agricultural research in Zambia. In: Scoones I and Thompson J (eds) Beyond Farmer First, pp 32–41. Intermediate Technology Publications, London, UK

Geertz C (1963) Agricultural Involution: The Process of Ecological Change in Indonesia. University of California Press, Berkeley, CA, USA

Gholz HL (ed) (1987) Agroforestry: Realities, Possibilities and Potentials. Martinus Nijhoff Publishers, Dordrecht, The Netherlands

Gladwin C. (1980) Cognitive strategies and adoption decisions: A case study of non-adoption of an agronomic recommendation (Mexico). In: Brokensha D, Warren DM and Werner O (eds) Indigenous Knowledge Systems and Development, pp 9–28. University Press of America, Lanham, MD, USA

Hatch JK (1976) The Corn Farmers of Motupe: A Study of Traditional Farming Practices in Northern Coastal Peru. Monograph, Land Tenure Centre, University of Wisconsin, Madison, WI, USA

Haverkort B, Gibbon D and Bayer W (1991) Research for the Future: An Overview of Exciting Research in Sub-Saharan Africa for the Development of Low-External Input and Sustainable Agriculture. SAREC-ILEIA, Leusden, The Netherlands

Hildebrand PE (1990) Farming systems research – extension. In: Jones JGW and Street PR (eds) Systems Theory Applied to Agriculture and the Food Chain, pp 131–143. Elsevier Science Publishing Co Inc, Amsterdam, The Netherlands

Howes M and Chambers R (1980) Indigenous technical knowledge: Analysis, implications and issues. In: Brokensha D, Warren DM and Werner O (eds) Indigenous Knowledge Systems and Development, pp 323–334. University Press of America, Lanham, MD, USA

Ison RJ (1993) Soft systems: A non-computer view of decision support. In: Stuth JW and Lyons BG (eds) Decision Support Systems for the Management of Grazing Lands, pp 83–118. Man and the Biosphere Series, volume 11, UNESCO, Paris, France

Johnson AW (1972) Individuality and experimentation in traditional agriculture. Human Ecology 1: 149–159

Kendon G, Walker DH, Robertson D, Haggith M, Sinclair FL and Muetzelfeldt RI (in press) Supporting customised reasoning in the agroforestry domain. New Review of Applied Expert Systems

Khush GS (1993) Breeding rice for sustainable agricultural systems. In: Buxton DR, Shibles R, Forsberg RA, Blad BL, Asay KH, Paulsen GM and Wilson RF (eds) International Crop Science 1. Based on the International Crop Science Congress sponsored by Iowa State University and the Crop Science Society of America, Ames, IA, 14–22 July. Crop Science Society of America, Inc, Madison, WI, USA

Macarenhas J, Shah P, Joseph S, Jayakaran R, Devavaram J, Ramachandran V, Fernandez A, Chambers R and Pretty J (eds) (1991) Participatory Rural Appraisal: Proceedings of the February 1991 Bangalore PRA Trainers Workshop. RRA notes 13. Sustainable Agriculture Programme, IIED, London, UK

Mathias-Mundy E, Muchena O, McKiernan G and Mundy P (1992) Indigenous Technical Knowledge of Private Tree Management. Bibliographies in Technology and Social Change Series no. 7. Technology and Social Change Program in collaboration with CIKARD and LEAD. Center for Indigenous Knowledge and Rural Development (CIKARD), Iowa State University Research Foundation, Ames, IA, USA

McKiernan G (1989) The CIKARD international documentation unit and library of indigenous knowledge for agriculture and rural development. In: Warren DM, Slikkerveer LJ, Titilola SO (eds) Indigenous Knowledge Systems: Implications for Agriculture and International Development. Studies in Technology and Social Change, no. 11. Iowa State University, Ames, Iowa, USA

Meehan P (1980) Science, ethnoscience, and agricultural knowledge utilisation. In: Brokensha D, Warren DM and Werner O (eds) Indigenous Knowledge Systems and Development, pp 377–386. University Press of America, Lanham, MD, USA

Millar D (1994) Experimenting farmers in Northern Ghana. In: Scoones I and Thompson J (eds) Beyond Farmer First, pp 160–165. Intermediate Technology Publications, London, UK

Nair PKR (1990) The prospects for agroforestry in the tropics. World Bank technical paper 131.

Posey DA (1985) Indigenous management of tropical forest ecosystems: The case of the Kayapo Indians of the Brazilian Amazon. Agroforestry Systems 3: 139–158

Raintree JB (1987) Design & Development Users Manual: An Introduction to Agroforestry Diagnosis and Design. ICRAF, Nairobi, Kenya

Raintree JB (1990) Theory and practice of agroforestry development and design. In: MacDicken KG and Vergara NT (eds) Agroforestry Classification and Management, pp 58–97. John Wiley and Sons, New York, NY, USA

Richards AJ (1939) Land, Labour and Diet in Northern Rhodesia. Oxford University Press for the International Institute of African Languages and Culture, Oxford, UK

Richards P (1985) Indigenous Agricultural Revolution. Hutchinson & Co, London, UK

Richards P (1994) Local knowledge formation and validation: The case of rice production in Sierra Leone. In: Scoones I and Thompson J (eds) Beyond Farmer First, pp 165–170. Intermediate Technology Publications, London, UK

Rocheleau DE (1987) The user perspective and the agroforestry research and action agenda. In: Gholz HL (ed) Agroforestry: Realities, Possibilities and Potentials, pp 59–87. Martinus Nijhoff Publishers, Dordrecht, The Netherlands

Scoones I and Thompson J (1994) Knowledge, power and agriculture – towards a theoretical understanding. In: Scoones I and Thompson J (eds) Beyond Farmer First, pp 16–32. Intermediate Technology Publications, London, UK

Simmonds NW (1985) Farming systems research: A review. World Bank technical paper no. 43

Slikkerveer LJ (1989) Changing values and attitudes of social and natural scientists towards indigenous peoples and their knowledge systems. In: Warren DM, Slikkerveer LJ and Titilola SO (eds) Indigenous Knowledge Systems: Implications for Agriculture and International Development. Studies in Technology and Social Change, no. 11. Iowa State University, Ames, IA, USA

Stolzenbach A (1994) Learning by improvization: Farmers' experimentation in Mali. In: Scoones I and Thompson J (eds) Beyond Farmer First, pp 1155–1160. Intermediate Technology Publications, London, UK

Thapa B, Sinclair FL and Walker DH (1995) Incorporation of indigenous knowledge and perspectives in agroforestry development. Part 2: Case-study on the impact of explicit representation of farmers' knowledge. Agroforestry Systems (this issue)

Van den Hoek A and Bekkering T (1990) Planning of agroforestry programmes in Java, Indonesia. In: Budd WW, Duchhard I, Hardesty LH and Steiner F (eds) Planning for Agroforestry. Developments in Landscape Management and Urban Planning, pp 79–100. Elsevier Science Publishing Co Inc, Amsterdam, The Netherlands

Walker DH, Sinclair FL and Kendon G (in press) A knowledge-based systems approach to agroforestry research and extension. AI Applications

Walker DH, Sinclair FL, Kendon G, Robertson D, Muetzelfeldt RI, Haggith M and Turner GS (1994) Agroforestry Knowledge Toolkit: Methodological Guidelines, Computer Software and Manual for AKT1 and AKT2, Supporting the Use of a Knowledge-Based Systems Approach in Agroforestry Research and Extension. School of Agricultural and Forest Sciences, University of Wales, Bangor, UK, 138 pp

Warren DM (1989) The impact of nineteenth century social science in establishing negative values and attitudes to indigenous knowledge systems. In: Warren DM, Slikkerveer LJ and Titilola SO (eds) Indigenous Knowledge Systems: Implications for Agriculture and International Development. Studies in Technology and Social Change. no. 11. Iowa State University, Ames, IA, USA

Warren DM (ed) (1991a) Indigenous agricultural knowledge systems and development. Agriculture and Human Values 8(1–2), 179 pp

Warren DM (1991b) Using Indigenous Knowledge in Agricultural Development. Discussion paper no. 127. The World Bank, Washington DC, USA

Warren DM and Cashman K (1988) Indigenous Knowledge for Sustainable Agriculture and Rural Development. Gatekeeper Series no. SA10, Sustainable Agriculture Programme. International Institute for Environment and Development, London, UK

Warren DM, Slikkerveer LJ and Titilola SO (eds) (1989) Indigenous Knowledge Systems: Implications for Agriculture and International Development. Studies in Technology and Social Change, no. 11. Iowa State University, Ames, IA, USA

Warren DM, Brokensha D and Slikerver LJ (eds) (1994) Indigenous Knowledge Systems: The Cultural Dimension of Development. Intermediate Technology Publications, London, UK, 640 pp

Winarto Y (1994) Encouraging knowledge exchange: Integrated pest management in Indonesia. In: Scoones I and Thompson J (eds) Beyond Farmer First, pp 150–155. Intermediate Technology Publications, London, UK

Agroforestry Systems **30**: 249–261, 1995.

Incorporation of indigenous knowledge and perspectives in agroforestry development

Part 2: Case-study on the impact of explicit representation of farmers' knowledge

B. THAPA[1], F. L. SINCLAIR[2] and D. H. WALKER[3]
[1] *Pakhribas Agricultural Centre, c/o BTCO, P.O. Box 106, Kathmandu, Nepal;* [2] *School of Agricultural and Forest Sciences, University of Wales, Bangor, Gwynedd LL57 2UW, UK;*
[3] *Division of Tropical Crops and Pastures, CSIRO, P.O. Aitkenvale, Queensland 4814, Australia*

Key words: Nepal, tree fodder, knowledge-based systems

Abstract. An explicit representation of farmers' knowledge about on-farm tree fodder resources was developed using knowledge-based systems software and methodology at a front-line agricultural research centre in the eastern mid-hills of Nepal. Sophisticated knowledge of tree-crop interactions and the feeding value of various tree fodders was revealed. Farmers' ecological knowledge was more developed in some areas than others and was largely complementary to that of agricultural researchers in Nepal and the scientific literature more generally. Evaluation of the combined knowledge of farmers and researchers related to interdisciplinary land use problems by the creation of explicit, encyclopaedic knowledge bases, offers the possibility of targeting research more precisely to gaps in knowledge constraining the productivity and sustainability of the farming system. While similar insight into the depth of farmers knowledge may have been gained by researchers adopting a less formal approach to knowledge acquisition, without an explicit, durable and manipulable record, subsequent analyses and continued use of the knowledge in the research and development process might not be achieved.

Introduction

In an accompanying paper [Walker et al., 1995b – this issue] it was argued that explicit representation of the ecological knowledge held by farmers could form an appropriate basis for planning, monitoring and evaluating agroforestry research and extension; and an overview was given of the AKT knowledge-based systems software and methodology developed to achieve this. The present paper documents the impact of the trial application of a knowledge-based systems approach to interdisciplinary research on tree fodder at a front-line agricultural research centre in Nepal.

Institutional context

Pakhribas Agricultural Centre (PAC) is located 25 km north-west of Dhankuta (87° 30′ E, 27° 19′ N), in the eastern hills of Nepal. Since 1982 it has held responsibility for agricultural research and development activities servicing the four districts of the Koshi Hills Area Rural Development Programme

(KHARDEP) now the Koshi Hills Development Programme (KHDP). After integration into the national agricultural development network of regional research farms and stations in 1988, PAC's command area was increased to a total of 11 districts in the Mechi, Koshi and Sagarmatha zones. The command area of PAC thus comprises areas characterised by strongly dissected relief ranging from 300 m to over 4000 m, a climate that varies from sub-tropical to alpine, and a heterogeneous farming community in terms of ethnic and socio-economic groups and farming activities [Gibbon and Shultz, 1989].

Problem context

Livestock play a critical role in crop cultivation and sustaining soil fertility in the hill farming systems of Nepal and are fed during the dry season on tree fodder. Land fragmentation, with consequent decrease in the size of farm holdings combined with decreasing access to dwindling common property forest resources that have been over-exploited, is causing farmers to incorporate an increasing number of fodder trees on their farmland [Carter and Gilmour, 1989]. The productivity of fodder trees is critical for maintaining livestock health and numbers and, therefore, the productivity and sustainability of people's livelihood system, but trees planted on crop terrace risers may also influence soil erosion through leaf drip effects and compete with staple food crops for light, water and nutrients. Farmers need, therefore, to achieve a balance between generating enough high quality fodder at appropriate times in the season and minimising the competitive effects of trees on staple food crops.

Research context

Research on tree fodder at PAC thus impinges on the work of four separately administered sections: the Forestry Section, concerned with the establishment and management of fodder trees; the Livestock Section, concerned with the improvement of the fodder supply to animals (the amount, feeding value and timing of tree fodder given to animals); the Veterinary Investigation and Analytical Services Section, responsible for fodder evaluation and the Agronomy Section. While there was no combined tree fodder research strategy in operation at PAC when the present work was initiated, information exchange and research planning was facilitated by a fodder working group comprising the four sections. The trial application of using a knowledge-based systems approach to construct an explicit representation of farmers' knowledge was initiated because it was clear from analyses of information flows [Sinclair et al., 1993] that there were opportunities for improving the effectiveness of fodder research at PAC by:

- making it more relevant to the priorities and existing knowledge of client farmers and

- making more effective use of previous research findings when new research was being planned.

Methods

The Agroforestry Knowledge Toolkits (AKT1 and AKT2) and associated methodology [Walker et al., 1994] support the creation and use of coherent and accessible knowledge bases, synthesising information elicited from farmers and development professionals [Walker et al., 1995a; Kendon et al., 1995]. This methodology as outlined in the accompanying paper [Walker et al., 1995b – this issue] addresses the need to record information:

- in suitably-sized chunks that represent useful knowledge without impairing the flexibility of combining the chunks variously to suit different purposes; and,
- in a form which allows knowledge to be expressed naturally but sufficiently unambiguously to be interpretable by others.

The AKT knowledge-based systems approach was applied at PAC to an investigation of farmers' knowledge about on-farm tree fodder resources. The study community was defined by the administrative boundary of Solma Village Development Committee comprising 3500 people in a collection of hamlets spread over an altitudinal range of 500 m to 2000 m. Sixty farmers selected randomly from the heads of households in the study community were interviewed using a questionnaire. A detailed tree inventory was undertaken on 30 of these farms. This was followed by the iterative creation and refinement of a knowledge base through semi-structured interviews with 40 purposively selected key informants, each interviewed an average of four times, for about one hour on each occasion. The purposive sample was stratified in terms of gender, altitude and ethnicity. Knowledge base creation took approximately two hours for every hour of interview material. It was undertaken on an iterative basis. Knowledge was entered into the knowledge base on the computer immediately after the first interview and the growing knowledge base was both modified and added to after each subsequent interview. Assessment of the content of the growing knowledge base itself provided the direction for further interviewing. A questionnaire survey was then used to the test the representativeness of the content of the knowledge base against the knowledge held by 50 randomly selected farmers from the community that had not previously been interviewed.

There was nothing particularly novel about the methods of knowledge elicitation, which followed a multi-method approach (see, for example, Rusten and Gold [1991] with strong emphasis on semi-structured interviewing guided by participant observation. Their use was, however, honed towards producing a complete and coherent record of local ecological knowledge by the iterative evaluation on computer of what had been collected, as it was collected. This was facilitated by:

- disaggregating knowledge into sets of unambiguous unitary statements using a restricted syntax (Table 1),
- considering the context of these unitary statements in terms of their conditionality and the local definition and classification of terms, and
- evaluating related sets of statements by constructing hierarchically structured node-and-link diagrams (Fig. 1).

Details of these methods of knowledge acquisition and how they are supported by use of the AKT software are published elsewhere [Walker et al., 1994; Walker et al., 1995a]. Statements in the knowledge base were tagged with their source so that people contributing knowledge could be credited with the knowledge they had contributed.

Results

The study demonstrated that farmers in the study area had knowledge of over 90 different tree species that were deliberately managed on farmland on the basis of sophisticated knowledge of tree-crop interactions and the amount and feeding value of fodder required within the farming system. Details of the knowledge that was acquired are reported elsewhere [Thapa et al., 1994; Thapa, 1994], the present focus is on the impact that the process of knowledge acquisition has had on research on tree fodder resources at PAC.

Systematic research on tree fodder is in its infancy in Nepal. This is partly because the basic information required to formulate research plans is scarce, so that a clear research strategy is lacking. The creation of a knowledge base, abstracting the knowledge available from a farming community, has made a significant contribution towards rectifying this situation. In the first place, it was evident that farmers were deliberately managing abundant natural regeneration of trees on crop terrace risers, obviating the need for the provision of nursery grown seedlings and schemes to motivate farmers to plant trees, which have dominated development activity in the forestry sector.

Tree-crop interactions

Investigation of farmers' knowledge about tree-crop interactions revealed that they understood that shade and splash erosion caused by leaf droplets (a process locally referred to as '*tapkan*') reduced crop yield, and that tree attributes such as leaf size, leaf texture, crown density, crown size, tree height and leaf inclination angle influenced shade and/or leaf drip effects. Because farmers viewed these as primary issues for consideration in managing aboveground interactions affecting crop yield, tree crown architecture was identified as an important area of consideration in planning tree fodder research.

Table 1. Some illustrative examples of natural language statements and their formal equivalents. The final knowledge base on tree fodder created through interaction with Solma farmers contained about 700 statements. The natural language statements are concise, unitary items of knowledge elicited from informants and the formal statements are the same knowledge represented using a restricted syntax of a formal grammar, fully specified in Walker et al. [1995a]. The formal representation both forces an unambiguous interpretation of the meaning of the statement and enables automated reasoning with the statement on computer. The formal statements, which may be conditional, include four fundamental elements: i) objects, which are physical items e.g. trees; ii) processes (or events), that are changes or fluxes, e.g. germination or soil erosion; iii) attributes, which are properties of objects or processes e.g. tree height or the rate of soil erosion; and iv) values, which are the measurable state of an attribute e.g. 5 m or 5 t ha^{-1} a^{-1}. The elements may be combined in four ways that involve i) associating attributes and values to objects or processes; ii) representing causal linkages; iii) representing other, user-specified linkages; and iv) making comparisons. A part construct is available to identify component parts of an object, and causation may be two-way, in which case an increase in *x* causing an increase in *y* is automatically equivalent to a decrease in *x* causing a decrease *y*; or one-way, in which case this equivalence does not necessarily hold.

Natural language statement	Formal statement
A decrease in crop stem strength causes an increase in the amount of crop lodging	att_value(part(crop, stem),strength, decrease) causes2way att_value(process(crop, lodging), amount, increase)
Utis [*Alnus nepalensis*] is a tall tree	att_value(Utis, height, tall).
Lopping causes a decrease in tree height	process(lopping) causes2way att_value(tree, height, decrease).
Trees cause an increase in the amount of shading if they are planted on crop terraces	Tree causes2way att_value(process(shading),amount, increase)if att_value(process(tree,planting),location, crop_terrace).
An increase in leaf maturity causes a decrease in fodder palatability	att_value(leaf,maturity,increase) causes1way att_value(fodder,palatability,decrease).
The growth rate of regrowth shoots after lopping is slow if the atmospheric temperature is low	att_value(process(regrowth_shoot,lopping), growth_rate, slow) if att_value(system, temperature, low).
Rato siris [*Albizia julibrissin*] fodder is toxic to animals	link(toxic_to, rato_siris_fodder, animal).
Badahar [*Artocarpus lakoocha*] fodder is more palatable than *gogun* [*Saurauia nepaulensis*] fodder	comparison(palatability, badahar_fodder, greater_than, gogun_fodder).
Shading influences crops	link(influence,shading,crop).
Nebharo [*Ficus roxburghii*] has big leaves	att_value(part(nebharo, leaf),size, big).

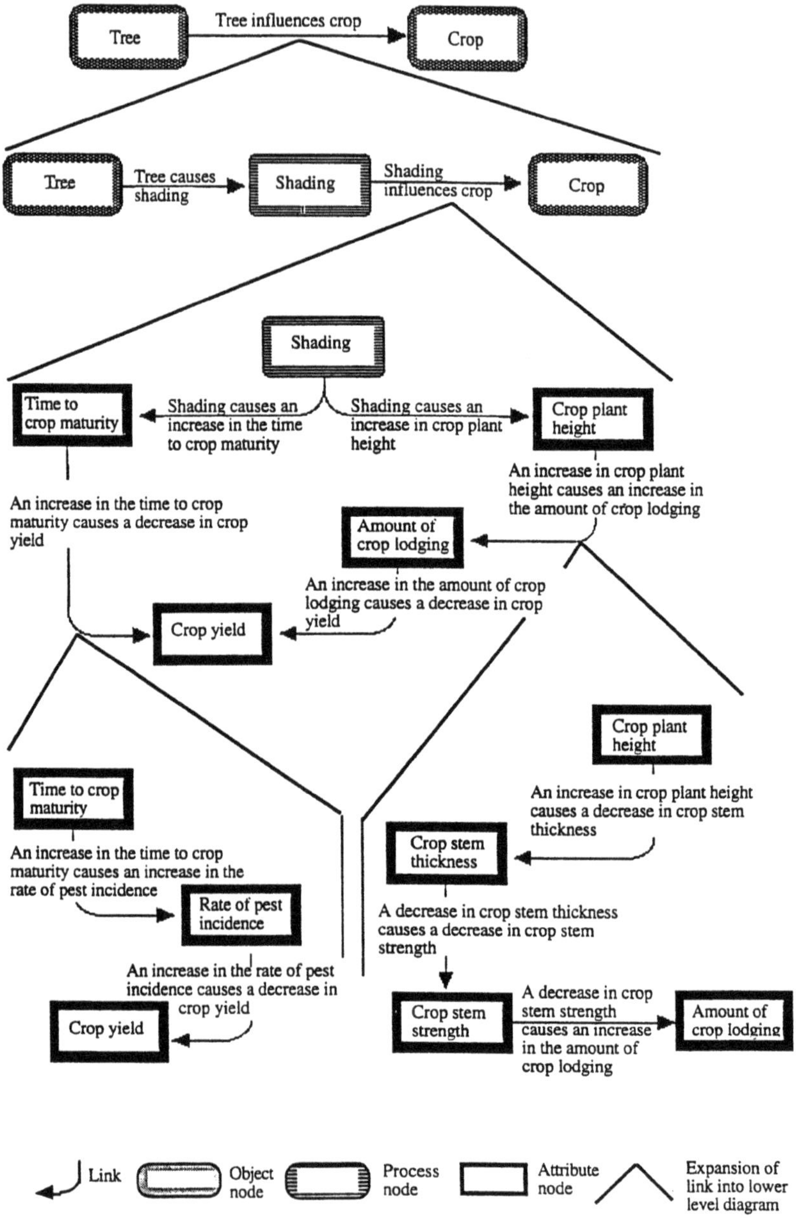

Fig. 1. Some statements about the effects of tree shading on crops synthesised into a set of hierarchically linked node-and-link diagrams using AKT(1) software [Walker et al., 1995a]. The boxes are nodes and the labelled arrows are the links, which also correspond to formal statements (Table 1). This is an illustrative example of the use of diagramming in developing a knowledge base. Diagrams, at any one hierarchical level, used in constructing the tree fodder knowledge base in the present study typically contained from 20 to 30 nodes and from 30 to 40 links [Thapa, 1994].

In the absence of such information previously, researchers have tended to follow a more conventional approach, emphasising survival rate, above-ground growth rate, and total foliage production as the most important parameters for selection of germplasm. Such research may be of little relevance to the target communities because these are not necessarily the parameters of most concern to the farmers who can, in fact, specify associative tree ideotypes [Huxley, 1985] suitable for integration in their farming systems, in terms of tree attributes affecting crop yield, fodder quality and resilience to lopping. The results of this study have already benefited researchers at PAC, modified their understanding of above-ground interactions and significantly improved the design of agroforestry research by including consideration of this knowledge when new research has been planned (L. Joshi, pers. comm.[1]).

A detailed analysis of the ways in which farmers and researchers differ in their understanding of processes, attributes and terminology in describing and rationalising aspects of tree fodder production and management practices, has important implications for future tree fodder research and extension programmes. For example, one of the key processes considered by farmers in managing tree-crop interactions termed '*tapkan*' concerns the erosive effect on soil and consequent reductions in crop yield caused by water droplets falling from tree leaves. This is an eclectic term with no exact scientific equivalent. The lack of understanding about *tapkan* by Nepali researchers indicated a gap in their knowledge and potential problems in interaction between the farmer and the researcher. Furthermore, the farmers' assertion that leaf size and texture affected the size of droplets falling from leaves and hence their erosive effect on soil, was in direct contradiction to the prevailing scientific literature that held, on the basis of controlled experimentation and knowledge of the surface tension properties of water, that droplet size was independent of canopy morphology [Brandt, 1989; Thornes, 1989]. Only recently have Hall and Calder [1993], using a new and expensive instrument (a disdrometer) capable of more accurate determination of droplet diameter, measured markedly different modification of rainfall drop size spectra under pine (*Pinus caribaea*), teak (*Tectona grandis*) and eucalypt (*Eucalyptus camaldulensis*) canopies. This new evidence suggests that the farmers' explanations may be valid scientifically and that their knowledge of the drop size modification by a wide range of tree species may be pertinent to controlling soil erosion on slopes and complementary to what is known scientifically. The knowledge articulated by farmers relating to *tapkan* can now be utilised by researchers in focusing future research and in developing detailed research plans appropriate to farmers' needs.

The study also revealed that the depth and spread of root systems was considered by farmers in managing below-ground interactions. Farmers classified many tree species as being either '*rukho*' (23 species) or '*malilo*' (17 species). *Rukho* trees (for example, *Ficus bengalensis*) were thought to exhibit competitive and/or allelopathic effects which inhibited crop germination and growth rate and so reduced crop yield whereas *malilo* trees (for example,

Albizia julibrissin) were thought to contribute to soil fertility through inputs from decomposing litter and have less competitive root systems, so enhancing crop yield. Since farmers viewed these as important issues for consideration in managing below-ground interactions, tree root architecture and leaf litter decomposition rates were identified as important areas of consideration in planning tree fodder research. Research on below-ground interactions in agroforestry has barely begun in Nepal, perhaps because of a combination of: forestry professionals not being aware of their character and so considering them of little or no immediate relevance; and, the inherent difficulty associated with studying roots.

However, analysis of the farmers' knowledge system suggests that research on below-ground interactions may have greater leverage than research on above-ground processes because farmers consider them important but know little about them, presumably because they are difficult for the farmers themselves to observe. Research on root architecture including intra-species and inter-species variation in rooting pattern of common *rukho* and *malilo* fodder tree species; the effects of tree fodder management techniques, particularly frequency of fodder lopping, on rooting pattern; and, the variation of root development in relation to different propagation or planting materials and its implications for resource sharing (in terms of niche differentiation and root competition between tree and crop components for soil water and nutrients); are areas immediately identifiable from the agroforestry literature as having potential for research to alleviate constraints within the system [see for example, Van Noordwijk and Purnomosdhi, this issue; Schroth, this issue].

The knowledge articulated by farmers on crop cultivar responses to shade also had direct implications for planning future crop improvement programmes. The study clearly suggested that there was considerable potential for selecting crop cultivars for different shade levels and environments, as determined by altitude and aspect, found in the mid-hills of Nepal. Considerable information on existing shade tolerant cultivars was also forthcoming. Unfortunately, most improved crop cultivars from government agricultural research stations in Nepal have been selected under full sunlight. This may have partly contributed to the low rate of crop cultivar adoption by farmers.

Fodder value

Scientists both in Nepal [Subba and Tamang, 1990; Panday and Osti, 1993] and elsewhere (M. Gill and P. Thorne, *pers. comm.*)[2] have been working to ascertain fodder quality through chemical analyses. Much research on nutritional value of tree fodder has been based on proximate analysis of fodder leaves, often emphasising crude protein and crude fibre content as the most important parameters for assessing fodder quality. More recently, a broader range of standard nutritive value assessment methods (which include measurements of digestibility and the content of anti-nutritive factors) have been used. However, results from these studies are far from conclusive and often contradictory between different laboratories [Mahato and Subba, 1988; Paterson, 1993], in some cases vindicating farmers preferences and in others finding no chemical explanations for them [Subba et al., 1992].

Analysis of the indigenous knowledge about tree fodder collected by Thapa [1994] was undertaken; it was found that farmers classified fodder trees, for various categories of livestock, and with various seasonal and age-dependent attributes, in terms of their nutritional value (*posilo* and *kam posilo* categories) and as either heating or cooling (*chiso* and *obano* categories; consistent with the earlier work elsewhere in Nepal of Rusten and Gold [1991]). Farmers assessed fodder quality in terms of its ability to satisfy animal appetite, its palatability, its effects on animal health and its effect on milk yield and the butterfat content of milk. Intra-species variability in fodder quality was consistently recognised for three previously undifferentiated species (*Ficus nemoralis*, *Bauhinia purpurea* and *F. roxburghii*) while some farmers recognised sub-species variants in a further four species (*F. lacor*, *F. semicordata*, *B. variegata* and *Saurauia nepaulensis*). Knowledge both of how to identify the sub-species and their fodder value was apparent (Table 2). Farmers also possessed detailed knowledge of how the timing, extent and technique of fodder lopping affected the amount and quality of fodder production from different species. This record of farmers' knowledge of fodder value has subsequently provided the framework for innovative research using indigenous knowledge to make sense of the results of laboratory-based nutritive

Table 2. Varietal characteristics of one of the species, *Dudhilo* (*Ficus nemoralis*) consistently recognised by Solma farmers as having distinct subspecies. Nepali words are explained in the text.

Farmers' variety	Varietal characteristics
sano pate dudhilo	– small leaf size, leaf shape is elongated with a pointed tip – sheds leaves early in the dry season (winter shedder) – rapid growth of regrowth shoots after lopping – can be lopped at least two or three times per year and so produce a large quantity of fodder – only *posilo* to and eaten preferentially by goats and sheep – more *chiso* than *thulo pate dudhilo*; and – produces more sticky milky latex than *thulo pate dudhilo* and hence causes leaves to stick in animal's throats more than *thulo pate dudhilo*
thulo pate dudhilo	– big leaf size, leaf shape is round with a flat tip – sheds leaves late in the dry season (spring shedder) – slow growth of regrowth shoots after lopping – can be lopped only one time per year for fodder, therefore, produces a smaller quantity of fodder than *sano pate dudhilo* – *posilo* and eaten preferentially by all classes of animals – more *obano* or less *chiso* than *sano pate dudhilo* – produces less sticky milky latex than *sano pate dudhilo* and hence less of a problem in respect of leaves sticking in animal's throats.

value assessment [Thorne et al., 1994]. The important point to note here is that the analysis was conceived and implemented in conjunction with livestock scientists after the indigenous knowledge about tree fodder had been recorded in a knowledge base by a researcher from the Forestry Section at PAC. Significantly, the knowledge base provided a sufficiently flexible, explicit and accessible record of the local knowledge system for an analysis to be undertaken that had not been conceived when the knowledge acquisition was planned and conducted.

Interesting correspondences were apparent between some terms for attributes used by farmers and different terms used similarly by researchers, for example 'bitterness' of leaves corresponded closely with 'tannin content' of leaves, indicating complementarity in the knowledge from the two sets of sources. In general the researchers' knowledge had a greater explanatory power with respect to what chemical constituents caused various effects on livestock but farmers knew the values of the attributes they recognised for many more species than did researchers for the attributes that they recognised.

Distribution of knowledge and its relationship to agricultural practice

Conceptual understanding of interactive processes such as shading and leaf droplet erosion were consistently held by farmers across gender, ethnic and altitudinal divides but detailed knowledge about species-specific tree attributes, which influenced the application of general concepts in practical management, varied amongst farmers, apparently associated with which particular species they had experience of cultivating [Thapa, 1994]. Access to a central repository of indigenous knowledge by less knowledgeable farmers, either directly or via the forestry extension service may, therefore, assist them in improving their practice. Comparison of indigenous ecological knowledge with farmer practice, by examining which types of tree were growing on which types of land, indicated that underlying ecological knowledge held by farmers did, at least to some extent, determine their practice. Practices known by farmers to have negative ecological consequences did, however, occur; for example, planting trees that were known to have a predisposition to cause soil erosion on crop terrace risers because of their high fodder value at critical times of the year. The ubiquitous persistence of such practices implies that farmers are forced to make trade-offs between short term productivity and longer term sustainability and indicates substantial potential for improvement of farming practice through research and extension targeted on these areas.

Conclusions

The case study has revealed the depth and consistency of indigenous ecological knowledge about agroforestry on one hillside in Nepal and its complementarity with existing scientific knowledge. Problems related to on-farm tree fodder production in Nepal clearly require interdisciplinary research. Farmers already combine sophisticated knowledge of fodder production by trees, their impact on associated soil and crops, and the feeding value of fodder for livestock, in arriving at complex trade-offs between these elements. The creation of explicit, encyclopaedic knowledge bases, combining interdisciplinary scientific knowledge with that of farmers, has the potential to have a profound impact on the research and development cycle both in terms of:

- targeting research at gaps in farmers' knowledge that constrain the productivity and sustainability of their farming system and
- marshalling the corporate knowledge of researchers related to a land use problem across traditional subject divisions.

While the same level of insight with respect to the nature and extent of farmers' knowledge might have been achieved through the application of less formal approaches to knowledge acquisition (although this seems doubtful) the conclusions drawn could not have been as easily scrutinised retrospectively. The research involved extented interaction with farming communities and, rather than simply developing an external ('etic' – Knight [1980]) description of landuse practice and constraints sought an 'emic' [Knight, 1980] explanation of system function, thereby revealing key issues in system management which had not been illuminated by previous Rapid Rural Appraisal type investigation at PAC [Chand and Gibbon, 1990]. Furthermore, the knowledge base remains a resource that can now be used in a variety of further research and planning activities.

Acknowledgements

Research on the formal representation of indigenous ecological knowledge reported here was funded by the UK Overseas Development Administration (ODA), Forestry Research Program. Development of the software and approach involved collaboration with Robert Muetzelfeldt of the Institute of Ecology and Resource Management of the University of Edinburgh; Dave Robertson, Mandy Haggith and Gill Kendon of the Department of Artificial Intelligence of the University of Edinburgh; Nishantha Jinadasa of the University of Sri Jayewardenepura, Sri Lanka (funded by the ODA South East Asian Development Division – SEADD); Gamini Hitinayake of the University of Peradeniya, Sri Lanka (also funded by SEADD); Felician Kilahama of the Department of Forestry and Beekeeping, Tanzania (funded by the Ford Foundation); Pornchai Preechapanya of the Royal Forest Department and University of Chiang Mai, Thailand (funded by the Ford Foundation, the Rockefeller Foundation and Winrock International) and Alison Southern of the University of Wales, Bangor (funded by a John Robert Jones scholarship and a Humphrey Griffith scholarship).

Notes

1. Mr Laxman Joshi is the Acting Head of Forestry Section, Pakhribas Agricultural Centre.
2. Dr Margaret Gill and Dr Peter Thorne are livestock specialists at the Natural Resources Institute, Chatham, UK.

References

Brandt CJ (1989) The size distribution of throughfall drops under vegetation canopies. Catena, 16: 507–524

Carter AS and Gilmour D (1989) Increase in tree cover on private farmland in Central Nepal. Mountain Research and Development 9(4): 381–391

Chand SP and Gibbon D (1990) Samuhik Bhraman: A rapid and appropriate method of prioritising and replanning agricultural research in Nepal. Technical paper 121, Pakhribas Agricultural Centre, BAPSO, Kathmandu, Nepal

Gibbon D and Schultz M (1989) Agricultural Systems in the Eastern hills of Nepal: Present situation and opportunities for innovative research and extension. Pakhribas Agricultural Centre, BAPSO, Kathmandu, Nepal

Hall RL and Calder IR (1993) Drop size modification by forest canopies: Measurements using a disdrometer. Journal of Geophysical Research 98: 18,465–18,470

Huxley PA (1985) The basis of selection, management and evaluation of multipurpose trees – an overview. In: Cannell MGR and Jackson JE (eds) Attributes of Trees as Crop Plants, pp 13–35. Institute of Terrestrial Ecology, NERC, UK

Kendon G, Walker DH, Robertson D, Haggith M, Sinclair FL and Muetzelfeldt RI (1995) Supporting customised reasoning in the agroforestry domain. New Review of Applied Expert Systems (in press)

Knight GG (1980) Ethnoscience and the African farmer: Rationale and strategy (Tanzania). In: Brokensha D, Warren DM and Werner O (eds) Indigenous Knowledge Systems and Development, pp 203–229. University Press of America, MD, USA

Mahato SN and Subba DB (1988) Nutritional evaluation of fodder at Pakhribas Agricultural Centre, Dhankuta. In: Robinson PJ (ed) Proceedings of the second meeting of the working group on fodder trees, forest fodder and leaf litter. FRIC occasional paper no. 2/88, pp 20–22. Forest Research and Information Centre, Kathmandu, Nepal

Panday SB and Osti NP (1993) Chemical composition of six fodder tree species. In: Mathema P (ed) Proceedings of the regional seminar on fodder trees, forest fodder and leaf litter. FRD occasional paper no. 1/93, pp 4–11. Forest Research Division, Kathmandu, Nepal

Paterson RT (1993) Use of Trees by Livestock: Anti-Nutritive Factors. Natural Resources Institute, Chatham, UK

Rusten EP and Gold MA (1991) Understanding an indigenous knowledge system for tree fodder via a multi-method on-farm research approach. Agroforestry Systems 15: 139–165

Schroth G (1995) Tree root characteristics as criteria for species selection and systems design in agroforestry. Agroforestry Systems (this issue)

Sinclair FL, Walker DH, Joshi L, Ambrose B and Thapa B (1993) Use of a knowledge based systems approach in the improvement of tree fodder resources on farmland in the eastern hills of Nepal. Pilot phase report. August 1993. School of Agricultural and Forest Sciences, University of Wales, Bangor, UK

Subba DB and Tamang PM (1990) Seasonal variation in the chemical composition of the leaves of Ficus species of fodder trees. In: Amatya SM and Philip MS (eds) Proceedings of the third meeting of the working group of fodder trees, forest fodder and leaf litter. FRIC occasional paper no. 2/90, pp 20–24. Forest Research and Information Centre, Kathmandu, Nepal

Subba DB, Tamang BB and Tamang PM (1992) Availability of protein from high altitude tree fodder during the dry season in the eastern hills of Nepal. Veterinary Review 7(2): 55–57

Thapa B (1994) Farmers' ecological knowledge about the management and use of farmland tree fodder resources in the mid-hills of eastern Nepal. PhD thesis, University of Wales, Bangor, UK, 271 pp

Thapa B, Sinclair FL and Walker DH (1994) Farmers' ecological knowledge about the management and use of farmland tree fodder resources in the eastern hills of Nepal: Implications for research and development. In: Sebiollotte M (ed) Systems Orientated Research in Agriculture and Rural Development: Papers, pp 654–660. CIRAD, Montpellier, France

Thorne P, Thapa B, Walker DH and Sinclair FL (1994) An investigation of tree fodder quality: Are novel, nutritive value assessment methods capable of meeting farmers' needs for information on feed quality. Project outline, Livestock Section, Natural Resources Institute, Chatham, UK

Thornes J (1989) Solutions to soil erosion. New Scientist (3 June): 45–49

Van Noordwijk M and Purnomosidhi P (1995) Root architecture in relation to tree-soil-crop interactions and shoot pruning in agroforestry. Agroforestry Systems (this issue)

Walker DH, Sinclair FL, Kendon G, Robertson D, Muetzelfeldt RI, Haggith M and Turner GS (1994) Agroforestry Knowledge Toolkit: Methodological Guidelines, Computer Software and Manual for AKT1 and AKT2, Supporting the Use of a Knowledge-Based Systems Approach in Agroforestry Research and Extension. School of Agricultural and Forest Sciences, University of Wales, Bangor, UK, 138 pp

Walker DH, Sinclair FL and Kendon G (1995a) A knowledge-based systems approach to agroforestry research and extension. AI Applications (in press)

Walker DH, Sinclair FL and Thapa B (1995b) Incorporation of indigenous knowledge and perpectives in agroforestry development. Part one: Review of methods and their application. Agroforestry Systems (this issue)

Agroforestry Systems **30**: 263–275, 1995.

The use and value of multiple methods to capture the diversity of endogenous agroforestry knowledge: an example from Rwanda*

C. DEN BIGGELAAR[1] and M. A. GOLD[2]
[1] *c/o ICRAF, United Nations Avenue, Gigiri, P.O. Box 30677, Nairobi, Kenya;* [2] *Department of Forestry, Michigan State University, East Lansing, MI 48824-1222, USA*

Key words: indigenous knowledge, knowledge ranking, participatory research, socioeconomic surveys, tree inventory

Abstract. To understand endogenous agroforestry knowledge and the processes behind the generation of such knowledge, multiple methods are needed. In this paper, the identification and characterization of farmer agroforestry experts will illustrate the necessity of multiple methods. Nonformal, participatory methods were employed to identify the research subjects (farmer agroforestry experts) and to gain insights into the emic conception and understanding of 'knowledge' and 'agroforestry'. An adaptation of the wealth ranking game was used to determine (1) the research subjects using locally defined criteria of 'being knowledgeable about agroforestry', and (2) how this knowledge is distributed within a community. Additional information, approaching the subject from an etic perspective, was gathered using formal surveys to collect socioeconomic data and to make an inventory of all trees and tree species.

Study results indicate that exclusive reliance on either emic or etic perspectives would not have yielded satisfactory results, as there is a differentiation in agroforestry knowledge and agroforestry systems between different groups of farmers. It is, therefore, recommended that future studies of endogenous knowledge systems combine qualitative and quantitative, participatory and formal data collection methods as they provide complementary and supplementary perspectives on a complex reality.

Introduction

According to Tindimubona [1990], science is 'A cumulative tradition of knowledge, an evergrowing body of reflections and ideas, experiences and actions of a great stream of thinkers and workers'. It is the tool with which people learn about their environment, its resources and problems, and how to control and utilize them productively and sustainably. As environments vary across the globe, one can therefore assume that 'science' differs also in these different environments. This is confirmed by Feyerabend [1988], who states that there are many different kinds of science, first world science being only one science among many others. Thus, if there are different kinds of science, each science tradition (or better, the people that adhere to that tradition) may

* An earlier version of this paper was published under the title 'Differentiating the Nature of Agroforestry Systems and Agroforestry Knowledge among Farmers in Southern Rwanda: Participatory and Formal Approaches as Complementary Methods' (*Culture and Agriculture* no. 50, Fall 1994 (in press).

arrive at different conclusions about what constitutes a problem or need, and what acceptable solutions will be. Science and technology are above all socially constructed entities [Bijker et al., 1987; Feyerabend, 1988; Knorr-Cetina, 1981]. 'It is social groups that give meaning to (science and) technology and problems are defined within the context of the meaning assigned by a social group or a combination of social groups. Because social groups define problems of technological development, there is flexibility in the way things are designed, not one best way' [Pinch and Bijker, 1987].

A good example of a socially constructed entity is agriculture. People in different environments have devised diverse solutions to problems and constraints they encountered in the natural and social environment, resulting in a multitude of agricultural systems. Farmer conceived and designed agroforestry systems were used as the object of this study. Farmers in the *secteurs*[1] of Maraba and Simbi (Maraba *Commune*, Butare) and Kibingo (Karama *Commune*, Gikongoro) in southern Rwanda were studied to ascertain differences in knowledge of agroforestry and the development of this knowledge over time through farmers' own efforts. An understanding of the dynamics behind these systems and their underlying logic and knowledge is essential for improving existing or developing new systems acceptable to the farmers [Rakotomanana, 1988]. The first part of this paper will report on how agroforestry and knowledge is understood from an emic (local) perspective. This information is then used to determine how knowledge of agroforestry is distributed within the community, and to identify local tree experts (i.e., the persons most knowledgeable about and most active in knowledge production of agroforestry). A variation of the wealth ranking game [Grandin, 1988] was used to identify these local tree experts. In the second part of the paper, the ranking game results will be compared with the results of a socioeconomic survey and inventory of tree species to obtain an objective, more scientific view of agroforestry (the etic perspective). These methods provide contradictory results due to the fact that the epistemological definition of (agroforestry) knowledge (developed in a time when people primarily relied on the gathering of tree products to meet their needs) no longer matches the present situation of trees purposively planted and managed on farms. Neither the reliance on tree experts as defined and identified according to local criteria as the best information sources, nor scientifically designed surveys are sufficient methods for the study of endogenous agroforestry knowledge by themselves. As will be shown in this paper, answering one of the questions posed by Scoones and Thompson [1994], multiple methods are necessary as they provide contrasting but complementary information on the nature of agroforestry systems and agroforestry knowledge. The findings also support the results of previous studies of indigenous knowledge [Rusten and Gold, 1991] detailing the necessity of multiple methods.

Methodology

The basic assumption guiding the research was that not every farmer would have the same knowledge of trees, largely as a result of differing socioeconomic and political positions within the community and differing biophysical conditions on her or his farm. In addition, farmers would also differ in their propensity and aptitude towards experimentation and innovation. As the goal of this study was to understand the processes of knowledge production, it was necessary to interact with the farmers most knowledgeable about and involved in generating and developing agroforestry knowledge and technologies, as defined from an emic perspective. This made a random selection of farmers unacceptable since it could have resulted in a sample of too few farmers experimenting with trees on their farms. An adaptation of the 'wealth-ranking game' [Grandin, 1988] provided a more suitable technique to determine the study's subjects.

The knowledge ranking game

An adaptation of the wealth ranking game was chosen for two reasons. First, taking opinions of a larger group of people into account was a more objective means of identifying local tree experts than relying solely on key informants. Second, it enabled researchers to learn the 'ways in which people construe the world of their experience, and describe it in categories inherent to those structurings' [Knorr-Cetina, 1981]. In this research, the game was *not* used to rank and group people according to their wealth but according to their knowledge of agroforestry. In addition to determining the subjects (farmers) for further in-depth study, the method provided an emic perspective on 'knowledge of trees', its distribution within a community, and a preliminary understanding of the logic of endogenous agroforestry systems in southern Rwanda. The method also provided some quantitative information on the characteristics used by the informants to describe the groups of farmers they identified.

The game was used with two modifications in this study. First, the ranking was performed with the goal of identifying the *individuals* most knowledgeable about agroforestry.[2] Second, the cards that were used did not have the names of heads of household, but of *individual male and female farmers*. In many societies it is the female farmers who collect and cultivate most tree crops. If only heads of household were identified on the cards, then the majority of these female farmers would be excluded since only a small percentage of rural households are headed by women in Rwanda (around 20%).

Operationalizing the ranking criterion. Operationalizing the ranking criterion 'knowledgeable about trees and tree cultivation' posed a challenge for two reasons. Since there is no equivalent term for 'knowledgeable' in Kinyarwanda, a descriptive equivalent had to be found. According to Kagame

[1958], Kinyarwanda uses the same word for knowledge and experience (*ubumenyi*), indicating that knowledge is gained through experience. Having *ubumenyi* implies having knowledge of (experience with) many different things or notions. Thus, in the area of agroforestry knowledge, having many different tree species on the farm meant that one has experience with tree cultivation and, by consequence, more knowledge of trees. Therefore, species richness on the farm was used as a proxy criterion for differentiating highly knowledgeable from less knowledgeable farmers. The Kinyarwanda translation of the ranking criterion 'being knowledgeable about trees' used was '*Abahinzi-borozi bafite ubumenyi bw'ibiti bwinshi*'.[3]

Choice of informants and sample size for the knowledge ranking procedure. After initial meetings in each *secteur* with farmers and authorities, the *conseillers* (aldermen representing the *secteur* in the *commune* council) were asked to prepare a list of all farmers (both male and female) in their respective *secteurs*, subdivided by *cellule* and *colline*. In total, there were 2159 households in the three *secteurs* (704 in Maraba, 746 in Simbi and 709 in Kibingo) spread over 63 *collines*. According to Grandin [1988], if the level of agreement within a community on the ranking criterion is high, a sample size of 3–5 informants per locale is sufficient. Given that 3–5 informants were needed for each *colline*, the sample size was determined to be 241 informants, or about 11% of all households. A systematic sample of every ninth name was taken from the lists prepared by the *conseillers*, alternating men and women so that the final sample contained 121 men and 120 women. These 241 informants ranked all male and female farmers of these 2159 households (i.e., heads of households and their spouses if applicable) in the three communities, with males ranking males and females ranking females.

Formal socioeconomic survey and inventory of trees

To obtain statistically analyzable data to confirm the perceived differences between tree experts and the general farm population, a formal socioeconomic survey and inventory of tree species was done on two samples of farmers drawn from the ranking game population. These two surveys provided an etic perspective to the understanding of local agroforestry systems. Upon completion of the ranking game, a series of 13 community interviews were held at the *cellule* level, during which participants were asked to chose 44 tree experts from among the group deemed most knowledgeable about agroforestry as determined with the ranking game. Thus, these tree experts were identified as experts by the community according to locally defined criteria of 'expertise', using a two-stage process of ranking game and community interviews. Sixteen percent of the tree experts thus chosen were female. A group of 70 farmers (of whom 29% was female) identified as less knowledgeable about agroforestry were randomly selected from the remaining individuals to serve as a comparison group. The socio-economic survey and tree inventory consisted of interviewer-administered questionnaires developed by the author in collaboration with researchers of the *Institut des Sciences Agronomiques du Rwanda*, and pretested with farmers in the research area not included in the two samples.

Results

Farmers' views of agroforestry

Lundgren [1982] defined agroforestry as a holistic approach to land use, based on the combination of trees and shrubs with crops, pastures or animals on the same land unit, either in sequence or all at the same time. In reality, as Rocheleau et al. [1989] wrote, most farmers cannot easily separate the holistic approach to managing a land *unit* from the integration of woody plants into agricultural and pastoral *landscapes*. Whereas formal agroforestry science is based on the systematic placement of trees relative to crops and pastures, rural people are often more concerned about the fit of the whole agroforestry practice, and trees in general, into the larger landscape. This landscape approach to resource management could also be observed in Rwanda.

Many outside observers (i.e., persons trained in and adhering to the Western scientific knowledge tradition) have noted a steady decrease in the amount of *forest* resources in Rwanda since the beginning of this century [Gatera, 1980; Nwafor, 1979; Runyinya, 1980]. However, research by, amongst others, Balasubramanian and Egli [1986], Hummel [1988], and Pinner [1988] has shown that the amount of *tree* resources in the landscape actually increased during this period as farmers were well aware of the importance and the numerous benefits trees provide. Indeed, the disappearance of forests was often cited by farmers interviewed during this research as a primary motivation for introducing certain species on the farm. Thus, while forests, natural woodlots and forest galleries have all but disappeared under the influence of increasing population and pressure on the land, farmers have transformed a formerly near treeless landscape [Loupias, 1903; Johanssen, 1915] into a landscape dotted with many trees. However, these trees are more dispersed over the landscape, integrated in farming and pastoral systems.

As a corollary to the increase in the number of trees on farms, farmers have generated extensive knowledge about the introduction and management of indigenous tree species on their land. In addition, they have acquired and adapted a wealth of knowledge of exotic species and silvicultural practices introduced since the 1930s, first by the Belgian colonial rulers and, since independence in 1962, by various development programs. These combined interventions in the landscape by both farmers and outsiders have led to very complex and multifaceted agroforestry systems. This is compounded by the fact that what constitutes a tree (*igiti*) in Rwandan epistemology is different from what is understood by the term in the Western world view. For Rwandan farmers, trees are all plants that are not grasses [Kagame, 1958]. The resulting agroforestry systems are more difficult to appreciate for researchers trained in the Western scientific tradition, as the inclusion of annuals and perennials, in addition to trees and shrubs, makes the systems more complex. Agroforestry in this research was defined as all *ibiti* found (either naturally seeded or purposively planted) and managed on the farm, regardless of their integration within space and/or time, or their economic or ecological interactions.

Indigenous conceptions of the nature of 'knowledge'

In Kinyarwanda, knowledge, experience and science are all encompassed within the word *ubumenyi*. Science (not to be confounded by science as understood in Western epistemology) is to be understood as the possession of knowledge as contrasted with ignorance or misunderstanding. According to Kagame [1958], *ubumenyi* implies having knowledge of a plurality of notions/objects because a simple knowledge limited to one object or a limited number of notions cannot be constituted under the term *ubumenyi*.

Consultants perceived *ubumenyi* as having four aspects, namely intelligence, experience, communication and putting information into practice. *Ubumenyi* manifests itself through the intelligence one has in one's heart.[4] *Ubumenyi* can be 'natural' (i.e., a gift of God), can result from experience, or can be obtained by being well-informed and having gone to school (although consultants considered knowledge learned in school 'artificial knowledge'). Whatever the source of knowledge, one needs to put the information into practice for it to be recognized as *ubumenyi*.

All four themes (intelligence, experience, communication and practice) have to be present for a person to be recognized as being knowledgeable about something by one's fellow farmers. Being knowledgeable about something means knowing its utility, holding its secrets and showing a sustained interest in it as revealed through one's actions. It depends not only on *what* one knows but equally on how one can explain that knowledge to another person and provide sufficient and reliable information about it so that the other person can attain the same benefits. In short, besides knowing facts, real knowledge is first a willingness to do something with these facts, second to put them into practice, and third to communicate your knowledge about these facts to others. These findings are both startling and interesting, as they are analogous to what Western societies ascribe to their scientists: knowing basic facts, putting these into practice in experiments, communicating the results and the knowledge gained from of these experiments in journals and in scientific meetings, and replicability (i.e., other scientists should be able to obtain the same results, all else being equal). However, the level at which Rwandan tree experts (as one category of indigenous scientists) operate and the methods they use are quite different from the one at which researchers trained in the Western knowledge tradition operate.

Knowledge ranking: a local perspective on the definition and identification of tree experts

Using species richness as the principal criterion to distinguish farmers, informants (n = 241) created five groups of farmers (I – most knowledge – through V – least knowledge) according to their perceived expertise of trees and tree cultivation (Table 1). The farmers identified as the most knowledgeable about agroforestry (Group I) all had many *ibiti*[5] on their farms. However, the explanatory variables farm size (mentioned 105 times) and

Table 1. Variables used by ranking game informants to characterize the various knowledge groups created by them (based on an analysis of descriptive statements).

	Most knowledge I	II	III	IV	Least knowledge V
Number of species					
≥ 15	6				
10–14	48	11	3		
5–9	41	38	23		
≤ 4	1	38	90	43	29
Only on the rugo			43	22	23
Soil quality					
Good	22	5	8	1	
Bad		3	6	2	2
Farm size					
Large	94	42	40	1	4
Medium	10	21	27		1
Small	1	12	57	34	39
No land				1	2
Animals					
Cattle	82	43	63	7	4
Small animals	20	28	25	17	11
No animals	7	10	17	12	17
Manure					
Abundant	42	3			
Yes-sufficient	18	6	8		
Little or none		1	9	4	7
Source of information/knowledge					
Self	13	2	1		
Extension workers	2				
Wealth					
Rich	39	1			
Poor				27	12

Note: Because of space limitations, not all characteristics have been included in this table. For a complete list, see Den Biggelaar (1994).

presence or absence of cattle (mentioned 109 times) were more often cited to characterize these farmers than species richness (mentioned 96 times). People with large farms have more land to plant trees, and can, therefore, have more trees in total and a greater species diversity. Similarly, cattle provide manure with which soil fertility can be maintained, and farmers in this group were identified as those having a sufficient (mentioned 18 times) to abundant (mentioned 42 times) manure supply. The ability to maintain soil fertility on at least part of the farm means that a harvest sufficient to feed the family (and perhaps some surplus) can be obtained from less land, thus freeing some land for the cultivation of tree crops. The most knowledgeable farmers were also identified as being richer than others, having woodlots, and continuing to plant more trees on their farms every year. Although few respondents mentioned where their fellow farmers had obtained their knowledge, for farmers in Group I 'self' was mentioned most frequently, meaning that these farmers were more likely to generate knowledge through their own trials and observation. This is also reflected in the statement that these farmers were actively engaged in searching for and obtaining planting materials.

For Groups II and III, species richness was the principal distinguishing variable, more so than in the differentiation of Groups I and II. In Group I, farmers had between five and 14 tree species, in Group II farmers had between one and nine species, while the majority of farmers in Group III fell in the category 'less than four species'. Simultaneous with decreasing species diversity was a decrease in farm size in these three groups, pointing to a relationship between farm size and tree diversity. Animal ownership was about equal in each of these groups, even though the number of statements referring to absence of animals increased from seven in Group I to 17 in Group III. About one-quarter of the statements on species diversity for the farmers in Group III mentioned that these farmers had no trees other than those on the *rugo* (i.e., the home compound). This was an indication that farmers in this group were increasingly being squeezed for land for crops and eliminating trees as a consequence. This was also supported by the observation made by the informants that these farmers need all their land for crops and therefore cannot cultivate trees.

The least knowledgeable farmers (Groups IV and V) were characterized as having only trees on the *rugo* (particularly those in Group V), or, in addition to trees on the *rugo*, having one or two species in their fields. They have small farms, with none or only a few small animals and little to no manure. Lack of land was an often heard reason for not having trees or only having few of them, in spite of the recognition of the importance of trees. The *rugo* is therefore an important location for agroforestry, since this is the only place left for cultivating trees for the majority of farmers (for example, according to the 1989 Communal Development Plan, 54% of households in Maraba *Commune* have farms of less than 0.25 ha).

Formal surveys: defining expertise from a scientific perspective

The results of the socio-economic and tree surveys showed that the main ranking criteria used by the informants during the game (land, animals, species diversity) were significantly different between tree experts and comparison farmers. Because land is the major production factor in Rwandan agriculture (albeit one that is becoming increasingly limited for many farmers) and because having experience with a diversity of species was considered the most important aspect of 'being knowledgeable about agroforestry', only land and species diversity will be considered in the analysis presented here. For a complete review of factors important to differentiate tree experts and comparison farmers, see Den Biggelaar [1994].

Species diversity. A total of 152 *ibiti* were found, 31 of which were exotics. This diversity of species is remarkable as most farmers said that there were no or only few trees on the land when they started cultivating it. Indeed, when older people were asked to compare the current vegetation with that of their youth, all stated that there are many more trees now in and around the farm and in the general landscape than when they were young. The only trees they remember being present on the farm were those planted on the *rugo* and between the bananas immediately surrounding the *rugo*. All species presently found on the farm, therefore, were purposively planted and managed by the farmers in the last two to three decades. The reason for planting and maintaining so many different tree species is to meet the diverse needs for tree products (fuelwood, poles, lumber, food, fruit, stakes, medicines). While wood needs could be satisfied with few species, for medical problems it is necessary to plant many different species to treat the many diseases and parasites from which people and animals suffer.

Diversity differences between experts and comparison farmers. The species inventory confirmed that tree experts have a significantly greater diversity ($p < 0.01$) of tree species on their farms (average 34 species) than comparison farmers (average 12 species) (Table 2). In part, this can be attributed to the way tree experts were identified. During the ranking game, emphasis was given to tree diversity as a proxy for 'being knowledgeable about trees and tree cultivation'. While not known during the ranking game, later discussions in community and focus group meetings confirmed that knowledge is gained through experience, and that experience with many different tree species is more important in gaining expertise and knowledge than just having many trees of a few species. It also corresponds to the definition of knowledge by Kagame [1958] (having experience with many different objects or notions).

While tree experts maintain a greater diversity of trees, the mean number of trees of each species on farms of tree experts and comparison farmers is about equal. High standard deviations in the number of trees per species are an indication of the variation in agroforestry systems between farms, and the

Table 2. Indigenous agroforestry systems: an overview of the number of trees, species and tree densities on farms of comparison farmers and tree experts.

	Comparison farmers				Tree experts			
	Kibingo (n = 35)	Maraba (n = 17)	Simbi (n = 18)	Total (n = 70)	Kibingo (n = 22)	Maraba (n = 11)	Simbi (n = 11)	Total (n = 44)
Mean no. trees species^{-1}	28	22	26	26	35[c,d]	22	14	27[b]
Standard deviation	93	52	84	82	122	69	35	98
Mean no. species on farm	11.4	12.8	11.3	11.7	38.7	33.9	29.6	34.4[a]
Mean no. trees on farm	319	282	294	304	1355	746	414	929
Mean farm size (ha)	0.24[c]	0.11	0.14	0.18[b]	2.13[c,d]	0.53	0.37	1.27[a,b]
Mean no. of trees ha^{-1}	1329	2564	2101	1689	636	1408	1119	731

[a] Between group difference significant at $p < 0.01$.
[b] Within group difference significant at $p < 0.01$.
[c] Difference between Kibingo and Maraba significant at $p < 0.05$.
[d] Difference between Kibingo and Simbi significant at $p < 0.05$.

individuality of farmers in designing a system meeting their particular needs and circumstances.

Farm size and tree density. The results of the socio-economic survey indicated that tree experts (n = 44) have an average farm size of 1.27 ha, which is significantly different (at $p < 0.01$) from the size of comparison group farms (n = 70) of 0.18 ha (Table 2). While tree experts had more land, a greater variety of tree species and a greater total number of trees, comparison farmers had an average tree density more than twice that found on tree expert farms: 1689 trees ha^{-1} (ranging from 1329 to 2564) compared to 731 trees ha^{-1} (ranging from 636 to 1408). Thus, there is an inverse correlation between size of land holding and tree density, indicating different strategies of resource use and different pathways to intensification. As high tree densities require farmers to have higher levels of management skills and greater knowledge of the various components and their interactions, the nature and content of the agroforestry knowledge between tree experts and comparison farmers may be divergent.

Conclusions

Tree experts identified in this study all had much experience derived from cultivating a diversity of species, which translated into their being the most knowledgeable about trees and tree cultivation in the opinion of informants. However, being the most knowledgeable about a diversity of tree species may not necessarily mean that they are also the persons having the most knowledge of managing these trees in highly integrated crop-tree systems. This study observed a differentiation in both agroforestry systems and agroforestry knowledge in the groups of tree experts and comparison farmers.

Comparison farmers understood the need for and importance of trees. But as they had little land available, they needed a greater awareness of component interactions, making their choice of species and species locations more critical. The smaller the farm, the more farmers in this group concentrated their efforts on fewer, less-competitive species, planted at a higher number of individuals of each species. This strategy resulted in very complex systems of low diversity, but with a high degree of integration of trees and crops within a small space. As agroforestry research concerns the study of the integration of and interaction between trees, crops and/or animals, future research on endogenous agroforestry systems should focus on the much more complex and integrated systems of the comparison farmer group to learn about farmers' knowledge about component interactions and the logic behind system designs.

The results of this research indicate that endogenous agroforestry knowledge is not homogenous but varies between individuals and groups of individuals, but with overlapping elements. These results confirm the observation by Scoones and Thompson [1994] that knowledge is manifold, discontinuous and dispersed, not singular, cohesive and systematized. Exclusive reliance on local definitions of 'knowledge' and 'being knowledgeable about agroforestry' for the identification of tree experts may, therefore, not provide satisfactory results. As a result of 'development' and other political, socioeconomic and demographic processes of change, the epistemological definition of knowledge defined by a society once largely dependent on gathering its tree products no longer appears applicable to the situation today in which people actively plant and manage trees on their farms. A more extensive kind of knowledge of trees is developing which includes knowledge about the planting and management of trees, their integration and interaction with crops and/or animals, as well as knowledge of their utility. The different strategies of agroforestry development farmers in this study pursued have resulted in multiple ways of knowing about trees and tree cultivation. To capture this diversity of agroforestry knowledge and agroforestry systems, it is recommended that future studies of endogenous knowledge combine qualitative and quantitative, participatory and formal data collection methods seeking information from both emic and etic perspectives, as they provide complementary and supplementary perspectives on a complex reality.

Acknowledgements

This research was made possible with grants from the Great Lakes Bean Program of the Centro Internacional de Agricultura Tropical (CIAT), the International Centre for Research in Agroforestry (ICRAF) with funds from the Rockefeller Foundation, and the Office of the Vice-President for Research and Graduate Studies at Michigan State University. Their support is greatly acknowledged. I also like to thank Laura DeLind and Anne Ferguson, all at Michigan State University, and the reviewers for their helpful comments.

Notes

1. Rwanda is administratively divided into *prefectures*, *communes*, *secteurs*, *cellules* and *collines*. These can be compared to provinces, districts, townships, neighborhoods and blocks.
2. Grandin used the game to determine wealth strata in a community, which could then be used as a sampling frame for formal surveys.
3. The exact translation of the ranking criterion is 'Cultivators/livestock keepers who have knowledge of many trees'.
4. Rwandans believe that intelligence is located in the heart, not in the head. As discussed in Appendix I-1 [Den Biggelaar, 1994], the heart's function is to centralize and orient sensations.
5. *Ibiti* is the plural form of *igiti*.

References

Balasubramanian V and Egli A (1986) The role of agroforestry in the farming systems in Rwanda with special reference to the BGM region. Agroforestry Systems 4(4): 271–290

Den Biggelaar CJPAM (1994) Farmer knowledge and experimentation with trees and tree cultivation in agroforestry systems in Rwanda. PhD dissertation, Michigan State University, East Lansing, MI, USA

Bijker W, Hughes TP and Pinch TJ (eds) (1987) The Social Construction of Technological Systems. The MIT Press, Cambridge, MA, USA

Commission Communale de Planification Maraba (1989) Eléments pour l'élaboration du plan de développement de la commune. Résultats du recensement mené en Août 1989. Désiderata de la population, monographie communale. Kigali, MININTER/Préfecture de Butare

Feyerabend P (1988) Against Method (Revised edition). Verso, London, UK/New York, USA

Gatera F (1980) Acroissement démographique et déforestation au Rwanda. Bulletin Agricole du Rwanda 13(1): 28–32

Grandin B (1988) Wealth Ranking in Smallholder Communities: A Field Manual. Intermediate Technology Development Group, Rugby, UK

Hummel L (1988) Boeren en Bomen in de Gemeente Muyaga (Rwanda): Farming Systems en Agroforestry. Agricultural University of Wageningen, Department of Silviculture, Wageningen, The Netherlands

Johanssen PF (1915) Ruanda: kleine Anfänge – grosse Aufgaben. Verlagshandlung der Anstalt Bethel, Bielefeld, Germany

Kagame A (1958) La Philosophie Bantu-Rwandaise de l'être. Académie Royale des Sciences Coloniales, Classe des Sciences Morales et Politiques, Bruxelles. Nouvelle Série In-8^0-VI (1957–1958)

Knorr-Cetina KD (1981) The Manufacture of Knowledge. Pergamon Press, Oxford, UK

Loupias (1903) Traditions et légendes des Batutsi. *Anthropos* III(1): 1–16

Lundgren B (1982) Introduction. Agroforestry Systems 1(1): 3–6

Nwafor JC (1979) Agricultural land use and associated problems in Rwanda. The Journal of Tropical Geography 48(June): 58–65

Pinch TJ and Bijker W (1987) The social construction of facts and artefacts. Or how the sociology of science and the sociology of technology might benefit each other. In: Bijker W, Hughes TP and Pinch TJ (eds) The Social Construction of Technological Systems, pp 18–50. The MIT Press, Cambridge, MA, USA

Pinner E (1988) Enquête sur l'agroforesterie dans la région BGM. Agricultural University of Wageningen, Department of Silviculture and Forest Ecology, Wageningen, The Netherlands/ FSR Project at ISAR-Karama Kigali, Rwanda

Rakotomanana JL (1988) Synthesis of the recommendations of the seminar. In: Agroforestry: The Efficiency of Trees in African Agrarian Production and Rural Landscapes. Proceedings of a seminar organized in Kigali, Rwanda, June 11–16, 1988, pp 47–48. Technical Centre for Agricultural and Rural Cooperation (CTA), Wageningen, The Netherlands

Rocheleau D, Wachira K, Malaret L and Wanhoji BM (1989) Local knowledge for agroforestry and native plants. In: Chambers R, Pacey A and Thrupp LA (eds) Farmer First: Farmer Innovation and Agricultural Research, pp 14–23. Intermediate Technology Publications, London, UK

Runyinya B (1980) Impact de l'homme sur les formations forestières des zones marginales au Rwanda. Bulletin Agricole du Rwanda 13(1): 14–23

Rusten EP and Gold MA (1991) Understanding an indigenous knowledge system for tree fodder via a multi-method on-farm research approach. Agroforestry Systems 15: 139–165

Scoones I and Thompson J (1994) Introduction. In: Scoones I and Thompson J (eds) Beyond Farmer First. Rural People's Knowledge, Agricultural Research and Extension Practice, pp 1–12. Intermediate Technology Publications, London, UK

Tindimubona AR (1990) Guest editorial: Science culture in Africa. Discovery and Innovation 2(4): 4–7

Agroforestry Systems **30**: 277–287, 1995.

Historical development of agroforestry in China

W. HSIUNG (XIONG), S. YANG and Q. TAO
Department of Forestry, Nanjing Forestry University, Nanjing 210037, P.R. China

Key words: agriculture, history, sericulture, slash-and-burn, small-farming economy

Abstract. In China both agriculture and agroforestry originated in forests and developed side by side from their very beginnings. As revealed by archaeological evidence from ancient times, ancestral Chinese inhabited forests where they sheltered themselves from external hazards and lived on the edible parts of plants and animals through hunting and gathering activities.

As early as the New Stone Age (7000–8000 years B.C.), fire was commonly used to burn the forests for slash-and-burn cultivation, which is a primitive form of agroforestry. Along with the rapid growth of population, the annexing of tribes, the collapse of clan society and the development of the slavery system, the nomadic mode of slash-and-burn farming evolved into settlement farming in the Xia Dynasty (2000–1600 B.C.). Peasants then engaged in settled cultivation.

During the Shang and West Zhou Dynasty (1600–800 B.C.), perpetual settlement farming encouraged the development of private land-ownership. Peasants planted trees in or around the crop fields and grew fruit plants, vegetables and farmed domestic animals in their home yards for self-sufficiency. Since then, various forms of agroforestry have gradually developed and laid the fundamental framework of the Chinese small-farming economy for more than 3000 years.

There has been a rapid growth of population in China over since the 1950s. At the same time, the area of arable land has decreased drastically and the environment has degraded rapidly as industrial development has taken place. The traditional labour-consuming and ineffective agroforestry management practices have not adapted to the current situation. In view of economic, ecological and social benefits, conversion of mono-biological production into a trinity system of agriculture, processing and marketing is suggested and planned experimentally. Such a management system, known as modern agroforestry, could be very beneficial to the development of modern China's rural economy and environmental conservation.

Introduction

The term 'agroforestry' appeared in the 1970's and is now in wide use. However agriculture in China was initiated in forests and its original development and practice has been closely related to and historically paralleled with forestry. Agroforestry may date back to the middle and late periods of the Old Stone Age. It has undergone embryonic and primitive states, through a traditional mode of evolution and then developed to the modern model of management by the mid 20th century [Agricultural Heritage Institute of China, 1984; Tang, 1985; Yan and Yi, 1991]. The whole period of agroforestry development extends over several hundred thousand years. Based on the occurrence and development of agriculture in China, the present paper analyzes the composition, structure, function and efficiency of agroforestry in view of the historical background of China's social economy.

Origin of agroforestry

Forests are the largest terrestrial ecosystem on the earth with favourable environmental conditions and plentiful biological resources. They serve as a cradle of human civilization and a base for the birth of agriculture and forestry. Food and living conditions are essentials of human existence and they are also recognized as indicators of social and economic development. Stoneware, charcoal, ashes and animal bones dating from the Old Stone Age onwards have been found by archaeologists [Archaeological Institute, 1984; Du, 1982]. These show that Chinese ancestors had long in the past begun to live in forests in which they took shelter and protected themselves against possible attacks. They hunted birds and animals for meat and gathered the seeds and fruits of plants as crops. There was an extremely slow historical process of evolution from eating raw and living prey to using fire for heating and cooking that took a hundred thousand years to complete [Agricultural Heritage Institute of China, 1984; Du, 1982].

The Old Stone Age is the first turning point in the human history of evolution. The activities of the ancestral Chinese people at this time progressed gradually from simple instinct for survival to intelligent thinking as demonstrated in the making of wooden and stone tools and the improvement of hunting and gathering. They gradually came to notice that some captured wild animals could be domesticated and that seeds falling into ashes from fires could germinate, grow, bloom and bear fruits and seeds. With the accumulation of such experiences, the idea of raising animals and cultivating plant crops was conceived. Consequently, a primitive mode of agriculture took shape from these early practices [Agricultural Heritage Institute of China, 1984; Du, 1982; Tang, 1985; Zhang, 1992]. Therefore, it can be said that agriculture originated on forested land and was closely related to forestry activities at its very beginning. The earliest primitive farming society was thus formed in China.

Primitive agroforestry

Around 10,000 years ago, ancient China entered a new historical era – the New Stone Age. Ancestral Chinese practised primitive farming activities [Archaeological Institute, 1984; Zhu, 1988]. In ancient China people were always united by blood and kinsmanship and engaged collectively in all activities. Some wise men who made contributions to the lives of their people were supported as chiefs or head of families or tribes. In some history books there are legends about Sui Ren who drilled wood to make fire; Fu Xi who wove nets for catching fish and animals; You Chao who constructed shelters from trees; Shen Nong who taught his people to farm; Xuan Yuan who built wheels and carts and Lei Zu who raised silkworms [Agricultural Heritage Institute of China, 1984; Du, 1982]. All of them made efforts to bring the society forward at that time and were deservingly deified and respected as representatives of wisdom for ancestral Chinese. In particular, Shen Nong advocated the use of fire to burn down forests for hunting animals and to open up clearings for cultivation. He also made efforts to improve wooden and stone farming tools and to grow cereal crops in the ash-enriched soil of burnt patches in forested land. Such practices marked the beginning of slash-and-burn cultivation. Unburnt forested areas, on the other hand, became places for gathering of wild fruits, hunting and breeding of animals. This is the earliest form of agriculture, so to speak, and a primitive form of agroforestry. At the beginning of the New Stone Age, this form of agroforestry was merely supplementary to hunting and gathering as a means of livelihood [Tang, 1985; Yan and Yi, 1991; Zhu, 1988].

Slash-and-burn cultivation was a big step forward in human economic life. It indicated that Chinese ancestors were no longer entirely dependent on the provisions offered by nature. They added some supplies of natural products through their wisdom and labour. However, they knew nothing about farming techniques such as fertilization, ploughing, weeding and irrigation. The growth and yield of crops mainly depended on the fertility of forested land. The people burned down forests and used the ashes as fertilizers [Agricultural Heritage Institute of China, 1984; Deng, 1986; Yan and Yi, 1991; Zhang, 1992]. They sowed extensively in the hope of obtaining a bumper harvest. As a result of continued slash-and-burn farming, soil erosion became severe, aggressive weeds grew luxuriously, and yield decreased rapidly. They were forced to move to other forested areas to repeat the slash-and-burn process, deserting the existing previously-cultivated land. This is known as shifting cultivation [Sharma, 1976]. After an interval of years, as fertility recovered through vegetational succession, they came back and started the clearing and cultivation all over again.

Slash-and-burn farming became popular in the New Stone Age. Its development, continuation, and rotation varied from one area to another, mainly depending on population, cultural development and the amount of forested land. In those isolated areas with less population but more forest resources, slash-and-burn farming lasted for a much longer time in various forms. Even today this farming system remains quite popular in the very underdeveloped remote areas occupied by minority nationalities in southwestern China. Tribes using it include the Du Long, Nu, Xi Wa and Yao. They practise different forms of slash-and-burn and sow extensively in the hope of obtaining large harvests. After an interval of a certain number of years (3–4 or 7–8 years, depending on the fertility of the land) they move in groups from one place to another. That is why they are also known as nomadic tribes. Such a farming system involves extensive destruction of forest and land resources, and the yield obtained from crops is low. These people still live in severe poverty [Tang, 1985; Yan and Yi, 1991; Yu and Shen, 1983].

Slash-and-burn used in silvicultural management has had a history of more than one thousand years. It has been widely employed in site preparation for reforestation and afforestation of Chinese fir (*Cunninghamia lanceolata*) in southern China. Farmers slash down all ground cover including small trees and shrubs in summer, burn them in autumn, level up the ground to mix the ashes and soil in winter and then either plant seedlings or cuttings of fir trees with crops in spring or grow crops only for the first one or two years before tree planting. In newly established fir stands, intercropping is also practised until crown closure. Such a short phase of food crop production used as a silvicultural measure to ensure the establishment of timber stands is similar to the 'taungya' system developed for management of teak (*Tectona grandis*) stands in Burma in the mid 1800s [King, 1987]. After 25–30 years when the stand approaches maturity and the trees reach timber size, harvesting is conducted. The stand is regenerated by sprouting from the stumps left behind. After another 25–30 years, the trees are again harvested and the stand is again vegetatively renewed. After the third rotation, the stand becomes much deteriorated and is no longer fit for vegetative regeneration. The harvested land is left deserted to the natural course of vegetational succession. After 40–50 years when the fertility is naturally restored, the process of slashing, burning, site preparation, tree-planting and intercropping starts all over again. Today slash-and-burn is still a major part of the traditional silvicultural management commonly practised in southern China [Sheng, 1983; Yu and Shen, 1983].

Traditional agroforestry

In the latter part of the New Stone Age (2500–2100 B.C.), the population increased, tribes annexed, the clan society collapsed and the slavery system appeared. This enhanced drastically the reformation and development of farming practices.

According to historical records, the first emperor of the Xia Dynasty (2000 B.C.) divided his territory into nine regions, dredged rivers across the country, distinguished soil types and documented produce, so laying the foundations for agriculture [Agricultural Heritage Institute of China, 1984; Du, 1982]. In some areas, the practice of exploiting virgin land evolved into farming on cultivated but deserted land. Slash-and-burn was no longer the main farming system except in less populated, but abundantly forested and underdeveloped areas, while in hilly areas and plains with larger populations, forests were burned down on a massive scale to be converted into arable land for growing crops. Instead of nomadic farming, the slave owners allotted portions of their land to each slave they controlled. The slaves no longer moved to forested land for cultivation but settled down in one place where they engaged in farming activities permanently. The forests on poor hilly land were preserved as pasture land or as hunting sites. Sometimes luxuriant woodlots on cultivated land were also preserved as 'community woods' for social gatherings and ritual sacrifices, in addition to the supply and allocation of timber and firewood as described in the Analects of Confucius [Yan and Yi, 1991; Zhang, 1992].

Settlement farming together with an increase in population in the Xia and Shan Dynasties (2000–1100 B.C.) became a motive force pushing agricultural development forward. By that time, people began to make and improve bronze and wooden agricultural implements, to use animal labour in farming, to grow different crops (such as rice, wheat, millet, sorghum, beans), to raise different domesticated animals (such as horse, cattle, sheep, chicken, pig, dog) and to cultivate fruit, vegetables, mulberry trees (*Morus alba*) and linen plants. There appeared the embryonic form of multiple management, a combination of farming, silviculture, animal husbandry and sericulture that laid the economic foundation of a small farming system for self-sufficiency. Gathering wild fruits and hunting wild animals became much less important activities and descended to a subordinate position [Agricultural Heritage Institute of China, 1984; Du, 1982; Tang, 1985].

During the West Zhou Dynasty (1100–700 B.C.), the slavery system gradually collapsed. A land-tenancy system, the 'nine squares', was widely adopted across the country particularly on the North China Plain as described by Mencius (400 B.C.) [Yan and Yi, 1991; Zhang, 1992]. A large square area of land was equally divided into nine small areas, of which the eight on the edges were allotted to serfs, while the central one belonged to the landlord and was managed by the eight serfs for him. The private ownership of land became gradually established and resulted in further development of small self sufficiency systems. With one serf family as a production unit, animal husbandry, sericulture and domestic handicrafts (such as brewing, silk reeling and weaving) were managed and developed in conjunction with food production for self-sufficiency. This was a major form of rural economy known as 'man-farming and woman-weaving' in old China [Tang, 1985; Wang, 1964]. Besides this, trees for timber and fruit were planted around farm houses as depicted in the Book of Songs (800 B.C.): 'trespass not on my lane, break not my willow trees; leap not over my fence, pick not my mulberry leaves; intrude not my garden, cut not my sandalwood trees'. Growing melons and vegetables, as well as raising domesticated animals, were also included in this comprehensive homestead management [Zhang, 1992].

During the period of Spring-Autumn (770–481 B.C.) and the War-States (490–220 B.C.), a comprehensive management system consisting of farming, silviculture, animal husbandry and fishery activity had further developed. 'To grow different cereals, food supply is assured, to plant mulberry trees and hemp and to rear domesticated animals, then farmers are affluent' (in Guan Zi, 400 B.C.). As an agricultural policy during these periods, comprehensive production for multiple purposes was advocated and encouraged as an effective way to prosperity. 'Those who can plant trees and raise silkworms are awarded with ounces of gold and hectolitres of grain' [Yan and Yi, 1991; Zhang, 1992]. Mencius (400 B.C.) highly praised the small farming system of multiple management and said 'if a family owns a certain area of land with mulberry trees planted around its house for breeding silkworms, domesticated animals raised in its yard for meat, and crop fields cultivated and managed properly for cereals, it will be prosperous and will not suffer from starvation' [Tang, 1985; Yan and Yi, 1991]. Meanwhile, forests and water resources were well protected and properly exploited. 'If the forests are timely felled, then an abundant supply of timber and fire wood is ensured, if the fishing net with relatively big holes is timely cast into the pond, then there is not shortage of fish and turtle for use' (Mencius) [Zhang, 1992]. This presents a very prosperous picture of the rural economy of multiple management and a vivid sketch of pastoral peace.

History next witnessed the unification and fall of the Qing Dynasty (221–207 B.C.) and the warring decade of the Chu and Han States (210–205 B.C.). By the apex of the Han Dynasty (about 100 B.C.), social security was stable and the total population grew steadily up to 40 million. Experience of agroforestry accumulated from the past was collated, revised and widely extended to serve as a guideline for agricultural policy making. As stated in Huai Nan Zi (about 100 B.C.): 'Ever since ancient times, those who want to become masters of their people should make the best use of time and land plus human labour for production. Consequently, crops grow flourishingly, domesticated animals are all thriving, trees are timely planted, farming fields are regularly worked on. Mulberry trees and hemp are grown on fertile soil, while on hilly and steep mountain slopes unsuitable for growing crops, trees are planted for timber and firewood. Therefore, different sites are properly used for appropriate production' [Agricultural Heritage Institute of China, 1984; Tang, 1985; Yan and Yi, 1991; Zhang, 1992]. Such timely and well arranged agroforestry activities resulted in enormous economic benefits.

Si Maqian (150 B.C.), a great historian in the Westen Han Dynasty, gave a detailed description of the growing of fruit trees, timber trees and bamboos in his book, Historic Records. Tree selection and management became a constituent part of agricultural activities at that time [Zhang, 1992].

With the long history of China's order and disorder, the development of agriculture also experienced a course of rise and fall. Changes of governments as well as dynasties were always accompanied by wars and confusion. Production was ruined and people suffered from starvation and homelessness. Afterwards when the political situation become steady, agriculture developed again and prospered. After the Han Dynasty (206 B.C.–220 A.D.), political security lasted for relatively long periods in such dynasties as the Jin (265–420), Tang (618–907), Song (960–1279), Yuan (1234–1368), Ming (1368–1661) and Qing (1661–1911). In each of these periods some advances were made in agricultural management and techniques. Double and triple interplanting and combination of crops and other products varied greatly, particularly in the Ming and Qiun Dynasties [Agricultural Heritage Institute of China, 1984; Du, 1982; Tang, 1985]. According to the analysis of Wang Xue Hu [1964], there exist 497 old agriculture books in China, of which 283 appeared in the Ming and Qing Dynasties, accounting for 57% of the total. There are 162 books related the management of forests, animal husbandry and fisheries which includes mulberry, tea, fruit, vegetables, flowers, silkworm and bees. This number makes up 32.6% of the total [Wang, 1964].

The development of sericulture merits special mention. Cocoons, tough silk pieces and spindles unearthed in 1921 from the relics of the Yang Shao and Long Shan Cultures (2300–2000 B.C.) and Chinese characters such as mulberry, silkworm and silk recognized on the inscriptions uncovered from the relics of the late Shang Dynasty (1200–1100 B.C.) show clearly that China has a very long history of mulberry planting, silkworm rearing and silk reeling and weaving. First started in the Yellow River areas and later spread southward to the Yangtze River regions, sericulture has developed steadily. Its economic significance is second only to the growth of cereal crops and is more important than animal husbandry. The laying of emphasis both on farming and sericulture has become an important guideline for making agricultural policies in different dynasties. In fact, sericulture is the earliest and most typical example of agroforestry in China. The Book of Songs (800 B.C.) mentions the techniques of pruning mulberry trees and collecting mulberry leaves for feeding silkworms [Zhang, 1992]. In the Fan Book of Agriculture (about 100 B.C.), there is an account of interplanting systems of mulberry with millet and hemp, using the excrement of silkworms as fertilizer [Du, 1982; Tang, 1985; Zhang, 1992]. Through continuous efforts and improvement by farmers of different times, sericulture together with animal husbandry or fisheries has become the major form of 'side-line'[1] production in the self-sufficient rural economy. The mulberry plantation is intergrown with crops and other plants. The hatching and rearing of silkworms is repeated several times a year. At the same time handicraft industries such as silk reeling, spinning and weaving have also been rapidly developed, marking the way to the production of an economic commodity. Early in 200 B.C. silk from China made its way to mid and west Asia and Europe along a route that became well-known as the 'Silk Road' and in China as 'Seres' [Agricultural Heritage Institute of China, 1984; Du, 1982; Yan and Yi, 1991; Zhang, 1992]. In the tenth century, crop fields dotted with mulberry patches and fishing ponds surrounded by mulberry fields became very popular in the Yangtze River region and the delta of the Pearl River where crops, sericulture, fisheries and animal husbandry were comprehensively managed [Mei, 1983]. Since silk is the major raw material for the textile industry, sericulture remains the major form of 'side-line' production in China.

The Ming and Qing Dynasties (14–19th centuries) witnessed the marked development of traditional agroforestry in China. The proliferation of various matching and combination modes of farming, silviculture, animal husbandry, fisheries and 'side-line' production was unprecedented. In the early half of the 20th century, however, both agriculture and agroforestry remained stagnant due to social disorder and upheaval. After the founding of New China in 1949 development resumed.

On a rough estimation, there exist in China more than 50 main types of traditional agroforestry. They can conveniently be grouped into five categories:

1. Agroforestry with crops dominant. This model is popular in the vast agricultural plains where trees are planted as individuals, or in clusters, patches, strips and belts in crop fields. In provinces such as Shangdon, Anhui and Henan, there are crop-*Paulownia fortunei*, crop-date (*Ziziphus jujuba*) and crop-persimmon (*Diospyros kaki*) intergrowing forms, while in the north, northwest and east of China, plantations such as '3-N'[2] shelter belts and forest networks have been built in the cultivated fields or along the sea coasts.
2. Agroforestry with forest dominant. In naturally or artificially forested areas such as timber stands, orchards, special tree plantations, watershed forests, scenic forests (such as forest parks, historical relics, temples) and conservation forests, various crops (cereals, vegetables, cash crops, medicinal plants and forages) are intergrown so as to enable multiple use of the stands without disturbance of their initial condition. They include ginseng (*Panax ginseng*) under forest in the northeast, tung-oil trees (*Vernicia fordii*), mountain-spice trees (*Litsea cubeba*), and gold threads (*Coptis chinensis*) in Chinese fir plantations, mushrooms in bamboo stands, tea under pine forests, coffee and pepper in rubber plantations, medicinal Morinda (*Morinda officinalis*), and obscured Homalomena (*Hamalomena occulta*) and villous Amomum (*Amomum villosum*) under tropical forests. Moreover, the undergrowth of forests is also utilised to raise poultry and other domesticated animals as well as wildlife, such as deer in the northwest forests, musk deer in the enclosed forest stands in Sichuan, sheep or goats in fir stands and chickens in bamboo plantations in Jiangsu.
3. Agroforestry with pasture-husbandry dominant. Trees and shrubs are planted in clusters or belts on grazing land to provide protection for herds and to ensure herbage production, as commonly seen in Inner Mongolia and the sparse forest areas in the south.
4. Agroforestry with fisheries dominant. Trees are planted along rivers, lakes, reservoirs and fish ponds not only for protecting dams against waves and preventing soil erosion, but also for helping the fisheries business. In the delta areas of the Yangtze and Pearl Rivers, fish ponds are usually surrounded by mulberry trees, fruit trees, tea plants, sugar cane and other plants. Fish are even raised in rice fields in some areas. Such practices have a long history and continue to develop.
5. Homestead and 'four-side'[3] management. Timber trees, fruit trees, vegetables and sometimes aquatic crops are grown and poultry, animals and fish are raised on a small scale around farm houses, near villages, as well as on road and stream sides so as to make the best use of land and water resources. Besides self-sufficiency, some products can be marketed as a part of the major income for families. Though their combination varies

greatly from place to place, such practices are very common in rural areas all over the country.

During the past several thousand years, agroforestry systems with crop production dominant in a combination of silvo-pastoral-fishery and 'side-line' activities was gradually formed and then strongly developed in the closed rural society of China where it has played an important role in the development of the self-sufficient economy. Such a system, though meeting self-sufficiency requirements, is small in production scale and low in output for the amount of labour required for its management in a market economy. The rapid growth of population, decrease in arable land, development of industry and deterioration of environment, particularly under the pressure of economic reform and open-door policies, have all conspired to poise this small farming system on the verge of collapse.

Modern agroforestry

In recent years Chinese scientists have become more concerned with the present status of agricultural systems in relation to the Chinese rural economy and social background and have put forward concepts such as 'large-scale agriculture', 'stereo (vertical) farming', 'comprehensive farming', 'complex farming' and 'ecological agriculture' as well as appropriate technical measures in the hope of reforming the production structure for better use of water and land resources, increasing the income of farmers and improving the rural economy [Deng, 1986; Hu, 1988; Yang, 1987]. Their endeavours have resulted in some advances in the development of modern agroforestry.

As a comprehensive management system agroforestry should include farming, silviculture, animal husbandry, fisheries, 'side-line' production, processing and marketing which are organized and operated in accordance with the modern market economy, theories, methods and techniques of system ecology and ecosystem engineering. Only in this way can natural resources (land, climate, and water) and social resources (labour, science and technology) be fully utilized and maintained so as to achieve large and sustained economic, ecological and social benefits. Therefore, modern agroforestry is based on biological production from which economic growth can be achieved through the processing industry and the marketing system, as shown below [Hsiung, 1991].

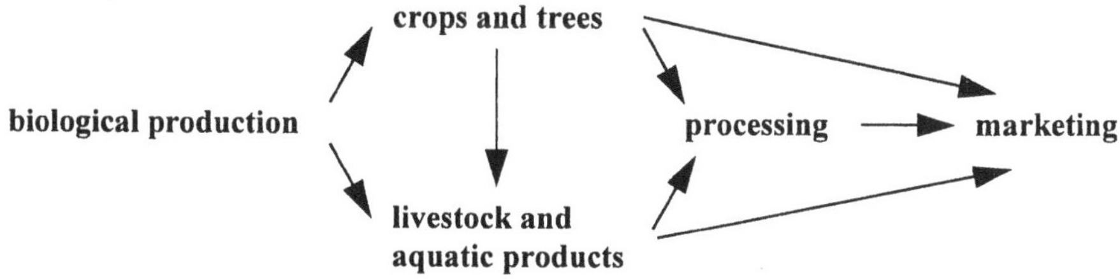

In this trinity system, biological production is achieved through material and energy transformation by crops, trees, animals and micro-organisms. The higher the rate of transformation, then the greater is the economic efficiency. Therefore, the proportion of the production of silviculture, animal husbandry, fisheries and 'side-lines' should be increased. As regards the growth of crops, the traditional duality of cereal-economic crops should be gradually changed into the trinity of cereal-forage economic crops so as to facilitate the development of animal husbandry.

A village or rural area is an agroecosystem which can be considered as a unit for the planning and design of modern agroforestry. Its natural conditions, socio-economic background, material supply, species diversity and technical level are important and should be carefully investigated and evaluated prior to design. In addition, attention should be paid to the proper protection and exploitation of natural resources.

Based on the conceptual framework mentioned above, model agroforestry systems can be transferred into tangible production systems with reference of the principles and methodology of ecosystem engineering [Hsiung and Zou, 1985; Hsiung, 1991; Qian, 1988; Zhu, 1988]. Trees, crops, domesticated animals, aquatic species and their intermediate processes should be correctly selected in view of their economic value and biological characteristics. These related components should be purposely integrated into a production system in an appropriate time sequence, spatial arrangement and numerical proportion so that they can fully perform their biological production role and create favourable conditions for successive components. According to the category and quantity of biological products (yields), processing systems and marketing systems can be correspondingly established. Thus, a whole rural economic system which combines agriculture, industry and trade as a trinity is formed. This is what is meant by modern agroforestry in China [Hsiung, 1991].

Experiments and practical trials have been conducted in recent years on variously labelled large-scale agriculture, comprehensive farming or ecological agriculture. Planning and design concerning watercourses, fields, forests, roads and village locations have been carried out with reference to local conditions. The production structure has been modified and organised to specialise in certain items such as cereal crops, cotton, chickens, ducks, pigs, cattle and aquatic products including their primary processing and marketing. Farmers operate them on a relatively large scale either as families or as responsible collectives and achieve remarkable economic profits. These integrated production systems belong to the category of modern agroforestry which, through persistent development and replenishment, are predicted to become an entirely new management system of the rural economy in China.

Notes

1. 'Side-line' refers to secondary productive activities such as silk-worm breeding, bee-keeping or poultry husbandry which complement the primary activity of crop production.
2. The '3-N' shelterbelt is a vast shelterbelt system which has gradually been established since the late 1970's in the North-East, North and North-West of China.
3. 'Four-side' management refers to the productive use of road-sides, river-sides, village-sides and crop field sides.

References

Agricultural Heritage Institute of China (ed) (1984) Agricultural History of China. Science Press, Beijing, China [in Chinese]

Archaeological Institute, Chinese Academy of Social Sciences (ed) (1984) Archaeological Discoveries and Studies in New China. Cultural Relics Press, Beijing, China [in Chinese]

Deng HH (1986) Neo-Agricultural Resolution. Guangxi People's Press, Nanjing, China [in Chinese]

Du SR (1982) History of China's Sciences and Technology. Science Press, Beijing, China [in Chinese]

Hsiung WY (1991) Ecosystem engineering and modern agroforestry. Journal of Ecology 10(1): 21–26 [in Chinese]

Hsiung WY and Zou JW (1985) On ecosystem engineering. Journal of Nanjing For Univ 1: 1–11 [in Chinese]

Hu ST (1988) Ecological Agriculture. Hubei Science and Technology Press, Wuhan, China [in Chinese]

King KFS (1987) The history of agroforestry. In: Steppler HA and Nair PKR (eds) Agroforestry: A Decade of Development, pp 3–11. International Centre for Research in Agroforestry (ICRAF), Nairobi, Kenya

Mei XH (1983) Approaches to ecosystems of mulberry plantations with crops, vegetables and animal husbandry. Journal of Ecology 2(4): 46–49

Qian XS (1988) System Engineering. Hunan Science and Technology Press, Chang Sha, China [in Chinese]

Sharma TC (1976) The Prehistoric Background of Shifting Cultivation. Proceedings of a Seminar on Shifting Cultivation in North-East India. India Council for Social Science Research, New Delhi, India

Sheng WT (1983) Site preparation for Chinese fir planting. In: Wu ZH (ed) Chinese Fir, pp 387–402. Chinese Forestry Press, Beijing, China [in Chinese]

Tang QG (1985) History of China's Agriculture. Agric Press, Beijing, China [in Chinese]

Wang YH (1964) Catalogue of Chinese Agriculture. Agric Press, Beijing, China [in Chinese]

Yan WI and Yi YH (1991) History of China's Agricultural Development. Tianjin Science and Technology Press, Tianjin, China

Yang TX (1987) Consideration of China's Agriculture. Office of Agricultural Modernization, Chinese Academy of Sciences, Shijiazhuang, China

Yu XT and Shen PQ (1983) Tree-crop interplanting. In: Wu ZH (ed) Chinese Fir, pp 431–441. Chinese Forestry Press, Beijing, China, [in Chinese]

Zhang JC (1992) Introduction to China's Traditional Forestry. Chinese Forestry Press, Beijing, China [in Chinese]

Zhu D (1988) Studies on Primitive Culture. Unite Book Store of Life, Study and Knowledge, Hongkong, China [in Chinese]

Zhu ZN (1988) On systematic optimal structure of agriculture. In: Xu GZ (ed) Application Examples of System Engineering, pp 49–55. Science Press, Beijing, China [in Chinese]

www.ingramcontent.com/pod-product-compliance
Ingram Content Group UK Ltd.
Pitfield, Milton Keynes, MK11 3LW, UK
UKHW020245290926
15470UKWH00002B/749
* 9 7 8 0 7 9 2 3 3 6 9 6 9 *